Communications in Asteroseismology

Volume 158
July, 2009

Proceedings of the

38th LIAC / HELAS-ESTA / BAG
Evolution and Pulsation of Massive Stars
on the Main Sequence and Close to it

Liège, Belgium, July 7-11 2008

Edited by
Arlette Noels, Conny Aerts, Josefina Montalbán,
Andrea Miglio and Maryline Briquet.

Austrian Academy
of Sciences Press

Vienna 2009 OAW

Communications in Asteroseismology

Editor-in-Chief: **Michel Breger**, michel.breger@univie.ac.at
Editorial Assistant: **Daniela Klotz**, klotz@astro.univie.ac.at
Layout & Production Manager: **Paul Beck**, paul.beck@univie.ac.at
English Language Editor: **Natalie Sas**, natalie.sas@ster.kuleuven.be

Coast editorial and production office
Türkenschanzstraße 17, A - 1180 Wien, Austria
http://www.univie.ac.at/tops/CoAst/
Comm.Astro@univie.ac.at

Cover Illustration

Conference logo of the 38th Liège International Astrophysical Colloquium
thanks to Anzio Giani and Anna Miglio

British Library Cataloguing in Publication data.
A Catalogue record for this book is available from the British Library.

Austrian Academy of Sciences Press
A-1011 Wien, Postfach 471, Postgasse 7/4
Tel. +43-1-515 81/DW 3402-3406, +43-1-512 9050
Fax +43-1-515 81/DW 3400
http://verlag.oeaw.ac.at, e-mail: verlag@oeaw.ac.at

38th Liège International Astrophysical Colloquium / HELAS-ESTA / BAG

38th Liège International Astrophysical Colloquium / HELAS-ESTA / BAG

Scientific Organizing Committee:

Conny Aerts, KU Leuven
Marcel Arnould, Univ. Libre de Bruxelles
Cesare Chiosi, Univ. of Padova
Franca D'Antona, INAF-Oss. Astr. Roma
Marc-Antoine, Dupret Paris-Meudon Obs.
Wojtek Dziembowski, Warsaw Univ.
Patrick Eggenberger, Univ. of Liège
Jadwiga Daszynska, Univ. of Wroclaw
Eric Gosset, Univ. of Liège
André Maeder, Geneva Obs.
Mario Monteiro, CAUP Porto
Arlette Noels, Univ. of Liège
Gregor Rauw, Univ. of Liège
Ian Roxburgh, QMU London
Hiromoto Shibahashi, Univ. of Tokyo
Sylvie Théado, Obs. Midi Pyrnes
Jean-Paul Zahn, Paris-Meudon Obs.

and members of the LOC

Local Organizing Committee:

Marcel Arnould (ULB)
Denise Caro
Angela Della Vecchia
Jean Marc Defise (CSL)
Alain Detal
Patrick Eggenberger
Maurice Gabriel
Eric Gosset
Mélanie Godart
Andrea Miglio
Josefina Montalbán
Arlette Noels
Jean-Yves Plesseria (CSL)
Gregor Rauw
Richard Scuflaire
Anne Thoul

Session 3 - Atmosphere, mass loss and stellar winds

Session 4 - Observed frequencies in pulsating massive stars

Session 5 - What can asteroseismology do to solve the problems ?

Session 6 - What about real stars ?

Special session : Future asteroseismic missions

Conclusions

Comm. in Asteroseismology
Vol. 158, 2009, 38th LIAC/HELAS-ESTA/BAG, 2008
A. Noels, C. Aerts, J. Montalbán, A. Miglio and M. Briquet., eds.

Preface

Arlette Noels-Grötsch

Institute of Astrophysics and Geophysics, University of Liège, Belgium

This special CoAst issue presents the proceedings of the colloquium

Evolution and Pulsation of Massive Stars
on the Main Sequence and Close to it

organized in Liège, Belgium, July 7-11 2008.

This colloquium was the 38th Liège International Astrophysical Colloquium (38thLIAC - http://www.ago.ulg.ac.be/APub/Colloques/Liac38); it was also held under the auspices of the European Helio- and Asteroseismology Network (HELAS - http://www.helas-eu.org/), the CoRoT Evolution and Seismic Tools Activity (CoRoT/ESTA - http://www.astro.up.pt/esta/) and the Belgian Asteroseismology Group (BAG - http://www.asteroseismology.be). We had the pleasure to welcome 102 participants from 16 countries from all over the world.

The main idea of this colloquium was first to confront the problems raised by "non standard" physics; second, to focus on the effects of these often missing physical processes on stellar evolution; and third, to analyse what asteroseismology can do to bring new light on these processes and their modelling in stars.

The targets were massive stars (O, B, WR) for which new and exciting results are now coming from asteroseismic interpretations of observed pulsations.

The first part of the conference was devoted to the internal structure of massive stars with session 1 dealing with the theoretical aspects of non standard physical processes while session 2 presented the challenges met in stellar modelling when introducing these physical processes. Session 3 was a "state of the art" in atmospheres, mass loss and stellar winds in massive stars. Session 4 presented the pulsation observations in massive stars while session 5 discussed the asteroseismic signatures of the physical processes involved in their internal structure. A final session 6 was essentially focused on the interpretation of the asteroseismic observations of ground-based, MOST and CoRoT targets. We also organized a special ESTA session on model comparison and a special session devoted to future asteroseismic missions (KEPLER, BRITE, PLATO, SIAMOIS, SONG).

After each presentation, participants were asked to reserve their questions for a panel discussion of about 30 minutes taking place after each session. The moderators were :

- Maurice Gabriel - ULg (Session 1)

- Georges Meynet - Geneva Obs.(Session 2)

- Eric Gosset and Gregor Rauw - ULg (Session 3)

- Joris De Ridder - KULeuven (Session 4)

- Arlette Noels - ULg (Session 5)

- Conny Aerts - KULeuven (Session 6)

- Jean-Marc Defise and Jean-Yves Plesseria - Centre Spatial de Liège (Special session on future asteroseismic missions)

- Mario Monteiro - Porto Univ. (Special ESTA session)

They all played a extremely important role in the good course of these very lively discussions. In this volume however, in an attempt to make its reading easier, we have redistributed the questions and answers according to the talks to which they belong.

All together, we heard 29 excellent invited talks, 21 no less excellent contributed talks and we had poster sessions twice everyday. I take this opportunity to heartily thank all the speakers, the authors of posters, the moderators of the panel discussions, all the participants; this colloquium was a great success, the ambiance was lively, friendly, curious, we all learned a lot from one another. Thanks to them all!

I am deeply grateful to Denise Caro who took in charge all the administrative tasks with a large, big smile. My thanks are also addressed to the Scientific Organizing Committee and the Local Organizing Committee who helped in more than one way to the nice course of this conference. I especially thank my friends and colleagues Josefina Montalban and Andrea Miglio who were at the origin of the whole story and who put tremendous efforts in carrying it through. Josefina, Andrea, I shall never forget!

We gratefully acknowledge the financial supports from our sponsors :

- HELAS - European Helio- and Asteroseismology Network

- FNRS - Fonds National de la Recherche Scientifique (Communauté française de Belgique)

- ULg - Université de Liège

- CSL - Centre Spatial de Liège

- AMOS - Advanced Mechanical and Optical Systems

Be them all heartily thanked!

Most of the theoretical research in asteroseismology done in Liège would not have been possible without the strong support from the Federal Science Policy Office (BELSPO) in the frame of the ESA/Prodex programs : CoRoT-Preparation to Exploitation and CoRoT-Data Exploitation. Many, many thanks!

A final word for these proceedings. Thanks to Michel Breger, Paul Beck and the CoAst staff; your journal is a wonderful opportunity to promote research in asteroseismology and to give to events such as this conference a well due scientific diffusion. All the authors have been (nearly...) ready in time, all the referees have acted with rapidity and great concern. A heavy work has been achieved by the editors : thanks to Conny Aerts, Maryline Briquet and, once more but not the last, thanks to Josefina and Andrea.

Arlette Noels-Grötsch

Comm. in Asteroseismology
Vol. 158, 2009, 38th LIAC/HELAS-ESTA/BAG, 2008
A. Noels, C. Aerts, J. Montalbán, A. Miglio and M. Briquet., eds.

List of participants

Conny **Aerts**, K.U. Leuven, Belgium, *conny@ster.kuleuven.be*
Georges **Alecian**, Observatoire de Paris - CNRS, France, *georges.alecian@obspm.fr*
Elio **Antonello**, INAF-O.A.Brera, Italy, *elio.antonello@brera.inaf.it*

Annie **Baglin**, LESIA Observatoire de Paris, France, *annie.baglin@obspm.fr*
Kevin **Belkacem**, Observatoire de Paris, France, *kevin.belkacem@obspm.fr*
Jonas **Blomme**, K.U. Leuven, Belgium, *jonasb@ster.kuleuven.be*
Ronny **Blomme**, Royal Observatory of Belgium, Belgium, *Ronny.Blomme@oma.be*
Mehdi-Pierre **Bouabid**, Observatoire Cote d'Azur, France, *bouabid@oca.eu*
Maryline **Briquet**, K.U. Leuven, Belgium, *maryline@ster.kuleuven.be*
Ines **Brott**, Utrecht University, The Netherlands, *brott@astro.uu.nl*

Matteo **Cantiello**, Utrecht University, The Netherlands, *M.Cantiello@astro.uu.nl*
Fabien **Carrier**, K.U. Leuven, Belgium, *fabien@ster.kuleuven.be*
Claude **Catala**, Observatoire de Paris, France, *Claude.Catala@obspm.fr*
Virginie **Chantry**, Université de Liège, Belgium, *chantry@astro.ulg.ac.be*
Andre-Nicolas **Chene**, Canadian Gemini Office, Canada, *andre-nicolas.chene@nrc-cnrc.gc.ca*
Cesare **Chiosi**, University of Padova, Italy, *cesare.chiosi@unipd.it*
Joergen **Christensen-Dalsgaard**, University of Aarhus, Denmark, *jcd@phys.au.dk*
Arthur **Cox**, Los Alamos National Laboratory, USA, *anc@lanl.gov*
Jan **Cuypers**, Royal Observatory of Belgium, Belgium, *Jan.Cuypers@oma.be*

Yassine **Damerdji**, Université de Liège, Belgium, *yassine.damerdji@ulg.ac.be*
Michaël **De Becker**, Université de Liège, Belgium, *debecker@astro.ulg.ac.be*
Philippe **de Meulenaer**, Université de Liège, Belgium, *pdemeulenaer@student.ulg.ac.be*
Selma **de Mink**, Utrecht University, The Netherlands, *SEdeMink@gmail.com*
Joris **De Ridder**, K.U. Leuven, Belgium, *joris@ster.kuleuven.be*
Jonas **Debosscher**, K.U. Leuven, Belgium, *jonas@ster.kuleuven.be*
Jean-Marc **Defise**, Centre Spatial de Liège, Belgium, *jmdefise@ulg.ac.be*
Pieter **Degroote**, K.U. Leuven, Belgium, *Pieter.Degroote@ster.kuleuven.be*
Tyl **Dermine**, Université Libre de Bruxelles, Belgium, *tdermine@astro.ulb.ac.be*
Maarten **Desmet**, K.U. Leuven, Belgium, *maarten.desmet@ster.kuleuven.be*
Pascual David **Diago Nebot**, Universitat de València, Spain, *Pascual.Diago@uv.es*
Boris **Dintrans**, Lab. d'Astrophysique de Toulouse - Tarbes, France, *dintrans@ast.obs-mip.fr*
Gulnur **Dogan**, University of Aarhus, Denmark, *gulnur@phys.au.dk*
Marc-Antoine **Dupret**, LESIA, Observatoire de Paris, France, *MA.Dupret@obspm.fr*
Wojciech **Dziembowski**, Warsaw University Observatory, Poland, *wd@astrouw.edu.pl*

Patrick **Eggenberger**, Université de Liège, Belgium, *eggenberger@astro.ulg.ac.be*
Maurice **Gabriel**, Université de Liège, Belgium, *gabriel@astro.ulg.ac.be*
Wolfgang **Glatzel**, Göttingen Univ., Germany, *wglatze@astro.physik.uni-goettingen.de*
Evert **Glebbeek**, Universiteit Utrecht, The Netherlands, *e.glebbeek@phys.uu.nl*
Mélanie **Godart**, Université de Liège, Belgium, *melanie.godart@ulg.ac.be*
Eric **Gosset**, Université de Liège, Belgium, *gosset@astro.ulg.ac.be*

Marie-Jo **Goupil**, LESIA Observatoire de Paris, France, *mariejo.goupil@obspm.fr*
Nicolas **Grevesse**, Université de Liège, Belgium, *Nicolas.Grevesse@ulg.ac.be*
David **Gruber**, Institute of Astronomy - University of Vienna, Austria, *gruber@astro.univie.ac.at*
Frank **Grundahl**, University of Aarhus, Denmark, *fgj@phys.au.dk*
Juan **Gutiérrez-Soto**, Observatoire de Paris-Meudon, France, *juan.gutierrez@obspm.fr*

Anne-Laure **Huat**, Observatoire de Paris, France, *anne-laure.huat@obspm.fr*
Mikolaj **Jerzykiewicz**, University of Wroclaw, Poland, *mjerz@astro.uni.wroc.pl*
Alain **Jorissen**, Université Libre de Bruxelles, Belgium, *ajorisse@astro.ulb.ac.be*
Christoffer **Karoff**, University of Aarhus, Denmark, *karoff@phys.au.dk*
Rainer **Kuschnig**, Institute for Astronomy University of Vienna, Austria, *kuschnig@astro.ubc.ca*

Norbert **Langer**, Univ. Of Utrecht, The Netherlands, *N.Langer@astro.uu.nl*
Yveline **Lebreton**, Observatoire de Paris, France, *Yveline.Lebreton@obspm.fr*
Laure **Lefevre**, Observatoire de Paris, France, *laure.lefevre@obspm.fr*
Jean-Christophe **Leyder**, Université de Liège, Belgium, *leyder@astro.ulg.ac.be*
Natacha **Linder**, Université de Liège, Belgium, *linder@astro.ulg.ac.be*
Alex **Lobel**, Royal Observatory of Belgium, Belgium, *alobel@sdf.lonestar.org*

André **Maeder**, Geneva Observatory, Switzerland, *andre.maeder@obs.unige.ch*
Pierre **Magain**, Université de Liège, Belgium, *Pierre.Magain@ulg.ac.be*
Laurent **Mahy**, Université de Liège, Belgium, *mahy@astro.ulg.ac.be*
Joao **Marques**, LESIA Observatoire de Paris, France, *joao.marques@obspm.fr*
Fabrice **Martins**, Université de Montpellier, France, *fabrice.martins@graal.univ-montp2.fr*
Georges **Meynet**, Geneva Observatory, Geneva University, Switzerland, *georges.meynet@obs.unige.ch*
Eric **Michel**, Observatoire de Paris, France, *Eric.Michel@obspm.fr*
Andrea **Miglio**, Université de Liège, Belgium, *a.miglio@ulg.ac.be*
Josefina **Montalban**, Université de Liège, Belgium, *j.montalban@ulg.ac.be*
Mario J.P.F.G. **Monteiro**, Centro de Astrofisica da Univ. do Porto, Portugal, *mjm@astro.up.pt*
Thierry **Morel**, Université de Liège, Belgium, *morel@astro.ulg.ac.be*
Benoit **Mosser**, Observatoire de Paris, France, *benoit.mosser@obspm.fr*
Andres **Moya**, Instituto de Astrofisica de Andalucia (CSIC), Spain, *moya@iaa.es*

Yaël **Nazé**, Université de Liège, Belgium, *naze@astro.ulg.ac.be*
Coralie **Neiner**, GEPI, Observatoire de Meudon, France, *coralie.neiner@obspm.fr*
Ewa **Niemczura**, Astronomical Institute, Wroclaw University, Poland, *eniem@astro.uni.wroc.pl*
Arlette **Noels-Grotsch**, Université de Liège, Belgium, *Arlette.Noels@ulg.ac.be*
Rhita-Maria **Ouazzani**, LESIA Observatoire de Paris, France, *Rhita-Maria.Ouazzani@obspm.fr*
Nesibe **Ozel**, LESIA Observatoire de Paris, France, *Nesibe.Ozel@obspm.fr*
Jean-Yves **Plesseria**, Centre Spatial de Liège, Belgium, *jyplesseria@ulg.ac.be*
Joachim **Puls**, University Observatory, Munich, Germany, *uh101aw@usm.uni-muenchen.de*
Gregor **Rauw**, Université de Liège, Belgium, *rauw@astro.ulg.ac.be*
Daniel **Reese**, University of Sheffield, United Kingdom, *d.reese@sheffield.ac.uk*
Pierre **Renson**, Université de Liège, Belgium, *renson@astro.ulg.ac.be*
Michel **Rieutord**, Lab. d'Astrophysique de Toulouse-Tarbes, France, *rieutord@ast.obs-mip.fr*
Pierre **Rochus**, Centre Spatial de Liège, Belgium, *prochus@ulg.ac.be*

Sophie **Saesen**, K.U. Leuven, Belgium, *sophie@ster.kuleuven.be*
Hideyuki **Saio**, Tohoku University, Japan, *saio@astr.tohoku.ac.jp*
Sébastien **Salmon**, Université de Liège, Belgium, *Sebastien.Salmon@student.ulg.ac.be*
Réza **Samadi**, LESIA Observatoire de Paris, France, *reza.samadi@obspm.fr*
Olivier **Schnurr**, University of Sheffield, United Kingdom, *O.Schnurr@sheffield.ac.uk*

Richard **Scuflaire**, Université de Liège, Belgium, *R.Scuflaire@ulg.ac.be*
Hiromoto **Shibahashi**, Univ. Of Tokyo, Japan, *shibahashi@dept.astron.s.u-tokyo.ac.jp*
Paul **Smeyers**, K.U. Leuven, Belgium, *Paul.Smeyers@ster.kuleuven.be*
Radoslaw **Smolec**, N. Copernicus Astronomical Center, Warsaw, Poland, *smolec@camk.edu.pl*
Juan Carlos **Suárez**, Instituto de Astrofsica de Andaluca (CSIC), Spain, *jcsuarez@iaa.es*

Suzanne **Talon**, Université de Montréal, Canada, *suzanne.talon@rqchp.qc.ca*
Sylvie **Théado**, Lab. d'Astrophysique de Toulouse-Tarbes, France, *stheado@ast.obs-mip.fr*
Anne **Thoul**, Université de Liège, Belgium, *Anne.Thoul@ulg.ac.be*
Katrien **Uytterhoeven**, Instituto de Astrofisica de Canarias, Spain, *katrien@iac.es*
Johan **van der Walt**, North-West University, South Africa, *johan.vanderwalt@nwu.ac.za*
Sophie **Van Eck**, Université Libre de Bruxelles, Belgium, *svaneck@astro.ulb.ac.be*
Sylvie **Vauclair**, Lab. d'Astroph. de Toulouse-Tarbes, France, *sylvie.vauclair@ast.obs-mip.fr*
Jean-Marie **Vreux**, Université de Liège, Belgium, *vreux@astro.ulg.ac.be*
Christoffel **Waelkens**, K.U. Leuven, Belgium, *Christoffel@ster.kuleuven.be*
Jean-Paul **Zahn**, Observatoire de Paris, France, *jean-paul.zahn@obspm.fr*

Introduction

Comm. in Asteroseismology
Vol. 158, 2009, 38th LIAC/HELAS-ESTA/BAG, 2008
A. Noels, C. Aerts, J. Montalbán, A. Miglio and M. Briquet., eds.

A colloquium for Arlette's... birthdayS

A. Baglin

Observatoire de Paris, LESIA, F-92195 Meudon, France

I do not intend to give here an overview of the scientific achievements of Arlette, which are very numerous, but only a few flashes on our common souvenirs.

Arlette is one of the students of Paul Ledoux, the father of the Liège stellar structure and stability group created in the early 60s. As such she has been educated as a theoretician, but always kept an eye on observations, trying to interpret them.

I met Arlette for the first time in 1967, in Paris, at the Institute of Astrophysics. She had been invited in our group, led by Evry Schatzman, to present the advancement of her work on secular stability of stellar models. She made a linear perturbation analysis of static models of low mass and looked at the secular solution, directly inspired by the work of Paul Ledoux. I remember that I had problems understanding what it meant exactly. I was puzzled by this secular instability with regards to evolution. So I asked Arlette about it, probably not very clearly, and I did not understand the answer...

This was for me the beginning of the cooperation with the Liège group. We in Paris, wanted to make computations of stellar structure models, a field which was in its enfancy. But in France we did not have sufficiently powerful machines to do that. Our bosses made an agreement (I must confess I do not know the content) in which we, the French, got access to the Computing Center of Liège University. At that time, information was fed to the computers using punched cards often very numerous, and packed in long and heavy cardboard boxes. So the boxes of the French people had to go to Liège... Each week, one of us (students of Schtazman) was driving to Liège, with all the boxes of the team members in his trunk. He was staying there the whole week, sleeping in the monacal room of the Institute in Cointe, and he was in charge of feeding the computer with all our programs and debugging if necessary. But often, as he was doing his own research in a very different domain, he could not understand the problems and we had to wait for his return to punch new cards for the next week!

Arlette likes to tackle controversial problems. I remember, for instance, her quite original contribution to the solar neutrino problem. Being a specialist of both secular and vibrational stability, she studied the possibility to mix the solar core through the vibrational instability of $g+$ low degree modes. And she found that this process should work but only below 1 solar mass.

Arlette likes broad subjects and new ideas. After looking at the Sun, she jumped to the highest possible masses. At that time people were puzzled by the discovery of a giant HII region by IUE. Was it excited by an extremely massive star or a cluster of smaller ones? She studied the vibrational instability of an elephant star (3000 solar masses!) and found a time scale of the instability of 145 years... very unlikely to be observed!

This was the beginning of a long period dedicated to the Wolf-Rayet stars looking for a scenario of their formation, explaining the proportion of the different types. She has been among the first, with the Liège group, to compute evolution of massive stars (60 to 100 solar mass) until the end of the He burning core. Once more she proposed the vibrational instability as a process to peal the star and produce the emergence of nuclear products at its surface. Already at that time she stressed the need for hydrodynamical descriptions to follow the instability... a long way to go.

Arlette has been and is very active in the organization of all the Liège colloquia (a series initiated by P. Swings in the late fifties) dedicated to stellar structure and evolution.

The first one was too early for Arlette to be involved in. In 1959 the 9th Liège Colloquium was entitled "Modèles d'étoiles et Evolution stellaire", and it was chaired by S. Rosseland. The 16th Liège Colloquium took place in 1969 and it was dealing with "Evolution stellaire avant la séquence principale". The 25th Liège Colloquium, in 1984, was organized in honour to Paul Ledoux and it was entitled "Theoretical problems in stellar stability and oscillations".

In 1995 the 32nd Liège Colloquium was chaired and published by Arlette with a title questioning the future of the domain : "Stellar evolution : What should be done?". The idea was to build common plans for the future at the very beginning of a new era... driven by the idea that observing stellar oscillations could indeed be possible and could open the way to seismology of stars as it was already developed for the Sun.

Let me cite again a sentence by Sir Arthur Eddington in 1926 in his founder book "The internal Constitution of the Stars" : *At first sight it would seem that the deep interior of the sun and stars is less accessible to scientific investigation than any other region of the universe. What appliance can pierce through the outer layers of a stars and test the conditions within?.* One has to add now: *Since 1980 we know that Seismology is such an appliance*, as written by Ian Roxburgh in the Eddington proposal.

Arlette decided to spend a lot of time with her colleagues of the Institute to build stellar models computed with high enough accuracy for seismology calculations. And now the CLES and LOSC codes play a major role in the ESTA comparison project.

When CoRoT was in great danger, our Ministery of Research asked us to prove that we were able to find partners interested in the mission. I immediately contacted Arlette and she reacted very quickly by forming the Belgium group of scientists supporting CoRoT, and applying for money. She has been very successful, and Belgium became a major partner. In 2001, the agreement between CNES and ESA/PRODEX for the Belgian contribution to CoRoT was signed.

Arlette is the coordinator of the Belgium contribution to CoRoT, and is a member of the Scientific Committee. On the instrumental side, Belgium has very successfully built the baffle (at CSL and Verhaert) and the mechanics of the equipment bay (SONACA). Belgium is supporting many students and post-doc researchers. So thanks to Arlette, the Belgium community is very active on the scientific interpretation of the CoRoT data.

For theoreticians like Arlette, nature is sometimes unfair! *"The terrible tragedy of science is the horrible murder of beautiful theories by ugly facts"*, attributed to Willy Fowler. As we just start to have a precise look on CoRoT data after 555 days in orbit, let's ask the question: "Will CoRoT unveil ugly facts?".

Ugly facts or wonderful ones, they will question and improve our knowledge of stellar evolution. A lot of fun for Arlette's future. . .

Session 1
Physics and uncertainties in massive stars on the main
sequence and close to it

Comm. in Asteroseismology
Vol. 158, 2009, 38th LIAC/HELAS-ESTA/BAG, 2008
A. Noels, C. Aerts, J. Montalbán, A. Miglio and M. Briquet., eds.

Rotation and waves in stars

S. Talon

Département de physique et RQCHP, Université de Montréal,
C.P. 6128, succ. centre-ville, Montréal (Québec) H3C 3J7- Canada

Abstract

In this review talk, I give an overview of the main physical mechanisms responsible for mixing inside rotating stars. As all these may act concurrently inside stars, a fair understanding of each is crucial to good stellar modeling. I will also discuss the role of internal waves, which may quite efficiently modify the star's differential rotation and thus, also play a major role in rotational mixing.

Rotational mixing

The need for extra mixing in stars is now widely recognized, and different rotational histories could well explain the variety of stellar behaviors. This requires self-consistent models that take into account the angular momentum evolution from the pre-main sequence on.

Shear instability

As solid body rotation corresponds to a lowest energy state, differential rotation gives rise to a pleiades of instabilities of which shear is expected to be the most efficient. For a detailed review of other well studied instabilities, see *e.g.* Talon (2007) and references therein.

Experiments show that turbulence develops even in the absence of the inflection points required for linear instability, and so, one has to study the non-linear regime. It may be tackled based on energy consideration. To extract energy from differential rotation, one must be able to overcome the density stratification characterized by the Brunt-Vaisala frequency N. By comparing kinetic and potential energies and assuming adiabaticity, one obtains the Richardson instability criterion

$$Ri \equiv \frac{N^2}{(du/dz)^2} < Ri_{\text{crit}} = \frac{1}{4} \qquad (1)$$

(see *e.g.* Chandrasekhar 1961). This is a *non-linear* condition for instability.

Dynamical shear instability: Along isobars, there is no restoring force, and thus, shear is unstable as soon as horizontal differential rotation is present. This *dynamical* instability leads to a large horizontal turbulent viscosity that reduces the *vertical* transport of elements (this is discussed at length in Talon et al. 2006).

Secular shear instability: In the direction of entropy stratification, both the thermal and the mean molecular weight stratifications hinder the growth of the instability. However, thermal diffusion K_T and horizontal shear D_h act as to reduce the stabilizing effect of the stratification. We obtain the *instability* criterion

$$\left(\frac{\Gamma}{\Gamma+1}\right) N_T^2 + \left(\frac{\Gamma_\mu}{\Gamma_\mu+1}\right) N_\mu^2 < \frac{1}{4}\left(\frac{du}{dz}\right)^2 \tag{2}$$

for eddies of size ℓ and velocity v, where $\Gamma = v\ell/K_T$ and $\Gamma_\mu = v\ell/D_h$ (Maeder 1995, Talon & Zahn 1997): there always exists an eddy that is small enough so that the instability criterion (2) will be satisfied. Turbulent diffusivity $D = v\ell$ corresponds to the largest eddy satisfying (2); this is a non-linear description, based on energy considerations.

Meridional circulation

The Eddington-Sweet meridional circulation is related to the thermal imbalance that is generally present in rotating stars (von Zeipel 1924), and which can be compensated for by a large scale advection of entropy S

$$\rho T \vec{u} \cdot \vec{\nabla} S = \vec{\nabla} \cdot \left(\chi \vec{\nabla} T\right) + \rho\varepsilon \tag{3}$$

(Vogt 1925; Eddington 1925). $\vec{u}$, the meridional circulation velocity, is calculated with the above equality. This large scale circulation also advects angular momentum and will thus modify the rotation profile. Feedback must be treated.

Following the pioneering work of Endal & Sofia (1976), models where first built by treating the advection of momentum as a purely diffusive process. A more refined description has been made by Zahn (1992) who, under the assumption of highly anisotropic turbulence driven by a horizontal dynamical shear instability, assumed the rotation state to become "shellular" (*i.e.* $\Omega = \Omega(P)$), simplifying the evaluation of feedback.

Angular momentum transport : If the only transport processes considered are meridional circulation and vertical turbulence ν_v, the distribution of angular momentum within the star obeys an advection-diffusion equation

$$\rho\frac{d(r^2\Omega)}{dt} = \frac{1}{5r^2}\frac{\partial(\rho r^4\Omega u)}{\partial r} + \frac{1}{r^4}\frac{\partial\left(\rho\nu_v r^4\frac{\partial\Omega}{\partial r}\right)}{\partial r}. \tag{4}$$

Note that the first term of this equation corresponds to an advective process. Neglecting the star's evolution, this equation admits a stationary solution in which advection is counterbalanced by turbulent diffusion, resulting in having a core rotating about 20% faster than the surface (Urpin et al. 1996, Talon et al. 1997).

If the star is not chemically homogeneous, meridional circulation will transform the vertical chemical stratification into horizontal variations of mean molecular weight that contribute to the heat flux via the equation of state. The new equilibrium requires an increase in differential rotation (compared to the homogeneous case) in order to maintain the asymptotic solution (Talon et al. 1997, Palacios et al. 2003).

Wind-driven circulation : If the star's surface is braked via a magnetic torque, as is the case of Pop I stars cooler than about 7000 K, the internal distribution of angular momentum is rapidly moved away from its equilibrium profile, with large shears developing in the outer regions. This is the regime of "wind-driven" circulation, and the "circulation-advection" model predicts strong mixing related to this state.

Chemical transport : In the case of chemical elements, the combined effect of meridional circulation and (strong) horizontal turbulence leads to (vertical) diffusion (Chaboyer

& Zahn 1992, Charbonneau 1992). This is the case because horizontal turbulence continually reduces horizontal chemical inhomogeneities, impeding efficient advection. The advection + horizontal diffusion + vertical diffusion equation can then be replaced by a purely diffusive equation, with vertical turbulent transport and an effective diffusivity depending on horizontal turbulent transport D_h and advective velocity u

$$\rho \frac{dc}{dt} = \rho \dot{c} + \frac{1}{r^2} \frac{\partial}{\partial r} \left[r^2 \rho V_{ip} c \right] + \frac{1}{r^2} \frac{\partial}{\partial r} \left[r^2 \rho \left(D_{\text{eff}} + D_v \right) \frac{\partial c}{\partial r} \right]. \tag{5}$$

The effective diffusivity is $D_{\text{eff}} = r^2 u^2 / (30 D_h)$ (Chaboyer & Zahn 1992); $\dot{c}$ is the nuclear production/destruction rate of the species, and V_{ip} is the atomic diffusion velocity.

Open problems in rotational mixing

This model of rotational mixing based on an advective formulation of meridional circulation and shear instabilities has been applied to a wide variety of stars. In massive stars, extensive studies have been conducted by the Geneva group, with great success in explaining problems such as He and N overabundances, the number ratio of red to blue super-giants or the Wolf-Rayet to O-type stars ratio (see e.g. Meynet 2007).

In low mass stars however, rotational mixing encounters two major problems. It shapes perfectly the hot side of the lithium dip, which corresponds to stars that suffer magnetic braking (Talon & Charbonnel 1998, Palacios et al. 2003) but fails for temperatures below $T_{\text{eff}} \lesssim 6700$ K. In the case of the Sun, complete self-consistent calculations predict that radial differential rotation remains large at the solar age[1] (Talon 1997, Matias & Zahn 1998). These results indicate that there must be another efficient transport process for angular momentum active when the surface convection zone becomes deep enough, and that is the reason why angular momentum transport by internal gravity waves has been examined.

Let us also mention, as limitations to the current model, the major uncertainty on the invoked horizontal turbulent diffusion coefficient D_h. Its magnitude itself remains quite uncertain, but furthermore, it does not take into account the Rossby radius. This is the scale at which Coriolis' acceleration becomes comparable to the vertical stratification and which, in the Earth's atmosphere, strongly limits horizontal homogenization. With typical values for H_P and N, it is of order $L_{\text{Rossby}} \simeq (50R)/(v_{\text{rot}} \sin \theta)$, with v_{rot} in km s^{-1}. This limit is thus relevant for stars rotating faster than about 100 km s^{-1}.

Another issue concerns the vertical transport of chemicals in the presence of strongly anisotropic turbulence. As a turbulent eddy is displaced vertically in the fluid, it looses some of its coherence by being mixed by horizontal turbulence (Vincent et al. 1996, Toqué et al. 2006). Talon et al. (2006) showed that such erosion is essential in the context of AmFm stars. However, such a reduction would degrade the good agreement obtained in massive stars. This might be linked to the Rossby radius, that could limit erosion in fast rotators.

Internal gravity waves

In astrophysics, internal gravity waves have been invoked as a source of mixing for elements (Press 1981, García López & Spruit 1991, Schatzman 1993, Montalbán 1994, Young et al. 2003), as a means to induce episodic mass-loss in Be stars (Ando 1986, Lee 2006) and as a way to synchronize binary stars from the surface to the core (Goldreich & Nicholson 1989). More recently, they have been invoked as an important process in the redistribution of angular momentum in single stars spun down by a magnetic torque (Schatzman 1993, Kumar & Quataert 1997, Zahn et al. 1997, Talon et al. 2002).

---

[1]This conclusion is also reached by the Yale group who model circulation as a diffusive process (Pinsonneault et al. 1989, Chaboyer et al. 1995).

Excitation of internal waves

In the single star context, waves are produced by the injection of kinetic energy from a turbulent region to an adjacent stable one. This is observed in numerical simulations of penetrative convection both in 2–D and 3–D (Hurlburt et al. 1986, Andersen 1994, Nordlund et al. 1996, Kiraga et al. 2000, Dintrans et al. 2005, Rogers & Glatzmeier 2005). They are excited by two different processes, namely convective overshooting in a stable region and Reynolds stresses in the convection zone itself. Both sources contribute to the excitation. We are still awaiting reliable prescriptions for these processes since current theoretical models (García López & Spruit 1991 for overshoot and Goldreich et al. 1994 for Reynolds stresses) are not in agreement with numerical simulations which, themselves, do not quite reproduce stellar conditions properly.

Momentum transport

Waves conserve their angular momentum as long as they are not damped. The non-local nature of this transport is what makes it so efficient. In stars, the most efficient damping process is heat diffusion by photon exchange, producing an attenuation factor proportional to the thermal diffusivity K, and inversely proportional to the wave's frequency ω and horizontal[2] wavelength $\sqrt{\ell(\ell+1)}/r^2$. For a wave of frequency ω, the local momentum flux is given by

$$\mathcal{F}_J(\omega, \ell, m, r) = \mathcal{F}_J(\omega, \ell, r_{cz}) \exp\left\{ -\int_r^{r_{cz}} \frac{[\ell(\ell+1)]^{3/2} \, KN^3}{(\omega - m\Delta\Omega)^4 \, r^3} dr \right\} \tag{6}$$

where a Doppler shift on the frequency is accounted for and $\Delta\Omega = \Omega(r) - \Omega_{cz}$. Locally, the total angular momentum luminosity $\mathcal{L}_J(r)$ is the sum of the contribution of all waves. The deposition of angular momentum is then given by the radial derivative of this luminosity. The evolution of angular momentum then follows

$$\rho \frac{d}{dt}\left[r^2\Omega\right] = \pm \frac{3}{8\pi} \frac{1}{r^2} \frac{\partial}{\partial r} \mathcal{L}_J(r). \tag{7}$$

The "+" ("−") sign in front of the angular momentum luminosity corresponds to a wave traveling inward (outward). This term should be added to the contribution of other transport processes. Let us finally mention that the flux of angular momentum is given by $\mathcal{F}_J = \frac{m}{\omega}\mathcal{F}_E$, where $\mathcal{F}_E$ is the energy flux, and thus, the angular momentum flux is dominated by the low frequency waves.

Wave-mean flow interaction and the SLO : We now examine the impact of IGWs on the radial distribution of angular momentum. For waves of low frequency and moderate to high ℓ, damping is strong enough so that most waves are damped very close to the base of the convection zone. These are the waves we will examine here.

Let us assume that prograde and retrograde waves are excited with the same amplitude. In solid body rotation, these are equally dissipated when traveling inward and have no impact on the distribution of angular momentum. In the presence of differential rotation, the situation is different. If the interior is rotating faster than the convection zone, the local frequency of prograde waves diminishes, which enhances their dissipation; the corresponding retrograde waves are then dissipated further inside. This produces an increase of the local differential rotation, and creates a double-peaked shear layer. In the presence of shear turbulence, this layer oscillates (Ringot 1998, Kumar et al. 1999). Viscosity is essential in this process and is responsible for the disappearance of the outermost peak as its slope gets steeper and the

---

[2]The vertical wavelength is much larger and thus, can be neglected.

peak is slowly absorbed by the convection zone.

Secular effects: If the radiative zone has the same rotation rate as the convection zone, over a complete SLO (shear layer oscillation) cycle, the magnitude of the prograde and retrograde peaks are on average equal, and waves of azimuthal number $+m$ and $-m$ have (on average) equal amplitudes after crossing the SLO. In the presence of differential rotation however, with the inner part rotating faster than the outer part as is the case of a solar type star that is spun down by a magnetic torque, the prograde peak is always larger than the retrograde peak and there is a net flux of negative angular momentum to be redistributed within the star's radiative zone. Talon et al. (2002) showed that this can spin down the interior of a star over long time-scales. Talon & Charbonnel (2005) later showed that this *asymmetric filtering* depends only on the difference of rotation rates at the base of the convection zone and at the base of the SLO ($\delta\Omega = \Omega_{cz} - \Omega_{SLO}$). In fact, the asymmetry remains even in the absence of a SLO.

The solar rotation: This description of IGWs has been applied to the solar rotation problem. Charbonnel & Talon (2005) showed that, at the age of the Sun, differential rotation indeed becomes quite small. In this simulation, low-degree waves penetrate all the way to the core and spin it down extremely efficiently, due to its small angular momentum ($\propto r^2$). Once the core has been spun down, the damping of retrograde waves, that carry the negative angular momentum, increases locally, creating a "slowness" front that propagates in a wave-like way from the core to the surface. As further braking takes place, a second front forms and propagates outward.

The Li dip: We mentioned earlier that rotational mixing could explain the hot side of the Li dip but destroy too much lithium in cooler stars. If the process that is required to explain the Sun's rotation profile becomes efficient in the center of the Li dip, there reducing rotational mixing, we may obtain a model that is valid for *all* main sequence stars. Talon & Charbonnel (2003) showed that the net angular momentum luminosity corresponding to IGWs does have the proper temperature dependency and thus could well explain the cool side of the lithium dip. Complete evolutionary calculations are under way to verify this proposition.

Waves in massive stars

The case of IGWs in massive stars has been examined by Pantillon et al. (2007). The first difference between these and the waves in low mass stars lies in that fact that now waves are produced in the core, and travel to the surface. While this may seem rather trivial, it is the source of a major difference in behavior since thermal diffusivity rises towards the surface, and this causes total wave damping, with no reflection for the low frequency waves that transport a significant amount of angular momentum.

Another major difference is the fact that the stars under consideration rotate much faster, so that the ratio $2\Omega/\omega$ is of the order or larger than one. This leads to wave confinement towards the equator, and renders the mathematical treatment much more difficult (this is discussed by Mathis et al. 2008). It also creates an asymmetry between the prograde and the corresponding retrograde horizontal wavelengths. This implies that, even in the absence of a pre-existing differential rotation, there exists an asymmetry in damping that will create local shears. Let us also mention that, as the horizontal wavelengths are much smaller in rapidly rotating stars, there may be only a limited angular momentum flux beyond the SLO. This remains to be evaluated properly.

References

Andersen, B.N. 1994, Solar Phys., 152, 241

Ando, H. 1986, A&A, 163, 97

Chaboyer, B., Demarque, P., & Pinsonneault, M.H. 1995, ApJ, 441, 865

Chaboyer B., & Zahn J.-P. 1992, A&A, 253, 173

Chandrasekhar, S. 1961, Hydrodynamic and Hydromagnetic Stability, Clarendon Press, Oxford, p.491

Charbonneau, P. 1992, A&A, 259, 134

Charbonnel, C., & Talon, S. 2005, Science, 309, 2189

Dintrans, B., Brandenburg, A., Nordlund, Å, & Stein R.F. 2005, A&A, 438, 365

Eddington, A.S. 1925, Obs., 48, 73

Endal, A.S., & Sofia, S. 1976, ApJ, 210, 184

García López, R.J., & Spruit, H.C. 1991, ApJ, 377, 268

Goldreich, P., & Nicholson, P.D. 1989, ApJ, 342, 1079

Goldreich, P., Murray, N., & Kumar, P. 1994, ApJ, 424, 466

Hurlburt, N.E., Toomre, J., & Massaguer, J.M. 1986, ApJ, 311, 563

Kiraga, M., Różyczka, M., Stepien, K., et al. 2000, Acta Astronomica, 50, 93

Kumar, P., & Quataert, E.J. 1997, ApJ, 475, L143

Kumar, P., Talon, S., & Zahn, J.-P. 1999, ApJ, 520, 859

Lee, U. 2006, MNRAS, 365, 677

Maeder, A. 1995, A&A, 299, 84

Mathis, S., Talon, S., Pantillon, F.P., & Zahn, J.-P. 2008, Solar Phys., 251, 101

Matias, J., & Zahn, J.-P. 1998, Sounding solar and stellar interiors, IAU Symposium 181, Nice, Eds. J. Provost & F.X. Schmider

Meynet, G. 2007, Stellar Nucleosynthesis: 50 years after BBFH, Eds. C. Charbonnel, J.-P. Zahn

Montalbán, J. 1994, A&A, 281, 421

Nordlund, A., Stein, R.F., & Brandenburg, A. 1996, Bull. Astron. Soc. of India, 24, 261

Palacios, A., Talon, S., Charbonnel, C., & Forestini, M. 2003, A&A, 399, 603

Pantillon, F.P., Talon, S., Charbonnel, C. 2007, A&A, 474, 155

Pinsonneault, M.H., Kawaler, S.D., Sofia, S., & Demarque, P. 1989, ApJ, 338, 424

Press, W.H. 1981, ApJ, 245, 286

Ringot, O. 1998, A&A, 335, 89

Rogers, T.M., & Glatzmeier, G.A. 2005, MNRAS, 364, 1135

Schatzman, E. 1993, A&A, 279, 431

Talon, S. 1997, "Hydrodynamique des étoiles en rotation", PhD Thesis, Université Paris VII

Talon, S. 2007, Stellar Nucleosynthesis: 50 years after BBFH, Eds. C. Charbonnel, & J.-P. Zahn

Talon, S., & Charbonnel, C. 1998, A&A, 335, 959

Talon, S., & Charbonnel, C. 2003, A&A, 405, 1025

Talon, S., & Charbonnel, C. 2005, A&A, 440, 981

Talon, S., Kumar, P., & Zahn, J.-P. 2002, ApJL, 574, 175

Talon, S., Richard, O., & Michaud, G. 2006, ApJ, 645, 634

Talon, S., & Zahn, J.-P. 1997, A&A, 317, 749

Talon, S., Zahn, J.-P., Maeder, A., & Meynet, G. 1997, A&A, 322, 209

Toqué, N., Lignières, F., & Vincent, A. 2006, GAFD, 100, 85

Urpin, V.A., Shalybkov, D.A., & Spruit, H.C. 1996, A&A, 306, 455

Vincent, A., Michaud, G., & Meneguzzi, M. 1996, Phys. Fluids, 8 (5) 1312

Vogt, H. 1925, Astron. Nachr., 223, 229

von Zeipel, H. 1924, MNRAS, 84, 665

Young, P.A., Knierman, K.A., Rigby, J.R., & Arnett, D. 2003, ApJ, 595, 1114

Zahn, J.-P. 1992, A&A, 265, 115

Zahn, J.-P., Talon, S., & Matias, J. 1997, A&A, 322, 320

DISCUSSION

Shibahashi: The movie you showed us concerning the angular momentum transfer indicates some eruptive events in addition to continuous transfer. Could you explain the physical reason of such eruptive events ?

Talon: The extraction of angular momentum from the core occurs through a front of "slowness that travels" from the center to the base of the convection zone. The magnitude of this front is proportional to surface extraction that is much stronger at the beginning. The reason for the creation of the front itself is that the core contains little angular momentum and so is most easily spun down. The following front propagation is related to a reduced local frequency (due to the Doppler shift).

Maeder: What damping of the gravity waves are you expecting from the horizontal turbulence ?

Talon: We expect it to be much smaller than thermal damping because it would act on the horizontal gradients which are much smoother than the vertical ones on which thermal diffusivity acts. In fact, vertical turbulence also acts on the vertical gradients, but it is weaker than K_T, and thus has a negligible impact on wave damping. Let me recall that we have $K_r^2 = (N^2/\omega^2 - 1)K_h^2$.

Comm. in Asteroseismology
Vol. 158, 2009, 38th LIAC/HELAS-ESTA/BAG, 2008
A. Noels, C. Aerts, J. Montalbán, A. Miglio and M. Briquet., eds.

The impact of magnetic fields on stellar structure and evolution

J.-P. Zahn

LUTH, Observatoire de Paris, 92195 Meudon, France

Abstract

Magnetic fields responsible for various activity phenomena have been detected at the surface of many stars. Presumably they are also present in the stars' deep interior. They channel stellar winds to large distance, thus increasing the loss of angular momentum. In stably stratified radiation zones, they tend to impose uniform rotation, and may thus interfere with the rotational mixing operating in these regions and influence the evolution of the host star. Moreover, certain magnetic configurations are known to be unstable, and they could produce some mixing of their own. We shall discuss these problems in the light of the latest developments.

Introduction

Let us briefly summarize what we know about the magnetism of stars. Magnetic fields are observed - more or less directly - at the surface of all solar-like low-mass stars, namely stars with a thick outer convection zone. These fields are variable on the scale of months or years, and they are highly structured; most probably they are generated through dynamo action driven by the convective motions. In contrast, among more massive stars that are deprived from such convection zones, only a small fraction, less than 5% according to Power et al. (2008), are hosting magnetic fields: these are the so-called Ap and Bp stars. These fields display much simpler topologies, and they seem unchanged over long time-scales (when compared to human life); that is why they are believed to be of fossil origin (Wade et al. 2007).

Since they are rooted in radiation zones, fossil fields evolve on a considerably longer time-scale than dynamo fields; the Ohmic decay time for a dipolar field is given by $R^2/2\pi\eta$, R being the radius and η the magnetic diffusivity. This time amounts to 10 Gyr for the Sun (Cowling 1957), and it scales with mass M and radius roughly as $M^{3/2}/R^{1/2}$. Therefore these fields are extremely long-lived, even if they are of high spherical degree. Unless they are destroyed by some instability, a possibility that we shall examine later on.

Impact of magnetic field on stellar structure

In order to have an impact on the overall structure of a star, the magnetic force, often called Lorentz force, ought to be able to compete with the two main forces that rule its hydrostatic equilibrium, namely gravity and pressure force. This would occur if the Alfvén velocity $V_A = B/\sqrt{4\pi\rho}$ equaled the sound speed; in the Sun that would require the field to be of order $2\,10^8$ G, which is considerably stronger than any field observed on the surface of main sequence stars.

However, one cannot rule out that such extremely strong fields are hiding in the deep interior of some stars. Presumably the only way to detect them would be through asteroseismology. The frequencies of the oscillation eigenmodes - either p or g-modes, depending on the restoring force (pressure or buoyancy) - are split and displaced by a magnetic field, much like what occurs under the effect of rotation, and this may be used to estimate the magnetic field. In massive stars, where only few eigenmodes are observed, moreover of low spherical degree, the task is rather difficult, except perhaps for the slowly pulsating B stars, where high order g-modes could bear the signature of a 100 kG field located near the boundary of the convective core (Hasan et al. 2005). The effect of a dipole field on the p-modes in a rotating star was considered in full detail by Bigot et al. (2000), with application to the rapidly rotating Ap stars.

Strong magnetic fields can suppress thermal convection

It is well-known that thermal convection can generate magnetic fields, but magnetic fields can also suppress the convective instability. This can be seen by examining the stability of a stratified fluid in presence of magnetic field. A perturbation described by its displacement $\vec{\xi} \propto \exp[st + i\vec{k} \cdot \vec{x}]$ has a growth-rate s that obeys the following dispersion relation, where we neglect all kinds of dissipation (thermal, Ohmic, viscous, in decreasing importance):

$$s^2 = \left(\frac{k_h}{k}\right)^2 N_t^2 - (\vec{k} \cdot \vec{V}_A)^2, \qquad N_t^2 = \left[\frac{g}{H_P}(\nabla - \nabla_{ad})\right], \qquad (1)$$

N_t being the buoyancy frequency. Here g is the local gravity, H_P the pressure scale height, $\vec{k}$ the wavenumber, k_h its horizontal component, and $\nabla = \partial \ln T / \partial \ln P$ designates as usual the logarithmic temperature gradient.

At first sight, it would seem that, to suppress convection, it requires a field strength of

$$\frac{B^2}{4\pi\rho} = V_A^2 \gtrsim N_t^2 R^2 \qquad (2)$$

according to (1), which translates into 10^7 G at the base of the solar convection zone, or 10^3 G near the surface. However that threshold is lowered by two orders of magnitude when diffusion is taken into account; the stability condition then becomes (Chandrasekhar 1960)

$$\frac{B^2}{4\pi\rho} = V_A^2 \gtrsim \frac{\eta}{K} N_t^2 R^2, \qquad (3)$$

with η and K being the Ohmic and thermal diffusivities ($\eta/K \approx 10^{-4}$ in the solar interior). The instability is of double-diffusive type, with heat diffusing much faster than the magnetic field.

In the sunspots we see a striking proof of that inhibition of thermal convection, by a field of kilogauss strength. That is also why we observe surface inhomogeneities in the Ap-Bp stars: due to the presence of magnetic field, there is no convection to smooth out such inhomogeneities, as in normal stars where convection exists in the thin superadiabatic layers due to the ionization of hydrogen and helium.

Magnetic fields amplify the loss of angular momentum by stellar winds

A well-established fact is that solar-like stars spin down as they evolve. At the ZAMS, some of such stars have equatorial velocities that reach 100 km/s, while that of the Sun is only 2 km/s. The only way stars can achieve this is by losing mass, through a wind whose mechanism has been elucidated by Parker (1958). In the absence of magnetic field, this is not a very efficient process, but the picture changes when one includes the magnetic field, as was pointed out by Schatzman (1962). The wind is then forced to rotate with the star up to a distance where the wind speed exceeds the Alfvén velocity: the lever arm is then no longer the radius R, but the so-called Alfvén radius, which in the present Sun is about $R_A \approx 15\,R$.

This mechanism also works for massive stars, where the magnetic field is likewise shaped by the wind, which is there driven by radiation; its dynamical behavior has been studied in detail in a series of papers by A. ud-Doula and S. Owocki (ud-Doula & Owocki 2002; Owocki & ud-Doula 2004; ud-Doula et al. 2006, 2008). They introduce a 'wind magnetic confinement parameter' $\eta* = B_{eq}^2/\dot{M}v_\infty$, which measures the ratio between magnetic field energy density (at the equator) and kinetic energy density of the wind (v_∞ is its terminal speed and $\dot{M}$ the mass loss rate). They show that for $\eta^* \gg$ the latitude dependent Alfvén radius $R_A(\theta)$ scales with the confinement parameter according to a rather simple law:

$$\left[\frac{R_A(\theta)}{R}\right]^{2q-2} - \left[\frac{R_A(\theta)}{R}\right]^{2q-3} = \eta^*[4 - 3\sin^2(\theta)], \qquad (4)$$

where θ is the colatitude, and q the magnetic exponent. For a dipole field $q = 3$ and therefore $R_A \propto \eta^{*1/4}$.

Magnetic fields inhibit rotational mixing

In the absence of magnetic field, the radiation zone of a rotating star undergoes mild mixing through internal motions - a combination of turbulence and large-scale flows - that are due to several causes. These are mainly the loss of angular momentum by a wind, eventually the coupling with an accretion disk, and the structural adjustments the star undergoes as it evolves (such as the moderate core contraction and envelope expansion on the main sequence).

Let us examine what occurs in the presence of magnetic field. One effect of such a field is the enhancement of the angular momentum loss, speeding up the rotational mixing. But the field also acts in the deep interior. When it is axisymmetric with respect to the rotation axis, it tends to render the rotation uniform along the field lines of the meridional field (Ferraro 1937). To estimate the strength of the axisymmetric field above which that rotational mixing is inhibited, we refer to the angular momentum transport equation, which has been horizontally averaged; we neglect the turbulent transport but include the torque exerted by the Lorentz force $\vec{j} \times \vec{B}/c$:

$$\rho\frac{d}{dt}\left(r^2\overline{\Omega}\right) = \frac{1}{5r^2}\partial_r\left(\rho r^4\overline{\Omega}U_2\right) + \overline{\vec{B}_p \cdot \vec{\nabla}(rB_\varphi/4\pi)}, \qquad (5)$$

with $\vec{B}_p$ and B_φ being the meridional and azimuthal components of the magnetic field. Note that it is only the leading term of the expansion in spherical harmonics of the vertical component of the circulation velocity, i.e. $u_r(r, \theta) = U_2(r)P_2(\cos\theta)$, that participates in the advection of angular momentum. Introducing the characteristic time t_{AM} for the angular momentum evolution, we see that the Lorentz torque balances the l.h.s. when

$$B_p B_\varphi \gtrsim B_{crit}^2 = 4\pi\rho\,\frac{R^2\Omega}{t_{AM}}. \qquad (6)$$

If we apply this result to a 15 $M_\odot$ star, taking $\rho = 0.07$ g/cm^3, $R = 4.5\,10^{11}$ cm, $R\,\Omega = 3\,10^7$ cm/s and setting somewhat arbitrarily $t_{AM} = 10^6$ yr, we find that the critical field strength for suppressing rotational mixing is of the order of $B_{crit} \approx 200$ Gauss. This crude estimate needs to be confirmed by more thorough calculations, for which the formalism has been derived by Mathis and Zahn (2005).

When the field is non-axisymmetric, as is often observed, uniform rotation tends to be enforced everywhere in the radiation zone. Little more is known about this oblique rotator model, as can be seen in Mestel (1999), and this is clearly a promising field for future research.

MHD instabilities in stellar radiation zones

Magnetic fields are liable to various instabilities, which could perhaps lead to mixing and may therefore have an impact on stellar evolution. These instabilities have been discussed by Spruit (1999) in a comprehensive review. He concluded that the instabilities most likely to play a role in stellar radiation zones are those studied by Tayler and his collaborators in the 1970's. These apply to axisymmetric fields and are linked to the classical pinch and kink instabilities.

Describing the fully non-linear regime of the instability turns out to be a difficult task. Spruit (2002) suggested that the instability should saturate when the turbulent magnetic diffusivity η_t, that supposedly is accompanying it, yields a zero growth rate. We had an opportunity to verify his prediction by examining the results of actual non-linear simulations performed with the 3-dimensional ASH code (the code is described in Clune et al. 1999, and Brun et al. 2004). The purpose of this calculation was to study the interaction of a fossil magnetic field with the solar tachocline (Brun & Zahn 2006), in order to check whether such a field could prevent this transition layer, between differential rotation in the convection zone and uniform rotation below, from expanding into the deep interior, as had been suggested by Gough and McIntyre (1998).

We started the simulations with a deeply buried poloidal field of dipole type and a uniformly rotating radiation zone, on the top of which we imposed the differential rotation of the convection zone. As time proceeds, the poloidal field diffuses upward and the differential rotation diffuses downward; their interaction produces a toroidal field which is antisymmetric with respect to the equatorial plane. Once the poloidal field reaches the convection zone, it imprints the differential rotation of that region on the radiative interior below (Ferraro's law). Since this is not observed, we concluded that such a fossil poloidal field as postulated by Gough and McIntyre does not exist in the Sun.

The bonus of carrying out these simulations with a 3-dimensional code was to capture the non-axisymmetric instabilities that affect the large scale magnetic field. We observed indeed the instabilities that had been described by Tayler and his collaborators. The first to appear is that of the initial poloidal field, with an azimuthal number $m \approx 40$. It is followed by that of the toroidal field, once that field had been generated through shearing the poloidal field by the differential rotation. The latter instability, of azimuthal order $m = 1$, is clearly the one studied by Pitts and Tayler (1985), although in this case it occurs in the presence of both the toroidal and poloidal field, a configuration deemed to be more stable than that of just a toroidal field.

This instability saturates when its energy reaches that of the fossil field, but it is not clear whether this is just a coincidence (see Fig. 1). An unexpected result was to find that the Ohmic decline of the poloidal field is not accelerated by the instability. The 3-dimensional perturbations associated with this instability behave as Alfvén waves, rather than as turbulence, and they do not produce the turbulent diffusion invoked by Spruit. We conclude therefore that they cannot either achieve any mixing of the stellar material.

Spruit made another interesting conjecture, namely that the instability could regenerate the toroidal field which triggers it, and would thus operate a dynamo, much as that driven by the convective motions in the convection zones. However the dynamo loop cannot work as he described it: it requires some type of α-effect to generate a mean electromotive force, as is well-known in dynamo theory (Parker 1955). As in the case of the Sun, the only way to check whether this α-effect is actually present is through numerical simulations.

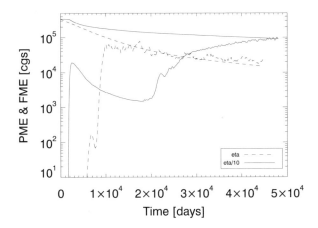

Figure 1: Temporal evolution of the energies of the mean poloidal field and of the instability field; these energies are summed over the computational domain. Note that the Ohmic decline of the poloidal field is not affected by the growth of the instability field, in spite of the fact that their energies become comparable: it demonstrates that the instability does not produce turbulent diffusion (courtesy A. S. Brun).

But so far these simulations have given conflicting results: Braithwaite (2006) claims to observe dynamo action, with field reversals, contrary to us, who don't see any regeneration of either the poloidal or the toroidal field (Zahn et al. 2007). The main difference between our simulations lies in the way the equations are solved. Our code is of pseudo-spectral type and this method achieves exponential convergence and machine accuracy in evaluating derivatives, which allows us to reach, with enhanced diffusivities, a magnetic Reynolds number of 10^5 for a resolution of 128x256x192. That should be more than sufficient to detect a dynamo, if it exists at all. Braithwaite uses instead a 6th order finite difference scheme, with a resolution of 64x64x33; his numerical diffusivity is tuned to ensure stability and it is not easy to infer from it the effective Reynolds number. Moreover, it appears that his simulations have not been run long enough beyond the transient phase to establish without any doubt that a dynamo is at work. Clearly further simulations are needed to settle this issue.

Conclusion and perspectives

Despite all the work that has been accomplished since the 1970's, notably by Leon Mestel, Roger Tayler and their collaborators, we still do not fully understand the origin of magnetism in massive stars and the role it plays in their evolution. But let me summarize the few results that are now firmly established.

Only a minority of massive stars host magnetic fields - less than 5% according to the latest survey. In those stars, the magnetic field is probably of fossil origin, and it is deeply rooted in the radiative interior. We don't know why magnetic stars are so few: presumably all stars initially had a fossil field, which they gathered from the interstellar medium during the process of formation, but it must have been destroyed later in most of them.

It has been known since the 1960's that magnetic fields enhance the loss of angular momentum through stellar winds, which explains why most magnetic stars are slow rotators. One would thus be tempted to conclude that rotational mixing is enhanced by this angular momentum loss; but it turns out on the contrary that the internal motions - either large scale circulations or turbulence - are frozen by the interior field. The field itself can become unstable, but the instability saturates in a regime that is wave-like and does not lead to turbulent diffusion. It is thus expected that magnetic stars will not show signs of internal mixing - we expect that soon observations will tell whether this prediction is correct.

The study of stellar magnetism is still in its infancy, but the situation is improving rapidly. The new generation of extremely powerful spectro-polarimeters is delivering much wanted observational constraints of high quality, while the steadily progressing computer resources allow numerical simulations with an increasing degree of realism. There will be much to discover and to explain for the young astrophysicists entering this exciting field!

References

Bigot, L., Provost, J., Berthomieu, G., et al. 2000, A&A, 356, 218

Braithwaite, J. 2006, A&A, 449, 451

Braithwaite, J., & Nordlund, A. 2006, A&A, 450, 1077

Brun, A. S., Miesch, M. S., & Toomre, J. 2004, ApJ, 614, 1073

Brun, A. S., & Zahn, J.-P. 2006, A&A, 457, 665

Chandrasekhar, S. 1961, Hydrodynamic and Hydromagnetic Stability (Oxford)

Clune, T. L., Elliott, J. R., Glatzmaier, G. A., et al. 1999, Parallel Comput., 25, 361

Cowling, T. G. 1957, Magnetohydrodynamics (Interscience, New York)

Decressin, T., Mathis, S., Palacios, et al. 2008, submitted to A&A

Ferraro, V. C. A. 1937, MNRAS, 97, 458

Gough, D.O., & McIntyre, M.E. 1998, Nature, 394, 755

Hasan, S. S., Zahn, J.-P., & Christensen-Dalsgaard, J. 2005, A&A, 444, L29

Mathis, S., & Zahn, J.-P. 2004, A&A, 425, 229

Mathis, S., & Zahn, J.-P. 2005, A&A, 440, 653

Mestel, L., 1999, Stellar Magnetism (Oxford Univ. Press)

Owocki, S., & ud-Doula, A. 2004, ApJ, 600, 1004

Parker, E. N. 1955, ApJ, 122, 293

Parker, E. N. 1958, ApJ, 128, 664

Pitts, E., & Tayler, R. J. 1985, MNRAS, 216, 139

Power, J., Wade, G. A., Aurière, M., et al. 2008, Contr. Astron. Obs. Skalnaté Pleso, 38, 44

Schatzman, E. 1962, Annales Ap, 25, 18

Spruit, H. C. 1999, A&A, 349, 189

Spruit, H. C. 2002, A&A, 381, 923

Talon, S., Zahn, J.-P., Maeder, A., & Meynet, G. 1997, A&A, 322, 209

ud-Doula, A., & Owocki, S. 2002, ApJ, 576, 413

ud-Doula, A., Owocki, S., & Townsend, H. D. 2008, MNRAS, 385, 97

ud-Doula, A., Townsend, H. D., & Owocki, S. 2006, ApJ, 640, L191

Wade, G. A., Silvester, J., Bale, K., et al. 2007, arXiv:0712.3614

Zahn, J.-P., Brun, A. S., & Mathis, S. 2007, A&A, 474, 145

DISCUSSION

Noels: While convection can create a magnetic field through a dynamo effect, a fossil magnetic field can freeze the convection in the outer layers of roAp stars, which is needed to excite pulsations. However some stars of about the same characteristics are stable, which could mean that convection is not frozen. What would be the general picture for A stars ?

Zahn: A horizontal field of equipartition strength is able to inhibit convection. In Ap stars this probably occurs in the upper convection zone.

Cox: What is a "good" overshoot for solar oscillations ?

Zahn: Implementing a smooth profile for the subadiabatic gradient below the solar convection zone somewhat improves the agreement with helioseismic data. But the main cause for the observed discrepancy is probably due to opacity, and thus to the assumed chemical composition.

Vauclair: Would you think that the same process could account for both the solid rotation in the solar interior and the confinement of the tachocline ?

Zahn: It is not clear which effect internal waves have on differential rotation, hence on the tachocline, since their present treatment involves horizontal averaging (to render the problem 1 dimensional).

Vauclair: Would you more favor the magnetic field and/or internal waves and/or something else for these two effects ?

Zahn: An explanation I find very promising has been proposed by E. Forgacs-Dajka and K. Petrovay. The cyclic poloidal dynamo field diffuses into the tachocline, thoroidal field is generated by the differential rotation, and the Laplace torque acts to reduce the differential rotation. The thickness of the tachocline is then determined by the turbulent magnetic diffusivity : $\eta \sim 10^{10}$ cm^2/s is compatible with the observed value.

Comm. in Asteroseismology
Vol. 158, 2009, 38th LIAC/HELAS-ESTA/BAG, 2008
A. Noels, C. Aerts, J. Montalbán, A. Miglio and M. Briquet., eds.

Gravitational settling and radiative accelerations

G. Alecian[1]

[1] LUTH, Observatoire de Paris, CNRS, Université Paris Diderot, 5 Place Jules Janssen, 92190 Meudon, France

Abstract

Atomic diffusion in a star modifies the concentration of elements in layers where mixing motions are weak or inexistent. How much local abundances could depart from what they would be if the star were entirely mixed (outside the core) will depend on the balance between gravitational and radiative accelerations. The latter are far from negligible as long as elements are not fully ionized. While gravity varies smoothly with radius inside the star, radiative accelerations have complex behavior and wide variations (roughly from 10^{-1} to 10^3 times the gravity) in a large portion of the star. According to their type and evolutionary stage, many stars may be significantly affected by atomic diffusion, through abnormal superficial composition and/or structural changes.

Introduction

Atomic diffusion is a slow microscopic process. Its effects in stars are often erased by macroscopic motions such as turbulence or large scale circulations that efficiently mix the stellar matter. However, this is not always the case. The most well-known situation is the one of Ap stars that present huge abundance anomalies for many metals. Since the pioneer work of Michaud (1970), these anomalies are understood to be caused by atomic diffusion in stable atmospheres[1]. Further studies have shown that most of the main sequence Chemically Peculiar stars (CP) are explained by atomic diffusion[2]. For instance, computations show that mercury can be enhanced by a factor of about 100 (in the atmosphere of a main sequence star with $T_{eff} = 10\,000$ K) in about 10 years, and iron by a factor of about 20 in about 1 century. Since computation of time-dependent diffusion cannot yet be carried out for atmospheres[3], these estimations have been made using a kinematic linear approximation from data provided by the numerical code of Alecian & Stift 2006. Diffusion time-scales are much larger in the interior of the star: iron stratification needs more than 10^6 years in layers around $T = 2.10^5$K. But even if it is a slow process, several numerical models have shown that enhancement of metals is compatible with the evolution of slowly rotating stars (see Turcotte & Richard 2003). Atomic diffusion may also act in other types of stars (even in solar type stars), but with less evident observational traces on abundances, since the effects are located in deeper layers than in CP stars (Turcotte et al. 1998). These effects might have consequences, for instance, on the evolution of those stars, or on their pulsations (see also Bourge & Alecian 2006a, Bourge et al. 2006b for β Cephei stars).

---

[1] Ap stars have stable atmospheres, because of their slow rotation, strong magnetic fields or tidal effect due to binarity.

[2] Also some White Dwarfs and Horizontal-Branch stars

[3] In optically thin medium, these computations are extremely heavy.

The diffusion velocity

At first approximation, the leading force in the diffusion process is often due to the momentum transfered from the radiation field to atoms during photoabsorption. This transfer of momentum gives rise to the radiative acceleration which acts in the direction opposite to that of gravity. A direct comparison between these two accelerations allows a first estimate of the diffusion velocity of ions, using to the usual approximate formula (a more elaborated expression can be found in Michaud & Proffitt 1993):

$$V_{Di} \approx - \left(D_{ip} + D_{turb} \right) \, \partial_r \ln \frac{n_i}{n_p} + D_{ip} \left[A_i \frac{m_p}{kT} \left(g_i^{rad} - g \right) \quad + \quad ... \right]. \tag{1}$$

This is the relative velocity of a trace ion (with mass $A_i m_p$, number density n_i) with respect to protons (n_p). D_{ip} and D_{turb} are respectively the ion-proton and turbulent diffusion coefficients, g_i^{rad} the radiative acceleration, and g the gravity. The first term corresponds to the pure diffusion, plus a diffusive term introduced by Schatzman (1969) to model the effect of turbulence. The second term (involving gravity and radiative acceleration) dominates the first one in radiative zones and outside stratification edges. For the total diffusion velocity of a given element, see for instance Alecian & Stift (2006).

Effect of mass-loss

In the previous section, one sees that the mixing process due to turbulence could be included in Eq.(1). But, the diffusion process is also sensitive to organized large scale motions such as meridional circulation and matter flow due to stellar mass loss (Michaud & Charland 1986, Alecian 1996). The latter can be modeled in spherical symmetry as a wind throughout the whole star, with velocity which may be easily estimated from the continuity equation. In the particular case of a given element A, the buildup of the stratification is then described using the equation of conservation of particles for the trace element A :

$$\partial_t n_A + \partial_z \left[n_A V_{D,A} \left(n_A, z \right) + n_A V_w \left(z \right) \right] = 0 \tag{2}$$

where n_A is the number density of A, $V_{D,A}$ its diffusion velocity, and V_w the wind velocity (the same for all particles). The effect of a wind can be drastic. For instance, a numerical solution of Eq.(2) for Am stars (Alecian, 1996) showed that superficial Ca can become clearly underabundant for an age $> 10^5$ years if the mass loss rate is smaller than $1.\,10^{-14}\,M_\odot$ per year, but significantly overabundant for a mass loss of $2.\,10^{-14}\,M_\odot$ per year, and for an age $> 10^6$ years.

Radiative acceleration vs. gravity

Computation of accurate radiative accelerations is often complicated by the huge amount of atomic data to handle, and because of specific physical complexities (Montmerle & Michaud 1976, Gonzalez et al. 1992, Alecian & Stift 2006) such as the momentum redistribution between ions of a given element, or the effects of magnetic fields. We will not discuss these aspects here; let us just remark that radiative accelerations are generally easier to evaluate in stellar interiors (optically thick) than in stellar atmospheres. A short review of the methods used to compute radiative accelerations in stellar interiors can be found in Alecian (2007).

The key points to understand the role of radiative accelerations are the following. For main sequence stars, gravity is about 10^4 cm sec^{-2} and increases slowly and monotonically toward the center of the star, while radiative accelerations have a wide variation domain (typically between 0 and 10^7) according to the depth, the local concentration of the element under consideration, and from one ion to the other. Therefore, the dominant term ($g_i^{rad} - g$)

in Eq.(1) may vary from -10^4 to $+10^7$, and then the diffusion velocity will vary in the same proportions! Another important point to emphasize is that there are very few circumstances when radiative acceleration is close to zero. The main one is when the medium is completely ionized, and this occurs very deep in the star ($T > 3.10^6 K$) for iron which is an important contributor to the opacity. Another circumstance is when an element is mainly in an ionization state with noble gas configuration, but apart for helium, this affects generally a limited zone in the star.

The case of a 7 $M_\odot$ main sequence star

To illustrate the effect of atomic diffusion for more massive stars than usual CP stars, we present in Fig.1 radiative accelerations obtained using, as input, a model for a 7 $M_\odot$ main sequence star ($T_{eff} = 19\,310$ K). This model was obtained by letting the star evolve until the age of $3.1\,10^7$ years, without taking into account the diffusion processes.

Radiative accelerations have been calculated with the SVP method (LeBlanc & Alecian 2004). One can notice that radiative acceleration of Fe exceeds gravity by more than a factor 100 around log T $\approx$ 5.4. This implies that strong accumulation of Fe could appear around that temperature, provided that atomic diffusion has enough time to proceed and if mixing motions are not too strong. At this stage, one cannot assert that this accumulation will really occur or not without carrying out more sophisticate self-consistent evolutionary calculations including large-scale circulations, mass loss and atomic diffusion. However, we have estimated the diffusion time-scale of iron to check whether, in the absence of mixing, such an accumulation is at least compatible with the age of this star. The diffusion time-scale has been obtained by calculating the time needed for iron atoms to go through a distance of one pressure scale height with the velocity given by Eq.(1). The diffusion time-scale presented in Fig.2 shows that accumulation of Fe can appear in about 10^5 years in layers with temperature around log T $\approx$ 5.4 (nearly at the position of the so-called iron-bump of opacity). This is significantly shorter than the age of the star. Note also that this diffusion time-scale is 100 times shorter than the time-scale for gravitational settling[4], which shows the importance of radiative acceleration.

Conclusion

We have recalled some basic aspects of atomic diffusion in stars and shown that accurate modeling must take it into account. Atomic diffusion is a physical process which exists in any non-uniform multicomponent gas. Because it is a slow process, its effects are often erased by mixing motions, but the latter are not always and not everywhere efficient enough in a star to ensure that atomic diffusion can be neglected. Beside the well known case of CP stars, where atomic diffusion is clearly identified as the process responsible for the very strong superficial abundance anomalies, models show that it can also affect evolution of other types of stars.

Several evolution codes include gravitational settling. However, we stress that to neglect radiative acceleration is not justified, as far as the medium is partially ionized and especially when iron is the dominant source of opacity. We can even assert that considering gravitational settling alone for metals should lead to wrong results in modeling stellar evolution. Computation of radiative accelerations is often very heavy and not easy to implement in existing codes, but this is not out of reach, especially when using the SVP approximation. This approximation has been implemented in CLES (*Code Liégeois d'Évolution Stellaire*) and TGEC (*Toulouse Geneva Evolution Code*) and those parts of the codes are currently under test. To be self-consistent, such computations also need codes that take into account the local change of opacity according to the local change in chemical composition. Of course,

---

[4] The diffusion time-scale for gravitational settling has been obtained by imposing $g_i^{rad} = 0$ in Eq.(1)

including atomic diffusion is not enough to model element stratifications in stars since it is generally in competition with large-scale motions, such as those induced by rotation or convection. To evaluate quantitatively the role of atomic diffusion, one needs to compare all these processes within a self-consistent modeling.

Acknowledgments. The Helio- and Asteroseismology Network (HELAS), financed by the European Commission, is acknowledged for financial support of accommodation and registration fee.

References

Alecian, G. 1996, A&A, 310, 872

Alecian, G. 2007, Stellar Evolution and Seismic Tools for Asteroseismology: Diffusive Processes in Stars and Seismic Analysis, Porto, November 20-23, 2006, Joint HELAS and CoRoT/ESTA Workshop, EAS Publications Series by EDP Science, Ed.C.W. Straka, Y. Lebreton and M. J.P.F.G. Monteiro., 26, 37

Alecian, G., & Stift, M.J. 2006, A&A, 454, 571

Bourge, P.-O., & Alecian, G. 2006a, ASP Conf. Ser. 349, 201

Bourge, P.-O., Alecian, G., Thoul, A., et al. 2006b, CoAst 147, 105

Gonzalez, J.-F., LeBlanc, F., Artru, M.-C., & Michaud, G. 1995, A&A, 297, 223

LeBlanc, F., Alecian, G. 2004, MNRAS, 352, 1329

Michaud, G. & Charland, Y. 1986, ApJ, 311, 326

Michaud, G. & Proffitt, C. R. 1993, in Inside the Stars, IAU COLLOQUIUM 137, Vienna, April 1992, ASP Conference Series, 40, ed. W. W. Weiss & A. Baglin (San Francisco: ASP), 246

Michaud, G. 1970, ApJ, 160, 641

Montmerle, T., & Michaud, G. 1976, ApJS, 31, 489

Schatzman, E. 1969, A&A, 3, 331

Turcotte, S., & Richard, O. 2003, Ap&SS, 284, Issue 1, 225

Turcotte, S., Richer, J., & Michaud, G. 1998, ApJ, 504, 559

DISCUSSION

Cox: I am wondering in more detail just how important the element settling and levitation is for real stars ?

Alecian: Atomic diffusion (due to gravitational settling and radiation accelerations) is known to modify superficial abundances of slowly rotating or/and magnetic stars (the well-known chemically peculiar stars) around 2 to 5 $M_\odot$. Atomic diffusion may also affect the structure of other stars at some stage of their evolution (in less spectacular way). But this still needs to be investigated.

Grevesse: How does one explain the "spots" predicted in the A stars by Oleg Kochukov ?

Alecian: Abundance patches (or spots) and rings have been observed since several decades in magnetic Ap stars. They are better mapped now. Atomic diffusion brings a natural explanation to those superficial structures since diffusion velocity is extremely sensitive to the orientation (and intensity) of the magnetic field.

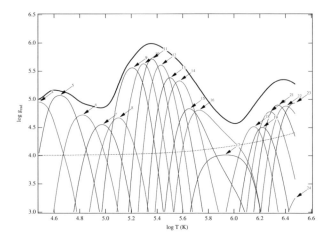

Figure 1: Radiative acceleration of Fe for a 7 $M_\odot$ main-sequence star. Arrows with the ion number indicate the detailed contribution of each ion to the total acceleration of iron (black heavy curve), log(accel. in cm sec^{-2}) vs. log(temperature in K). The dotted line is the gravity.

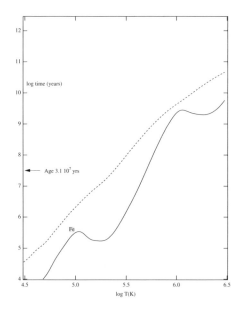

Figure 2: Diffusion time-scale of Fe inside a star (for the same stellar model as Fig.1), log(time in years) vs. log(temperature in K). The arrow indicates the age of the star for this model. The dashed line indicates the diffusion time-scale for gravitational settling.

Comm. in Asteroseismology
Vol. 158, 2009, 38th LIAC/HELAS-ESTA/BAG, 2008
A. Noels, C. Aerts, J. Montalbán, A. Miglio and M. Briquet., eds.

Semiconvection

H. Shibahashi

Department of Astronomy, University of Tokyo, Tokyo 113-0033, Japan

Abstract

During the main sequence evolution of a star more massive than $10\,M_\odot$, a superadiabatic but still convectively stable zone appears just outside the convective core. This is mainly because the adiabatic temperature gradient is low due to high radiation pressure in such massive stars. I will review the treatments of such zones. I will point out that such zones appear in the atmospheres of white dwarfs in a certain effective temperature range, and predict a possible instability of white dwarfs having such a superadiabatic but convectively stable layer.

Introduction

Massive stars during the main sequence stage produce energy by the CNO cycles, of which temperature dependence is so high that the energy generation rate increases sharply near the stellar center and drops rapidly outward as temperature decreases. As a consequence the temperature gradient required for radiative energy transport, $\nabla_{rad} \equiv (d\ln T/d\ln p)_{rad} = 3p\kappa L_r/(16\pi acGT^4 M_r)$, in the nuclear burning core becomes steeper than the 'adiabatic temperature gradient', $\nabla_{ad} \equiv (\partial\ln T/\partial\ln p)_S$, and then the nuclear burning core of these stars is fully convective. T, p, S, κ, M_r and L_r are the temperature, pressure, entropy, opacity, mass and luminosity at the considered radius, G is the gravitational constant, a is the radiation density constant, and c is the speed of light. As hydrogen is depleted in the central region, the convective core recedes with time and a shell with varying chemical composition, which we call hereafter the μ-gradient zone, is left between the outer edge of the present convective core and that of the convective core formed at the initial main sequence state. Energy is transported by radiation in the μ-gradient zone. In fact, the temperature gradient in the μ-gradient zone of stars less than $\sim 10\,M_\odot$ is lower than the adiabatic temperature gradient.

However, the situation is not so simple for more massive stars. Because of the contribution of radiation pressure to the equation of state, the adiabatic temperature gradient in the μ-gradient zone of such massive stars is so low that the temperature gradient there is superadiabatic. In a pioneering evolutionary study, Schwarzschild & Härm (1958) treated this zone as a semiconvective zone, in which the chemical composition was adjusted by a partial mixing to maintain convective neutrality, according to their expression, by a criterion

$$\nabla = \nabla_{ad}, \tag{1}$$

where $\nabla \equiv d\ln T/d\ln p$, but all energy was assumed to be transported by radiation ($\nabla = \nabla_{ad} = \nabla_{rad}$). Sakashita & Hayashi (1959, 1961) pointed out that the μ-gradient should be taken into account as a criterion for convective stability, and treated the μ-gradient zone of such massive stars as a partially mixed zone so that the temperature gradient satisfied

$$\nabla = \nabla_{ad} + \beta(4 - 3\beta)^{-1}\nabla_\mu, \tag{2}$$

where $\nabla_\mu = d\ln\mu/d\ln p$ and $\beta \equiv p_{gas}/(p_{gas} + p_{rad})$ is the ratio of gas pressure to total pressure. Here μ denotes the mean molecular weight.

The treatment of the μ-gradient zone in such very massive stars is important because it affects their later evolution significantly. There has been considerable controversy as to whether partial or complete mixing is achieved in that zone and which criterion should be used if mixing occurs indeed (e.g., Gabriel, 1969, 1970, Sothers 1970, Simpson 1971, Stothers & Chin 1975, Sreenivasan & Wilson 1978, Stevenson 1979, Langer et al. 1985, Spruit 1992, Canuto 2000). It should be noted that such zones are also present in stars of about $1\,M_\odot$ when a small convective core expands due to the growing importance of the CNO cycle (Gabriel & Noels 1977).

Criteria for convection

Suppose that a fluid parcel lifts up from its equilibrium position in a gravitationally stratified layer maintaining the pressure equilibrium with its surroundings. Let the pressure and density of the parcel at its equilibrium position be p_0 and ρ_0. The corresponding quantities of the outside environment at the parcel's equilibrium level are naturally identical to those of the parcel. Let us denote the displacement of the parcel in the vertical direction by ξ. At the height of the displacement position, the pressure and density of the environment fluid are $p_0(1+\xi d\ln p/dz)$ and $\rho_0(1+\xi d\ln \rho/dz)$, respectively, where z denotes the vertical coordinate. The pressure of the displaced fluid parcel is immediately adjusted with its environment. If the thermal time scale of the layer is much longer than the time required for dynamical motion, the density of the parcel varies adiabatically with motion; $\rho_0(1 + \xi\Gamma_1^{-1}d\ln p/dz)$, where $\Gamma_1 \equiv (\partial\ln p/\partial\ln\rho)_S$. If the parcel's density is lower than the surroundings, the parcel lifts up further and this is the onset of convection. Therefore, a criterion for occurrence of convection is

$$A \equiv d\ln\rho/dz - \Gamma_1^{-1}d\ln p/dz > 0. \tag{3}$$

This expression was derived by Karl Schwarzschild (1906) and the quantity A defined above is known as the Schwarzschild discriminant.

With the help of thermodynamic relations, in the fully ionized region, the Schwarzschild discriminant is written as (Ledoux 1947)

$$A = -d\ln p/dz \left[-(\partial\ln\rho/\partial\ln T)_p (\nabla - \nabla_{ad}) - \nabla_\mu \right]. \tag{4}$$

Hence, if the pressure is approximated by ideal gas plus radiation pressure, the criterion for convective instability is expressed as

$$(\nabla - \nabla_{ad}) - \beta(4 - 3\beta)^{-1}\nabla_\mu > 0. \tag{5}$$

This expression is widely known as the Ledoux criterion. In the case of a chemically homogeneous layer, $\nabla_\mu = 0$, the criterion is reduced to $\nabla - \nabla_{ad} > 0$, and the condition for convective neutrality is given by equation (1). On the other hand, in the presence of $\nabla_\mu \neq 0$, for the onset of convection, the temperature gradient is required not only to be superadiabatic but also to overcome the gradient of chemical composition, which tends to stabilize the layer, and the condition for convective neutrality is given by equation (2).

Since the convective core is well mixed, $\nabla_\mu = 0$ there. Hence, at the edge of the convective core, $\nabla - \nabla_{ad} = 0$. Evolving off the ZAMS, a μ-gradient zone appears where $\nabla_\mu > 0$ outside of the core. While the μ-gradient zone just outside of the convective core of stars less massive than $10\,M_\odot$ is subadiabatic ($\nabla < \nabla_{ad}$), such a zone for stars more massive than $10\,M_\odot$ is superadiabatic ($\nabla > \nabla_{ad}$) though the layer is nevertheless convectively stable in the sense of equation (5). The problem is how to treat this superadiabatic but convectively stable layer in stellar evolution modelling.

Linear overstability

Even when the medium is convectively stable in the sense of equation (5), overstability (vibrational instability with growing amplitude) may occur in a medium of varying molecular weight if the temperature gradient is superadiabatic ($\nabla > \nabla_{ad}$). This was first pointed out by the local analysis of Veronis (1965) in the case of a water layer with a gradient of the salt content, and by Kato (1966) in the context of astrophysics.

Local analysis

Let us consider a plane-parallel layer of fluid with a varying molecular weight in hydrostatic and radiative equilibrium under a constant gravitational field, $\mathbf{g} = -g\mathbf{e}_z$, in the Boussinesq approximation (density is treated as a constant in the equation of continuity and in the equation of motion except for the buoyancy term). If we assume the spatial and temporal dependence of perturbation of the form $f' \propto \exp(i\mathbf{k} \cdot \mathbf{x} + st)$, we get a characteristic equation for growth rate s, which becomes a cubic equation in the case of no nuclear energy generation, no viscosity and constant radiative conductivity (Kato 1966):

$$s^3 + s^2 Kk^2/(c_p\rho) + sk_h^2 N^2/k^2 + Kk_h^2 gH_p^{-1}\nabla_\mu/(c_p\rho) = 0, \tag{6}$$

where K is the radiative conductivity, c_p is the specific heat at constant pressure, $H_p \equiv -dz/d\ln p$ is the pressure scale height, $k_h^2 = k_x^2 + h_y^2$, $k^2 = k_h^2 + k_z^2$, and $N^2 = -gA$ is the squared Brunt-Väisälä frequency.

For weak nonadiabaticity, which is usually the case, the roots of equation (6) are given by

$$s_{1,2} \simeq \pm iNk_h/k + \beta^{-1}(4 - 3\beta)Kk^2 gH_p^{-1}(\nabla - \nabla_{ad})/(2c_p\rho N^2) \tag{7}$$

and

$$s_3 = -Kk^2 gH_p^{-1}\nabla_\mu/(c_p\rho N^2). \tag{8}$$

Equation (7) indicates that even when the layer is dynamically stable ($N^2 > 0$, i.e., $A < 0$) the layer is unstable in the sense that the amplitude of oscillatory motion grows with time if the temperature gradient is superadiabatic ($\nabla > \nabla_{ad}$). The physical reason for this overstability is that the radiative heat exchange brings about an asymmetry in the oscillatory motion in such a way that an oscillatory element overshoots its equilibrium position with an increasing velocity. (Equation (8) indicates the thermal instability in the case of an adverse μ-gradient, which I do not discuss anymore here.)

Linear analyses cannot determine whether the instability grows nonlinearly enough to mix up the material. However, if we assume this is the case, equation (1) is the local condition for the neutrality of overstability. In this sense, Schwarzschild and Härm's (1958) treatment can be regarded as that they assumed a partial mixing to satisfy the neutrality of overstability. It should be reminded, however, that this is not the necessary condition for stability but the sufficient condition, because stability cannot be determined locally and can be settled only by global stability analysis of a whole star as to nonradial g modes. In fact, the subadiabatic ($\nabla < \nabla_{ad}$) radiative zone works for damping as seen in equation (7).

Global analysis

The investigation of nonradial oscillations of stars has been remarkably developed in the 1970s and 1980s (see, e.g., Ledoux 1974, Unno et al. 1989). One of the important characteristics of nonradial oscillations is that eigenfunctions have large amplitudes at certain localized regions. If we rewrite the temporal dependence s by introducing the frequency $\sigma \equiv -i\,s$ in equations (6), a dispersion relation concerning the vertical wavenumber k_z and the real part of frequency σ_R is obtained; $k_z^2 = (N^2/\sigma_R^2 - 1)k_h^2$. From this relation, it is obvious that the perturbation is propagative vertically (radially in a spherical geometry) only where the squared Brunt-Väisälä frequency is higher than the squared frequency and in other areas the perturbation is evanescent.

Paying attention to these characteristics, Shibahashi & Osaki (1976), Scuflaire et al. (1976), and Gabriel & Noels (1976) have carefully studied the vibrational stability of gravity modes in massive stars of M $\gtrsim$ 10 M$_\odot$. Evolutionary models used in stability analysis were computed with Ledoux's criterion, equation (5), for convective stability and no partial mixing was assumed to occurThis was done to examine the possible overstability that may cause mixing.

The squared Brunt-Väisälä frequency N^2 is zero to high precision in the convective core. Outside the μ-gradient zone, —that is, in the radiative region extended outside the convective core formed at the initial zero-age state, N^2 varies gradually with stellar evolution keeping its ratio to GM/R^3 to be of the order of $\sim$ 5 (see, e.g., Unno et al. 1989). On the other hand, in the μ-gradient zone, $N^2/(GM/R^3)$ becomes higher and higher after evolving off from the zero-age along with the development of ∇_μ.

While the peak value of N^2 in the μ-gradient zone is still lower than $\sim$ 5 GM/R^3, the μ-gradient zone and the outer radiative zone form a joined wave propagation zone for g modes. In such a situation, the amplitudes of g modes are of the same order of magnitude both in the subadiabatic, outer radiative zone and in the superadiabatic, μ-gradient zone, and the radiative dissipation in the outer radiative zone overcomes the destabilizing effect. However, once the peak value of N^2 in the μ-gradient zone exceeds $\sim$ 5 GM/R^3, the μ-gradient zone left behind by the receding convective core in massive stars behaves like a single potential well that traps gravity waves only there for modes of $\sigma_R^2 \gtrsim$ 5 GM/R^3. Hence, the necessary condition for global overstability is that N^2 in the superadiabatic μ-gradient zone is higher than $\sim$ 5 GM/R^3. By approximating $\nabla_\mu \gg |\nabla - \nabla_{ad}|$, this condition can be rewritten as

$$\nabla_\mu \gtrsim 5 \, (r/R)^3 (M_r/M)^{-1} (H_p/r). \tag{9}$$

In such a case, the favorably well-trapped modes are oscillations with frequencies higher than $\sim$ (5 GM/R^3)$^{1/2}$ with the spherical degree $l \gtrsim R/d$, where d is the thickness of the μ-gradient zone and R denotes the stellar radius. All the forementioned research groups have shown by a global stability analysis that radiative dissipation in the outer, subadiabatic region is negligibly small for these modes and that such gravity modes well-trapped in the μ-gradient zones are in fact overstable due to the mechanism discussed in the previous subsection. The e-folding time for unstable modes is typically 10^3 - 10^4 yr, which is shorter than the evolutionary time of the stars. Then overstability is expected to grow to a finite amplitude.

Outcome of nonlinear development

Description of mixing as a diffusion process

The growing waves are supposed to be eventually disrupted when the amplitudes reach some certain critical limits, resulting in efficient material mixing, which may be described with a diffusion equation:

$$\partial \rho X/\partial t = r^{-2}\partial/\partial r[r^2 D(\partial \rho X/\partial r)], \tag{10}$$

where X denotes the mass fraction of hydrogen and D denotes a diffusion coefficient which must be a function of the growth rate as well as the physical quantities of the layer considered. The problem then turns to how to evaluate the diffusion coefficient. Stevenson (1979) evaluated $D \sim K/(c_p\rho)\beta^{-1}(4 - 3\beta)(\nabla - \nabla_{ad})/\nabla_\mu$ by assuming that the most unstable modes transfer energy to modes of the shorter wavelength which eventually disrupt by shear instability. Langer et al. (1983) derived D in another way by introducing the concept of a mixing length and an effective velocity, of which product gives D. They estimated the mixing length and the effective velocity from the wavelengths and the growth rates of linear unstable modes, respectively, and got $D \sim K/(c_p\rho)\beta^{-1}(4-3\beta)(\nabla - \nabla_{ad})gH_p^{-1}/N^2$. Gabriel (1995) argued that the mixing induced by semiconvection may be described with a concept analogous to turbulent diffusion, and derived an expression for D, which is similar to Langer et al. All the results quoted above imply that material mixing proceeds efficiently. It should be noted, however, that Spruit (1992) assumes that overstable convective flux is carried in thin plumes and concludes that the mixing rate is much smaller.

It should be noted that all the estimates of diffusion coefficient D were so far made with an implicit assumption that the linearly unstable modes were well trapped in the μ-gradient zone. Since the expression for D is proportional to $\nabla - \nabla_{ad}$, one might think that the mixing process would continue till the condition $\nabla = \nabla_{ad}$ (equation (1)) is satisfied. As pointed out in the previous section, however, the radiative dissipation occurring in the outer radiative zone would damp the modes if N^2 in the superadiabatic μ-gradient zone would become lower than that in the adjacent, outer, chemically homogeneous radiative zone. Hence, it is likely that the mixing process stops once either $\nabla = \nabla_{ad}$ or $\nabla_\mu \simeq 5 \, (r/R)^3 (M_r/M)^{-1}(H_p/r)$ (cf. equation (9)) is satisfied.

Numerical simulation

Merryfield (1995) carried out a 2D numerical simulation of hydrodynamics of a superadiabatic, gravitationally stratified layer with a varying molecular weight, aiming at modeling conditions of the μ-gradient zone in a 30 $M_\odot$. He showed that the outcomes are dramatically different depending on how strongly the instability is driven. In the case of large superadiabaticity, the initial disturbances evolve into large amplitudes and break to lead up to the material mixing. On the other hand, in the case of small superadiabaticity, such overturning does not occur and, instead, horizontally propagating motions with large wavelengths are organized. It is not yet clear why these numerical results are different from any of the theories described in the above. The Prandtl number and the molecular diffusion coefficient adopted in the numerical simulation are much larger than the stellar case, and these differences might be the cause.

Semiconvection in white dwarfs

Semiconvection appears not only in main sequence stars but also in white dwarfs. In terms of surface composition, the majority of white dwarfs are divided into two major subgroups: those with hydrogen atmospheres (DAs), which constitute about 80% of all white dwarfs, and those with helium atmospheres (DOs and DBs). The ratio of the memberships of these two subgroups is sensitive to the effective temperature. There are few white dwarfs showing helium lines between 45 000 K and 30 000 K, and this exclusion zone is known as the "DB gap." Shibahashi (2005) proposed the following model to explain the DB gap: Since the temperatures of the hotter and the cooler ends of the DB gap coincide with the effective temperatures where the He II/III and the He I/II convection zone show up respectively, convective mixing is suspected to be the cause of appearance of DOs and DBs outside the DB gap. Inversely, chemical separation due to gravitational settling in the convectively stable atmosphere is suspected to be the cause of the presence of the DB gap. If one adopts this scenario, the potentially DB white dwarfs near the cool end of the DB gap are expected to have a superadiabatic layer which is nonetheless convectively stable due to a steep chemical gradient. Such stars are vibrationally unstable due to radiative heat exchange, and then a new type of white dwarf variables are predicted near the cool end of the DB gap. A search for this new class of pulsating white dwarfs has been carried out (Kurtz et al. 2008).

References

Canuto, V. M. 2000, ApJ, 534, L1

Gabriel, M. 1969, A&A, 1, 321

Gabriel, M. 1970, A&A, 6, 124

Gabriel, M. 1995, in Stellar Evolution: What Should Be Done, Proc. of the 32nd Liège Int. Astrophys. Colloq., ed. A. Noels, D. Fraipont-Caro, M. Gabriel, N. Grevesse, and P. Demarque (Liège, Université de Liège), 95

Gabriel, M., & Noels, A. 1976, A&A, 53, 149

Gabriel, M., & Noels, A. 1977, A&A, 54, 631

Kato, S. 1966, PASJ, 18, 374

Kurtz, D. W., Shibahashi, H., Dhillon, V. S., et al. 2008, MNRAS, 389, 1771

Langer, N., El Eid, M. F., & Fricke, K. J. 1985, A&A, 145, 179

Langer, N., Sugimoto, D., & Fricke, K. J. 1983, A&A, 126, 207

Ledoux, P. 1947, ApJ, 105, 305

Ledoux, P. 1974, in Stellar Instability and Evolution, IAU Symp. 59, ed. P. Ledoux, A. Noels, and A. W. Rogers (Dordrecht, Reidel), 135

Merryfield, W. J. 1995, ApJ, 444, 318

Sakashita, S., & Hayashi, C. 1959, Prog. Theor. Phys., Kyoto, 22, 830

Sakashita, S., & Hayashi, C. 1961, Prog. Theor. Phys., Kyoto, 26, 942

Schwarzschild, K. 1906, Gött. Nach., 1, 41

Schwarzschild, M., & Härm, R. 1958, ApJ, 128, 348

Scuflaire, R., Noels, A., Gabriel, A., & Boury, A. 1976, Astrophys. Space Sci., 39, 463

Shibahashi, H. 2005, EAS Publication Series, 17, 143

Shibahashi, H. 2007, AIP Conference Proceedings, 948, 35

Shibahashi, H., & Osaki, Y. 1976, PASJ, 28, 199

Simpson, E. E. 1971, ApJ, 165, 295

Spruit, H. C. 1992, A&A, 253, 131

Sreenivasan, S. R., & Wilson, W. J. F. 1978, Astrophys. Space Sci., 53, 193

Steevenson, D. J. 1979, MNRAS, 187, 129

Stothers, R. 1970, MNRAS, 151, 65

Stothers, R., & Chin, C.-W. 1975, ApJ, 198, 407

Unno, W., Osaki, Y., Ando, H., et al. 1989, Nonradial Oscillations of Stars (Tokyo: Univ. of Tokyo Press)

Veronis, J. 1965, J. Marine Re., 23, 1

Comm. in Asteroseismology
Vol. 158, 2009, 38th LIAC/HELAS-ESTA/BAG, 2008
A. Noels, C. Aerts, J. Montalbán, A. Miglio and M. Briquet., eds.

Overshooting

B. Dintrans

LATT, Université de Toulouse, CNRS, 14 av. Edouard Belin, 31400 Toulouse, France

Abstract

Overshooting occurs in stars when convective elements penetrate into adjacent radiative zones. In the Sun, it leads to the so-called 'tachocline' at the base of the outer convection zone. This region is becoming a key ingredient of the standard solar dynamo model as strong toroidal magnetic fields may be generated there. However, this overshoot is not predicted by the mixing-length theory of convection where convective elements must stop at the border of a convectively unstable region. I will review the main properties of this convective overshoot in stellar interiors, with in particular its subtle dependence with the thermal diffusion, and will present the most recent results obtained from 2-D and 3-D direct numerical simulations (DNS) of penetrative convection.

Introduction

In the standard model of thermal convection based on the mixing-length phenomenology (Böhm-Vitense 1958), the penetration of convective elements into surrounding stable layers is not possible. Indeed, the mixing-length velocity v of a convective blob is related to its temperature contrast δT by

$$v^2 \propto \frac{\delta T}{T} g\ell \text{ with } \delta T \propto \nabla - \nabla_{ad}, \tag{1}$$

where g and ℓ denote the gravity and the mixing length, respectively, and ∇ and ∇_{ad} the usual temperature 'nabla'. Following Schwarzschild's criterion, the interface that separates a convective zone from a radiative one corresponds to $\nabla = \nabla_{ad}$. Therefore the blob must stop here. However, the mixing-length formalism is based on an acceleration budget and setting the blob acceleration to zero does not mean that its velocity itself is zero, i.e., the mixing-length theory neglects *inertia*. In fact, the blob penetrates into the stable layer over a small distance, then it is slowed down by the buoyancy braking and finally it stops. This small penetration is not anecdotal because it has important consequences in stellar physics:
- it leads to different evolutionary tracks in the HR diagram as massive stars live longer on the main sequence due to the additional mixing induced by the overshooting above the convective core (Rosvick & Vandenberg 1998, Perryman et al. 1998).
- in the standard $\alpha - \Omega$ model of the solar dynamo, the regeneration of poloidal magnetic fields into toroidal ones takes place in the tachocline where vertical and latitudinal differential rotation is expected to be most efficient (Parker 1955, Brandenburg & Subramanian 2005).
- helioseismology showed that the solar core rotation is almost rigid (Brown et al. 1989). Internal gravity waves propagating in stellar radiative zones are suspected to play the main role in this angular momentum redistribution (Zahn et al. 1997, Talon & Charbonnel 2003). The most wide-spread excitation model involves penetrative convection from neighboring

convection zones as strong downdrafts extend a substantial distance into the adjacent stable zones so that internal gravity waves can be randomly generated (Dintrans et al. 2005).

Modelling of overshooting

Assuming that the temperature gradient in the overshoot layer is close to the adiabatic one of the next convection zone, Roxburgh (1978) investigated the overshoot of convective cores and derived what is now called the 'Roxburgh's integral constraint':

$$\int_0^{r_c} (L - L_{rad}) d\left(\frac{1}{T}\right) > 0, \tag{2}$$

where the radius r_c corresponds to the edge of the overshoot region, while L and L_{rad} denote the total and radiative luminosity, respectively. The overshoot layer corresponds in this relation to the region where $L < L_{rad}$ (i.e., negative contribution) while the convection zone means $L > L_{rad}$ (i.e., positive contribution). However, this relation gives an upper limit of the overshoot extent because it does not take into account the dissipation acting on the penetrating elements. Indeed, the viscous dissipation $\mathcal{D}_\nu$ enters in the heat equation and affects the blob dynamics such that the correct relation reads now (Roxburgh 1989):

$$\int_0^{r_c} (L - L_{rad}) d\left(\frac{1}{T}\right) = \int_0^{r_c} 4\pi r^2 \frac{\mathcal{D}_\nu}{T} dr \text{ with } \mathcal{D}_\nu \sim \nu \left(\frac{\partial u_i}{\partial x_j}\right)^2, \tag{3}$$

where ν denotes the kinematic viscosity. As the viscous dissipation is always positive, this term acts to decrease the negative part of the integrand in the left-hand side and thus the radius r_c.

Following Roxburgh's investigation, Zahn (1991) showed that the penetration extent also depends on the value of the Peclet number Pe associated with the convective elements, that is,

$$Pe = \frac{v\ell}{\chi}, \tag{4}$$

where χ denotes the thermal diffusivity. Small Peclet's numbers mean that the radiative diffusion is stronger than the advection and the convective blob rapidly loses its identity compared to the surrounding medium. As a consequence, it does not feel anymore the buoyancy braking and travels over long distances in the radiative zone (Zahn called this regime 'overshoot'). On the other hand, a blob with a large Peclet number does not thermalize rapidly with its surrounding (i.e., no radiative losses) and keep its identity: it is then strongly slowed down by the buoyant force and its penetration is weaker (the so-called 'penetration' regime). Penetration leads to an adiabatic stratification below the convection zone because of the induced entropy mixing, whereas overshoot has no significant influence on the local stratification.

Assuming that the velocity W of the penetrating motions obeys the usual mixing-length scaling of thermal convection, Zahn derived the following relations for both the subadiabatic extent of a convective region and the width of the thermal boundary layer:

$$L_p \propto f^{1/2} W^{3/2} \text{ and } L_{bound} \propto \left(\frac{\chi H_p}{g}\right)^{1/2}, \tag{5}$$

where f denotes a filling factor of convective plumes (i.e., fraction of the area occupied by the plumes) and H_p the pressure scale-height at the bottom of the convection zone. Applied to the Sun, these two scalings lead to a penetration extent of about $0.2 - 0.3 H_p$ ($\sim 15\,000$ km) and an overshoot layer $\simeq 1$ km. These scalings however seem to overestimate the amount of convective penetration inferred by helioseismology as the study of the oscillations in the asymptotic laws of solar acoustic modes rather predicts an upper limit of $\sim 0.1 H_p$ for the extent of the whole convective overshooting (Roxburgh & Vorontsov 1994; Christensen-Dalsgaard et al. 1995; see also Aerts et al. 2003 for the case of a massive star).

Numerical simulations of overshooting

The bad news

The rapid development of supercomputers has lead several groups to undertake direct numerical simulations of convective overshooting. However, many numerical difficulties appear when one tries to simulate stellar convection:

- the molecular viscosity in stellar interiors is small, typically $\nu \sim 1$ cm^2/s (Parker 1979). Granulation cells in the upper convection zone of the Sun correspond to a typical size and velocity of about $L = 1000$ km and $U = 1$ km/s, respectively. It leads to Reynolds' number of about $Re = UL/\nu \sim 10^{11}$, meaning that solar convection is a highly turbulent and nonlinear phenomenon.

- we know from Kolmogorov's theory of turbulence (Kolmogorov 1941) that the ratio between the largest scale (i.e., the injection scale L) and the smallest one (i.e., the dissipation scale ℓ_d) behaves as $L/\ell_d \sim Re^{3/4}$ such that the solar dissipation scale is about one centimeter. In other words, one needs (at least) 10^8 gridpoints in each spatial direction to well reproduce both the injection of energy in the largest structures and its viscous dissipation in the smallest ones (the turbulent cascade scenario). As this number of gridpoints is clearly beyond the scope of today's supercomputers, several numerical techniques have been developed to overcome this restriction: (i) instead of doing Direct Numerical Simulations (DNS), one can perform Large-Eddy Simulations (LES) where the small-scale dynamics is modelled using a sub-grid scale approach (Lesieur 2008); (ii) one can also artificially increase the dissipation scale by adding a large-scale viscosity in the whole computational domain, the price to pay being that motions are essentially laminar ($Re \sim 10^2 \cdots 10^4$).

- finally, another problem concerns the different time scales that coexist in a convection zone. Indeed, the local turnover time scale of convection is roughly given by:

$$\tau \sim H_p/U_{RMS}, \tag{6}$$

where U_{RMS} is the typical (turbulent) RMS velocity of convective eddies. Applied to the solar convection zone, it leads to turnover time scales of about a day at the surface to several months at the bottom (Spruit 1974). As a consequence, simulations should reproduce motions whose dynamical time scales can vary by two orders of magnitude in the computational domain. Moreover, the thermal time scale is also much longer than these turnover time scales by orders of magnitude as, e.g., the solar thermal flux expressed in dynamical units is very small at the bottom of the convection zone ($F_\odot/(\rho c_s^3) \sim 10^{-11}$). This ratio is in essence equal to the ratio of the dynamical to thermal time scales, meaning that realistic simulations should be integrated on a very long time to make sure that a thermal-relaxed state is reached. It is not possible in practice to do that with a decent spatial resolution and the thermal fluxes are commonly overestimated by orders of magnitude in numerical simulations.

DNS of convective overshooting in local cartesian boxes

Following Hurlburt et al. (1984), the majority of DNS done so far use polytropic solutions for the internal structure of the star. Indeed, polytropes with $\rho \propto T^m$ (m being the polytropic index) allow to easily specify the strength of the convective instability as the hydrostatic equilibrium in that case implies

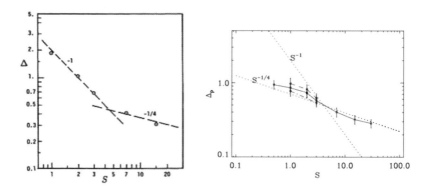

Figure 1: Penetration extent Δ v.s. stability parameter S. *Left*: in the 2-D case where both the penetration and overshoot regimes exist (from Hurlburt et al. 1994). *Right*: in the 3-D case where only the overshoot regime remains (from Brummell et al. 2002).

$$\frac{dT}{dz} = -\frac{g}{(m+1)R_*},\tag{7}$$

where g is the downward-directed gravity, z the altitude and R_* the perfect gas constant. For a monatomic gas with an adiabatic index $\gamma = 5/3$, this relation shows that the temperature gradient is equal to the adiabatic one $(= -g/c_p)$ when the polytropic index is $m_{ad} = 1/(\gamma - 1) = 3/2$. As a consequence, a stable layer corresponds to $m > m_{ad}$ and a convectively-unstable one to $m < m_{ad}$.

Hurlburt et al. (1986, 1994) used this model to study the overshooting in 2-D simulations of compressible convection for a solar case, that is, a radiative zone with polytropic index $m_3 > 3/2$ located above a convective zone with polytropic index $m_2 < 3/2$. In particular, they studied the influence of the relative stability parameter S on the penetration extent Δ, where $S = -(m_{ad} - m_3)/(m_{ad} - m_2)$, and found the following scalings (Fig. 1, left):

$$\begin{cases} \Delta \propto S^{-1} \text{ for small S-values: penetration regime,} \\[2mm] \Delta \propto S^{-1/4} \text{ for large S-values: overshoot regime.} \end{cases}\tag{8}$$

However, these scaling laws have not been fully confirmed in the 3-D case by Brummell et al. (2002) as only the overshoot regime $\Delta \propto S^{-1/4}$ is recovered (Fig.1, right). The reason simply lies in the smaller filling factor f for the convective plumes compared to the 2-D case (Eq. 5): the spaced downdrafts that penetrate into the stably stratified layer are not enough to maintain an adiabatic stratification below the convection zone and only the thermal boundary layer exists.

Other interesting 2-D and 3-D simulations of penetrative convection are the ones done by Roxburgh & Simmons (1993) and Singh et al. (1998). Indeed, a more realistic model of the star interior has been considered in the 2-D case through a temperature-dependent radiative conductivity. The main result is that the viscous dissipation contribution in the RHS of Roxburgh's integral constraint (3) decreases with the Prandtl number. It suggests that in the astrophysical limit of very small Prandtl numbers ($Pr \sim 10^{-6}$), the stable and unstable contributions in the LHS of Roxburgh's integral strictly balance each other and are sufficient to obtain a good estimate for the overshoot extent (Roxburgh & Simmons 1993). In the 3-D case (Singh et al. 1998), a sub-grid scale modelling of the small scales led to the confirmation

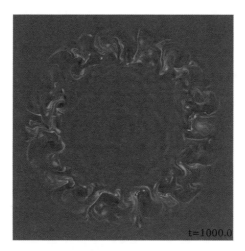

Figure 2: Vorticity isocontours in a 2-D star-in-a-box simulation of penetrative convection for a solar-like star (in preparation by Dintrans & Brandenburg).

of the scaling relationships between this extent and both the imposed bottom flux and the vertical velocity of convective downdrafts (see Eq. 5).

Global DNS

In the last decade, global direct simulations of overshooting have greatly improved following two numerical approaches:
- the spherical shape of the star is reproduced using a pseudo-spectral code where physical fields are projected onto spherical harmonics. Browning et al. (2004) used this method with the 3-D anelastic ASH code to study the core overshooting in a massive A-type star of $2M_\odot$, where only the inner 30% by radius of the star was considered (convective core and some of its surrounding radiative envelope). Although the achievement of the thermal relaxation in these simulations remains questioning, they found that the convective core adopts a prolate shape while overshooting is larger at the equator than at the poles, yielding an overall spherical shape to the whole central region.
- the star is embedded in a topologically Cartesian domain and the sphericity is reproduced through radial-dependent forcing or damping profiles. This novel approach has been developed by Dintrans & Brandenburg, in preparation and first results in 2-D are promising as both the penetration and overshoot regimes are observed, whereas internal gravity waves are efficiently excited in the central radiative zone (Fig. 2).

Conclusion

Convective overshooting is today included in stellar evolution codes and evolutionary tracks of massive stars reproduce satisfactorily the observations of clusters when an overshooting above convective cores is assumed. Theoretical studies showed that the overshoot region is composed of two nested layers whose size depends on the value of the Peclet number of penetrating motions. Numerical studies in 2-D and 3-D of penetrative convection confirmed this picture, except that only the overshoot layer (i.e., the thermal boundary one) exists in 3-D due to the weak filling factor of convective plumes. These local and global direct numerical simulations are however still quite far from realistic values of stellar interiors. Flows are less turbulent than in reality and the overshoot regime dominates. Further progress is clearly needed on the numerical side of this problem (e.g., development of large-eddy simulations), in its theoretical modelling (e.g., turbulent closure models; Kupka & Montgomery 2002) and in the observational signatures of overshooting from asteroseismology.

References

Aerts, C., Thoul, A., Daszyńska, J., et al. 2003, Science, 300, 1926

Böhm-Vitense, E. 1958, ZA, 46, 108

Brandenburg, A., & Subramanian, K. 2005, PhR, 417, 1

Brown, T. M., Christensen-Dalsgaard, J., Dziembowski, W.A., et al. 1989, SoPh, 343, 526

Browning, M. K., Brun, A. S., & Toomre, J. 2004, ApJ, 601, 512

Brummell, N. H., Clune, T. L., & Toomre, J. 2002, ApJ, 570, 825

Christensen-Dalsgaard, J., Monteiro, M. J. P. F. G., & Thompson, M. J. 1995, MNRAS, 276, 283

Dintrans, B., Brandenburg, A., Nordlund, Å., et al. 2005, A&A, 438, 365

Hurlburt, N. E., Toomre, J., & Massaguer, J. M. 1984, ApJ, 282, 557

Hurlburt, N. E., Toomre, J., & Massaguer, J. M. 1986, ApJ, 311, 563

Hurlburt, N. E., Toomre, J., Massaguer, J. M., & Zahn, J.-P. 1994, ApJ, 421, 245

Kolmogorov, A. N. 1941, DoSSR, 30, 301

Kupka, F., & Montgomery, M. H. 2002, MNRAS, 330, L6

Lesieur, M. 2008, Turbulence in Fluids (Springer-Verlag)

Parker, E. N. 1955, ApJ, 122, 293

Parker, E. N. 1979, Cosmical Magnetic Fields, (Oxford University Press)

Perryman, M. A. C., Brown, A. G. A., Lebreton, Y., et al. 1998, A&A, 331, 81

Rosvick, J. M., & Vandenberg, D. A. 1998, AJ, 115, 1516

Roxburgh, I. W. 1978, A&A, 65, 281

Roxburgh, I. W. 1989, A&A, 211, 361

Roxburgh, I. W. & Simmons, J. 1993, A&A, 277, 93

Roxburgh, I. W., & Vorontsov, S. V. 1994, MNRAS, 268, 880

Singh, H. P., Roxburgh, I. W., & Chan, K. L. 1998, A&A, 340, 178

Spruit, H. C. 1974, SoPh, 34, 277

Talon, S., & Charbonnel, C. 2003, A&A, 405, 1025

Zahn, J.-P. 1991, A&A, 252, 179

Zahn, J.-P., Talon, S., & Matias, J. 1997, A&A, 322, 320

Comm. in Asteroseismology
Vol. 158, 2009, 38th LIAC/HELAS-ESTA/BAG, 2008
A. Noels, C. Aerts, J. Montalbán, A. Miglio and M. Briquet., eds.

Thermohaline convection in main sequence stars

S. Vauclair[1]

[1] Laboratoire d'Astronomie de Toulouse-Tarbes, Université de Toulouse, 14 Avenue Edouard Belin, 31400 Toulouse, France

Abstract

Thermohaline convection is a well-known process in oceanography, which has long been put aside in stellar physics. In the ocean, it occurs when warm salted layers sit on top of cool and less salted ones. Then the salted water rapidly diffuses downwards even in the presence of stabilizing temperature gradients, due to double diffusion between the falling blobs and their surroundings. A similar process may occur in stars in case of inverse μ-gradients in a thermally stabilized medium. This process has important consequences in stellar physics.

What is thermohaline convection?

When warm salted layers sit on top of cool unsalted ones, they rapidly diffuse downwards even when the stabilizing effect of the temperature gradient exceeds the destabilizing effect of the salt gradient. This is due to the different diffusivities of heat and salt. When a warm salted blob falls down in cool fresh water, the heat diffuses out more quickly than the salt. The blob goes on falling due to its weight until it mixes with the surroundings. This leads to the so-called "salt fingers".

The condition for the salt fingers to develop is related to the density variations induced by temperature and salinity perturbations. Two important characteristic numbers are defined:
- the density anomaly ratio

$$R_\rho = \alpha \nabla T / \beta \nabla S \qquad (1)$$

where $\alpha = -(\frac{1}{\rho}\frac{\partial \rho}{\partial T})_{S,P}$ and $\beta = (\frac{1}{\rho}\frac{\partial \rho}{\partial S})_{T,P}$ while ∇T and ∇S are the average temperature and salinity gradients in the considered zone
- the so-called "Lewis number"

$$\tau = \kappa_S/\kappa_T = \tau_T/\tau_S \qquad (2)$$

where κ_S and κ_T are the saline and thermal diffusivities while τ_S and τ_T are the saline and thermal diffusion time scales.

The density gradient is unstable and overturns into dynamical convection for $R_\rho < 1$ while the salt fingers, which represent another kind of convection with a larger mixing time scale, grow for $R_\rho \geq 1$. On the other hand they cannot form if R_ρ is larger than the ratio of the thermal to saline diffusivities τ^{-1} as in this case the salinity difference between the blobs and the surroundings is not large enough to overcome buoyancy.

Salt fingers can grow if the following condition is satisfied:

$$1 \leq R_\rho \leq \tau^{-1} \qquad (3)$$

In the ocean, τ is typically 0.01.

The stellar case

Thermohaline convection may occur in stellar radiative zones in the presence of inverse μ-gradients. In this case $\nabla_\mu = \mathrm{d}\ln\mu/\mathrm{d}\ln P$ plays the role of the salinity gradient while the difference $\nabla_{ad} - \nabla$ (where ∇_{ad} and ∇ are the usual adiabatic and local (radiative) gradients $\mathrm{d}\ln T/\mathrm{d}\ln P$) plays the role of the temperature gradient.

When the destabilizing effect of the μ-gradient is more important than the stabilizing effect of the temperature gradient, the medium is dynamically unstable. This situation is described by the so-called Ledoux criterion:

$$\nabla_{crit} = \frac{\phi}{\delta}\nabla_\mu + \nabla_{ad} - \nabla < 0 \tag{4}$$

where $\phi = -(\partial \ln \rho / \partial \ln \mu)_T$ and $\delta = (\partial \ln \rho / \partial \ln T)_\mu$ When ∇_{crit} vanishes, marginal stability is achieved and thermohaline convection may begin as a "secular process", namely on a thermal time scale (short compared to the stellar lifetime!).

As for the ocean case, salt fingers will form if the following condition is verified :

$$1 \le |\frac{\delta(\nabla_{ad} - \nabla)}{\phi(\nabla_\mu)}| \le \tau^{-1} \tag{5}$$

with $\tau = D_\mu/D_T = \tau_T/\tau_\mu$ where D_T and D_μ are the thermal and molecular diffusion coefficients while τ_T and τ_μ are the corresponding time scales.

In stars the value of the τ ratio is typically 10^{-10} if D_μ is the molecular (or "microscopic") diffusion coefficient but it can go up by many orders of magnitude when the shear flow instabilities which induce mixing between the edges of the fingers and the surroundings are taken into account.

The effects of thermohaline convection as a mixing process in stars are far from trivial. Many detailed studies in the water case have been published, for example Gargett & Ruddick 2003) who gave precise comparisons between numerical simulations and laboratory experiments. However the stellar case is different as mixing then occurs in a compressible fluid.

As pointed out by Charbonnel & Zahn 2007a, two different parametrisation recipes, as given by Ulrich 1972 or Kippenhahn et al. 1980 can differ by two orders of magnitude. This clearly illustrates that treating thermohaline convection as a simple diffusion process, using a simple diffusion coefficient, may lead to wrong results.

Vauclair 2004 gave a discussion of these parameterisation procedures. The basic problem here concerns the vertical shear flow instability which occurs between the fingers and the inter-fingers medium, which is not taken into account in the Ulrich procedure. This instability leads to local turbulence which increases the mixing at the edge of the fingers. Consequently, a process of self-destruction appears for the blobs, so that fingers eventually reach a regime where they cannot form anymore: this effect is taken into account in the Kippenhahn et al. procedure.

All this becomes much more complicated if one wants to take into account the competition between thermohaline convection and other macroscopic motions. This question has recently been addressed by Canuto et al. (2008).

Some stellar examples

Thermohaline convection has to be taken into account any time when inverted μ-gradients are built. Situations with ^{4}He enhancement in stellar outer layers were discussed several times in the past, either due to mass transfer (Stothers & Simon 1969), or due to helium diffusion in a stellar wind (Vauclair 1975). The observations, which show that helium is enhanced by a factor two in average, prove that thermohaline mixing must be important in this case, otherwise the helium overabundance would be much larger. However, thermohaline mixing must not be completely efficient, otherwise no helium enhancement would be left at all. This may be due to the presence of a magnetic field. Extensions of these kinds of studies have been discussed for roAp stars by Balmforth et al. 2001. Another interesting case is that of ^{3}He burning regions, specially in giant stars (Abrams & Iben 1970, Ulrich 1971), Eggleton et al. 2006, Charbonnel & Zahn 2007b.

Thermohaline convection should also be taken into account in computations of iron accumulation inside stars. Detailed computations of such iron accumulation, due to atomic diffusion processes, have been performed in several frameworks. Charpinet et al. 1996 found that, in the case of sdB stars, iron accumulation could help triggering stellar oscillations due to the iron-related kappa-mechanism. Richard et al. 2001 found that such an iron accumulation can lead to a special iron-induced convective zone which may drastically change the internal stellar structure during main sequence evolution. However, as the iron accumulation leads to an inverted μ-gradient, the effects of thermohaline convection must be tested in all these computations before reaching a definitive conclusion (in preparation by Théado and Vauclair).

Other cases where thermohaline convection must occur are related to the accretion of metal-rich matter onto main sequence stars. This could happen during the process of planetary formation and migration in the early times of planetary systems. The question arose whether the overmetallicity observed in exoplanet host stars may be due to accretion of planetary material. Vauclair 2004 discussed the effect of thermohaline convection in this situation. It seems clear that, due to this process, most of the accreted material must fall below the "normal" (schwartzschild criterion) convective zone. Several results now converge to reach the conclusion that the observed overmetallicity is primordial.

Another case concerns carbon enhanced main sequence stars (CEMPs). These stars with abundance anomalies are supposed to have suffer some accretion of material coming from an AGB companion. Stancliffe et al.2007 pointed out that in case of accretion of metal-rich matter, this material would subsequently fall down inside the star due to thermohaline convection. In a more recent paper (Thompson et al. 2008), we suggested that, between the stellar birth and the time when the AGB sends its processed material onto it, the main sequence star had time to suffer helium and heavy element diffusion below its convective zone, thereby creating a stabilizing μ-gradient. In the presence of this diffusion-induced μ-gradient, outside matter may accumulate in the convection zone until the overall μ-gradient becomes flat. This effect can save the whole accretion process in this case. New computations have been done by Stancliffe & Glebbeek 2008 reaching similar conclusions.

In summary, thermohaline convection, which has long been forgotten in astrophysics, has to be taken into account in several important cases during stellar evolution. Its competition with other processes like meridional circulation, magnetic fields, etc. will also have to be computed. It may have important consequences, including for asteroseismology as it modifies the accumulation of elements related to the triggering of acoustic waves.

References

Abrams, Z., & Iben, I., Jr. 1970, ApJ, 162, L125

Balmforth, N. J., Cunha, M. S., Dolez, N., et al. 2001, MNRAS, 23, 362

Canuto, V.M., Cheng, Y., & Howard, A.M. 2008, Geophys. Res. Let., 35, 2, L02613

Charbonnel, C., & Zahn, J.-P. 2007b, A&A, 476, L29

Charbonnel, C., & Zahn, J.-P. 2007a, A&A, 467, L15

Charpinet, S., Fontaine, G., Brassard, P., & Dorman, B. 1996, ApJ, 471, L103

Eggleton, P. P.,Dearborn, D. S. P., & Lattanzio, J.C. 2006, Science, 314, 1580

Gargett, A., & Ruddick, B. 2003, ed., Progress in Oceanography, vol 56

Kippenhahn, R., Ruschenplatt, G., & Thomas, H.C. 1980, A&A, 91, 175

Richard, O., Michaud, G., & Richer, J. 2001, ApJ, 558, 377

Stancliffe, R. J., Glenneck, E., Izzard, R.G., & Pold, O.R. 2007, A&A, 464, L57

Stancliffe, R. J., & Glebbeek, E. 2008, MNRAS., 389, 1828

Stothers, R., & Simon, N.R. 1969, ApJ, 157, 673

Thompson, I.B., Ivans, I.I., & Bisterzo, S., et al. 2008, ApJ, 667, 556

Ulrich, R. K. 1971, ApJ, 168, 57

Ulrich, R. K. 1972, ApJ, 172, 165

Vauclair, S. 1975, A&A, 45, 233

Vauclair, S., Dolez,N., & Gough, D.O. 1991, A&A, 252, 618

Vauclair, S. 2004, ApJ, 605, 874

Comm. in Asteroseismology
Vol. 158, 2009, 38th LIAC/HELAS-ESTA/BAG, 2008
A. Noels, C. Aerts, J. Montalbán, A. Miglio and M. Briquet., eds.

Rotational mixing in Magellanic clouds B stars - Theory versus observation

I. Brott[1], I. Hunter[2], A. de Koter[1,4], N. Langer[1], D. Lennon[3] and P. Dufton[2]

[1] Sterrenkundig Instituut Utrecht, Universiteit Utrecht, Princetonplein 5, 3584CC Utrecht, The Netherlands
[2] Astrophysics Research Centre, School of Mathematics & Physics, The Queen's University of Belfast, Belfast, BT71NN, Northern Ireland,UK
[3] Space Telescope Science Inst., 3700 San Martin Drive, Baltimore, MD 21218, USA
[4] Astronomical Institute Anton Pannekoek, University of Amsterdam, Kruislaan 403, 1098SJ Amsterdam, The Netherlands

Abstract

We have used VLT FLAMES data to constrain the physics of rotational mixing in stellar evolution models. We have simulated a population of single stars and found two groups of observed stars that cannot be explained: (1) a group of fast rotating stars which do not show evidence for rotational mixing and (2) a group of slow rotators with strong N enrichment. Binary effects and fossil magnetic fields may be considered to explain those two groups. We suggest that the element boron could be used to distinguish between rotational mixing and the binary scenario. Our single star population simulations quantify the expected amount of boron in fast and slow rotators and allow a comparison with measured nitrogen and boron abundances in B-stars.

Introduction

Rotational mixing is an important process in massive stars. It can affect surface abundances and the internal structure of the star. The VLT FLAMES Survey (Evans et al. 2005) produced nitrogen abundances for a large sample of B stars, providing valuable empirical constraints for this mixing process. Nitrogen is an easily observed tracer element for rotational mixing. It is produced in the stellar center and can be mixed gradually to the surface over the main sequence lifetime. So, in fast rotating stars one expects N enhancements towards the end of the main sequence, while slow rotators should not show any N enhancement. To compare predictions from our stellar evolution models to the observations from Hunter et al. (2008a) (hereafter FLAMES data), we simulate a population of single stars to which we apply the same selection criteria. We limit ourselves here to an analysis of main sequence (MS) stars in the Magellanic Clouds.

The Model Grid and Population Simulation

As input for the population simulation we have calculated a grid of more than 500 evolutionary sequences until core hydrogen exhaustion. The models have been calculated with the stellar evolution code BEC (Langer 1991; Heger et al. 2000). The models include rotational mixing, angular momentum transport by magnetic fields and theoretical mass loss rates of Vink et al. (2001). Rotational mixing and core overshooting have been calibrated at an evolutionary

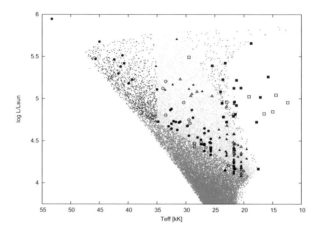

Figure 1: Result of a population synthesis calculation for 10^6 stars between 5 and 50 $M_\odot$ for LMC composition in the HR diagram. Overplotted are the observed FLAMES data. Circles are stars with $\log g > 3.7$ dex (i.e. early MS stars), triangles are stars with $3.2 \leqslant \log g \leqslant 3.7$ dex (i.e. late MS stars), squares symbolize stars with $\log g < 3.2$ dex. Open/filled symbols represent probable binaries/single stars, respectively. The colored regions are explained in the text. Only the yellow region is considered in the analysis below.

model representing the average mass and rotation velocity of the LMC sample (Hunter et al. 2008b). The mass and initial velocity range of the model grid have been chosen to cover most of the FLAMES data, i.e $5 - 50\,M_\odot$, $0 - 500$ km/s. It covers three metallicities, using chemical compositions for SMC, LMC and Galactic evolution models (in preparation by Brott and collaborators). As input chemical composition for our stellar models we used C,N,O abundances from H II regions (Kurt & Dufour 1998) and Mg, Si and Fe from unenriched B-stars. The abundances for remaining elements have been taken from Asplund et al. (2005) and lowered by 0.4 and 0.7 dex for the LMC and SMC, respectively.

By comparing isochrones to the LMC and SMC FLAMES data we find a wide age spread, consistent with a constant star formation rate for both observation sets. This is in agreement with most of the observed B-stars being field stars (Hunter et al. 2008c). We assume a Salpeter initial mass function. Since the initial velocity distribution of the stars is unknown, we assume that the observed distribution is close to the initial one. Mass loss is generally small for MS B-stars at LMC metallicity, justifying this approximation. For each simulated star an age, initial mass, initial rotational velocity and a random inclination is drawn from the appropriate distribution functions.

Fig. 1 shows an HR diagram of the observed and simulated data points of the LMC. In order that our simulations are consistent with the FLAMES observational dataset, we have excluded simulations that fall below the FLAMES magnitude limit (green) or that have $T_{eff} > 35$ kK. The analysis of the hotter O-type sample (blue) by Mokiem et al. (2007) has not been considered as no Nitrogen abundance estimates were available (blue). We have excluded post-MS stars, which we define as having $\log g < 3.2$ dex (red) (Hunter et al. 2008b). This gravity limit is appropriate for models with initial masses of $13\,M_\odot$ but may be too high for higher mass models due to the large overshooting parameter ($\alpha = 0.335$) in the models. We note that the observed points that we have excluded are all slow rotators and hence will not significantly affect our conclusions. In the next section we compare all remaining data points (yellow) to the FLAMES data.

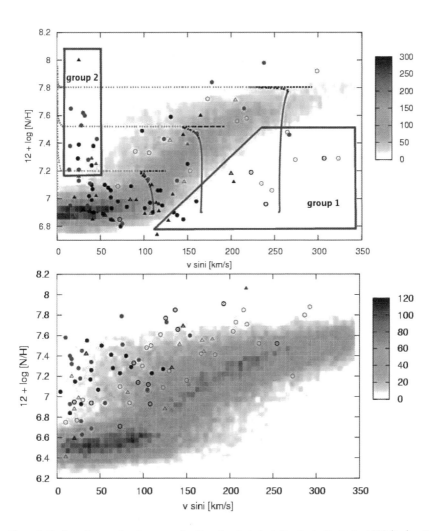

Figure 2: Surface nitrogen abundance as a function of projected rotational velocity for the LMC (top) and SMC (bottom). Observations and a set of 13 M$_\odot$ evolution tracks for different initial rotation velocities are compared. Red (dark gray) symbols are lower MS objects ($\log g > 3.7$), the corresponding parts on the tracks are marked in the same color. Black symbols are upper MS objects ($3.2 \leqslant \log g \leqslant 3.7$), the corresponding part on the track is marked in the same color. Triangles show possible binaries, circles are single stars and open symbols are upper limits to the nitrogen abundance. The shading in the background gives the number of simulated stars in this field, as shown by the bars on the right.

Nitrogen in the LMC and SMC

We have performed a simulation of 10^6 stars between 5 and 50 $M_\odot$ for both Magellanic Clouds. Only data passing the observational selection criteria (LMC $\sim$ 4.1%, SMC $\sim$ 4.9%) were kept. For better visibility we have binned the data, shown in the density plots in Fig. 2. The evolutionary tracks in Fig. 2 top (LMC) show that nitrogen is mixed gradually to the surface during the MS evolution. The population simulation for single stars shows for both the LMC (top) and SMC (bottom) that, while in fast rotating stars a significant enrichment in N is expected, slow rotators show almost no enhancement. While this is true for most of the LMC FLAMES data, there are two groups that do not agree with the simulation results. At high $v \sin i$ and low N, a group of mainly upper limits is populating the diagram (group 1). The simulation cannot populate this region, because fast rotating and non-enriched stars are very young and close to the Zero Age Main Sequence (ZAMS). Thereby they are too faint to pass the survey magnitude limit. These stars could have been spun up in a binary system, but only two objects in this group are identified radial velocity variables. Group 2 (low $v \sin i$ and highly N enriched) is also in clear contradiction to the predictions. These stars cannot all be pole on stars, but are intrinsically slow rotators. Morel et al. (2008) have shown for a Galactic sample of slowly rotating N-rich stars that a large fraction of them is magnetic. This suggests that internal magnetic fields might give rise to the abundance anomalies of this group.   About 50% of the FLAMES sample are consistent with our single-star models with rotational mixing (see Maeder et. al, this volume), however neither set of models can explain groups 1 and 2. In fact the binary models of Langer et al. (2008) show promise of accounting for these groups and, to add further ambiguity to the interpretation, can explain the so-called normal stars as well.

In the SMC (Fig. 2, bottom) almost all observed stars seem to have higher nitrogen enrichments than predicted by the models, even though the baseline is in agreement with the most unenriched stars in the sample. However, almost all faster rotators ($v \sin i > 100$ km/s) have upper limits, hence are consistent with the predictions. As suggested by Hunter et al. 2008a the enriched group at low velocities might be the analogue of group 2 in the LMC. In both cases, the group consists mainly of high log g objects.

Boron to Discriminate between a Binary and Single Star Scenario

The true binary fraction in the sample is of key importance to understand the groups that are in disagreement with the single star simulation. The element boron can only exist in the coolest outer layers of the star. It will be gradually destroyed by rotational mixing as shown in Fig. 3. In a mass transfer scenario practically boron free material is dumped onto the mass gainer. Thus, in a pure binary sample, boron is either almost at its initial abundance or it is significantly depleted by mass transfer (Fliegner et al. 1996). Boron abundances therefore will help identify effects of single versus binary stars in diagnostics such as Fig. 2, even though Boron measurements in the Magellanic Clouds are still very challenging at present.

Acknowledgments. Thanks to Peter Anders for many helpful discussions.

References

Asplund, M., Grevesse, N., & Sauval, A. 2005, ASPC, 336, 25

Evans, C.J., Smartt, S.J., Lee, J-K., et al. 2005, A&A, 437, 467

Fliegner, J., et al. 1996, A&A, 308, L13

Heger, A., Langer, N., & Woosley, S.E. 2000, ApJ, 528, 368

Hunter, I., Brott, I., Langer, N., et al. 2008a, A&A, submitted

Hunter, I., Brott, I., Lennon, D.J., et al. 2008b, ApJ, 676, L29

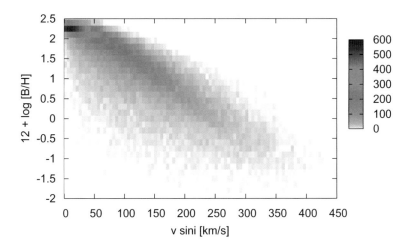

Figure 3: Predicted boron vs. rotational velocity for a population of single stars in the LMC. The trend reflects the transport of this element to hot deep layers, through rotational mixing, where it is subsequently destroyed.

Hunter, I., Lennon, D.J., Dufton, P.L., et al. 2008c, A&A, 479, 541

Kurt, C.M. & Dufour, R.J. 1998,RMxAC, 202

Langer, N. 1991, A&A, 252, 669

Langer, N., Cantiello, M., Yoon, S.-C., et al. 2008, IAUS, 250, 167

Maeder, A., Meynet, G., Ekstrom, S., et al. 2009, CoAst, 158, 72

Mokiem, M.R., de Koter, A., Evans, C.J., et al. 2007, A&A, 465, 1003

Morel, T., Hubrig, S., & Briquet, M. 2008 A&A, 481, 452

Vink, J.S., de Koter, A., & Lamers, H.J.G.L.M. 2001, A&A, 369, 574

DISCUSSION

Maeder: The zero of your sample stars does not coincide with the zero of your model simulations of rotating stars for the SMC. Thus, we can wonder how meaningful is the comparison.
Brott: Since most of these stars are upper MS stars, we can *not* assume that the average N abundance is representing the initial compositions. Using CNO abundances from HII regions should give us a better approximation of the initial conditions, since they are not yet polluted by the remnants of rotation while the stars might be. The fact that the HII-region base line and the base line from the stars do not agree shows that there is a factor which we still do not understand. It might be related to the HII regions measurements or to the stellar sample.
Maeder: The relation between the N enhancements and the vsini depends very much on the masses and one has to be very careful to make the comparisons for narrow ranges of masses, ages and metallicities.
Brott: We have carefully looked at the mass range that has to be considered, by comparing the observed sample to isochrones.

Meynet: In your plot $(12 + \log [N/H])$ versus vsini for the SMC, there is no observed stars with high vsini and low [N/H] as there seems to be in the LMC sample. Have you an explanation for that?

Brott: The single star population synthesis does not predict any high vsini/low [N/H] stars because this part of the track is below the magnitude limit. For the observations, this might be an observational bias because it is much harder to measure abundances, especially for fast rotators, in the SMC compared to the LMC.

Comm. in Asteroseismology
Vol. 158, 2009, 38th LIAC/HELAS-ESTA/BAG, 2008
A. Noels, C. Aerts, J. Montalbán, A. Miglio and M. Briquet., eds.

On the origin of microturbulence in hot stars

M. Cantiello[1], N. Langer[1], I. Brott[1], A. de Koter[1,2] S. N. Shore[3], J. S. Vink[4], A. Voegler[1]
and S.-C. Yoon[5]

[1] Astronomical Institute, Utrecht University
Princetonplein 5, 3584 CC, Utrecht, The Netherlands
[2] Astronomical Institute Anton Pannekoek, University of Amsterdam, Kruislaan 403, 1098 SJ,
Amsterdam, The Netherlands
[3] Dipartmento di Fisica "Enrico Fermi", Università di Pisa, via Buonarroti 2, Pisa 56127 and INFN -
Sezione di Pisa, Italy
[4] Armagh Observatory, College Hill, Armagh, BT61 9DG, Northern Ireland (UK)
[5] Department of Astronomy & Astrophysics, University of California, Santa Cruz, High Street, Santa
Cruz, CA 95064, USA

Abstract

We present results from the first extensive study of convection zones in the envelopes of hot
massive stars, which are caused by opacity peaks associated with iron and helium ionization.
These convective regions can be located very close to the stellar surface. The region in the
Hertzsprung-Russel diagram in which we predict the convection zones and the strength of
this convection is in good agreement with the occurrence and strength of microturbulence in
massive stars. We further argue that convection close to the surface may trigger clumping at
the base of the stellar wind of hot massive stars.

Introduction

With the introduction of the so-called iron peak in stellar opacities (Iglesias et al. 1992)
a convection zone appears in the envelope of sufficiently luminous massive main sequence
models (Stothers & Chin 1993). This convective region contains little mass and is usually
not discussed in the context of stellar evolution calculations. Here, we mainly focus on
the question whether the occurrence of sub-surface convection might be correlated with
observable small scale velocity fields at the stellar surface and in the stellar wind. A similar
idea has been used to explain microturbulence in low mass stars (Edmunds 1978), in which
envelope convection zones are extended and reach the stellar photosphere. While Edmunds
(1978) concludes that the same mechanism can *not* explain microturbulent velocities in O
and B stars, the iron-peak induced sub-photospheric convection zones in these stars were not
yet discovered at that time. In fact, we demonstrate below that sub-surface convection may
not only cause photospheric velocity fields which are observable, but possibly even induce
clumping at the base of the stellar wind.

Method

We performed a systematic study of sub-surface convection by calculating models of hot stars
with a hydrodynamic stellar evolution code (see for example Yoon et al. 2006). The Ledoux
criterion is used to determine which regions of the star are unstable to convection, and the

convective velocity is calculated according to the Mixing Length Theory (Böhm-Vitense 1958) with mixing length parameter $\alpha = 1.5$. The opacities in our code are extracted from the OPAL tables (Iglesias & Rogers 1996). We calculated a grid of non-rotating stellar evolutionary models for initial masses between $5\,M_\odot$ and $100\,M_\odot$, at metallicities of Z=0.02, Z=0.008 and Z=0.004, roughly corresponding to the Galaxy, the LMC and the SMC, respectively.

Sub-surface convective regions

The iron convective region in the envelope of hot stars corresponds to a peak in the opacity at $\log T \simeq 5.3$. The appearance and properties of this subsurface convective zone have been studied. In particular we used our grid of models to map on the HR diagram the average convective velocity in the upper part of the iron convection zone $\langle v_c \rangle$. Our main findings are:

- For given luminosity and metallicity, $\langle v_c \rangle$ increases with decreasing surface temperature. The convection zones are located deeper inside the star (in radius, not in mass), and the resulting larger pressure scale height leads to larger velocities. At solar metallicity and $10^5\,L_\odot$ (i.e. roughly at $20\,M_\odot$) the velocities increase from less than $1\ \mathrm{km\,s}^{-1}$ at the ZAMS to more than $5\,\mathrm{km\,s}^{-1}$ in the supergiant regime, where $\langle v_c \rangle = 2.5\,\mathrm{km\,s}^{-1}$ is achieved at $T_\mathrm{eff} \simeq 30\,000\,\mathrm{K}$. At SMC metallicity, the iron convection zone is absent at the ZAMS for $L < 10^5\,L_\odot$, and a level of $\langle v_c \rangle = 2.5\,\mathrm{km\,s}^{-1}$ is only reached at $T_\mathrm{eff} \simeq 20\,000\,\mathrm{K}$.

- For given effective temperature and metallicity, the iron zone convective velocity increases with increasing luminosity, as a larger flux demanded to be convectively transported requires faster convective motions. We found threshold luminosities below which iron convection zones do not occur, i.e., below about $10^{3.2}\,L_\odot$, $10^{3.9}\,L_\odot$, and $10^{4.2}\,L_\odot$ for the Galaxy, LMC and SMC, respectively.

- The iron convection zones become weaker for smaller metallicities, since due to the lower opacity, more of the flux can be transported by radiation. The threshold luminosity for the occurrence of the iron convection zone quoted above for $Z = 0.02$ is ten times lower than that for $Z = 0.004$. And above the threshold, for a given point in the HR diagram, the convective velocities are always larger at larger metallicity.

Microturbulence

The microturbulent velocity ξ is the non-thermal component of the photospheric velocity field which has a correlation length smaller than the size of the line forming region. The non-thermal velocity component with a larger correlation length is referred to as macroturbulence (e.g. Aerts et al. 2009).

The convective cells in the upper part of a convection zone can generate acoustic and gravity waves propagating outward. The problem of sound waves generated by turbulent motions has been discussed by Lighthill (1952) and extended to a stratified atmosphere by Stein (1967) and Goldreich & Kumar (1990). In the presence of stratification, gravity acts as a restoring force and allows the excitation of gravity waves as well. For both, acoustic and gravity waves, the most important parameter determining the emitted kinetic energy flux is the velocity of the convective motions. This is the reason why we used $\langle v_c \rangle$ as the crucial parameter determining the efficiency of subsurface convection. Lighthill (1952) and Stein (1967) showed that convection excites acoustic and gravity waves, with a maximum emission for waves with $\lambda \sim H_{P,c}$, the pressure scale height at the top of the convective region. They calculated the amount of convective flux F_c that is going into acoustic waves, $F_{\mathrm{ac}} \sim F_c\,M_c^5$, and gravity waves, $F_g \sim F_c M_c$, where we take $F_c \sim \rho_c \langle v_c \rangle^3$ and M_c is the Mach number for

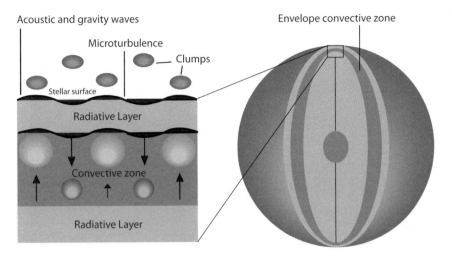

Figure 1: Schematic representation of the physical processes connected to subsurface convection. Acoustic and gravity waves emitted in the convective zone travel through the radiative layer and reach the surface, inducing density and velocity fluctuations. In this picture microturbulence and clumping at the base of the wind are a consequence of the presence of subsurface convection.

convection calculated at the top of the convective region. Since convection in our models is highly subsonic, gravity waves are expected to extract more energy than acoustic waves from the convective region. These waves can propagate outward, steepen and become dissipative in the region of line formation. Here, they may induce density and velocity fluctuations (Fig.1). The energy associated with the induced velocity fluctuations must be comparable or smaller than the energy in the waves above the convective zone. For microturbulence to be excited by this process, it is required that $E_g \geq E_\xi$ or

$$\frac{E_g}{E_\xi} \sim M_c \left(\frac{\rho_c}{\rho_s}\right) \left(\frac{\langle v_c\rangle}{\xi}\right)^2 \geq 1, \tag{1}$$

where ρ_c is the density at the top of the convective region, ρ_s is the surface density, ξ is the microturbulent velocity, $E_\xi \sim \rho_s \xi^2$ is the energy per cubic centimeter associated with the microturbulent velocity field and $E_g \sim M_c \rho_c \langle v_c\rangle^2$ is the amount of convective energy per cubic centimeter which is transported outward by gravity waves. In the ratio (1) we only consider energy per cubic centimeter since the volume of the line forming region is comparable to the volume of the upper part of the convective zone. We evaluated the ratio (1) using the data calculated from our models and a value for the microturbulent velocity ξ of $10\,\mathrm{km\,s^{-1}}$. This value has been chosen according to a set of microturbulent velocity determinations in massive stars by Trundle et al. (2007) and Hunter et al. (2008), which has been obtained in the context of the ESO FLAMES Survey of Massive Stars (Evans et al. 2005), and has been analyzed in a coherent way. This provides a relatively large database, even after restricting the data set to slow rotators, i.e. $v_{rot} \sin i < 80\,\mathrm{km\,s^{-1}}$. The error in the microturbulent velocity measurements is about 5 $\mathrm{km\,s^{-1}}$, which justifies our choice of $\xi = 10\,\mathrm{km\,s^{-1}}$ in the evaluation of the ratio (1). Fig. 2 shows that the process of excitation of microturbulence through sub-surface convection is energetically possible. Moreover, the region where sub-surface convection is efficiently generating gravity waves corresponds very well with the location of stars in which a microturbulent velocity field is clearly present. Using

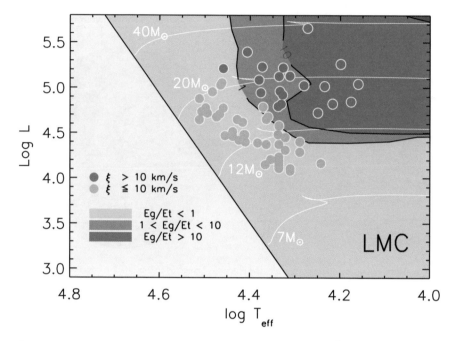

Figure 2: This HR diagram shows values of the ratio E_g/E_ξ for $\xi = 10\,\mathrm{km\,s^{-1}}$ at LMC metallicity. We estimated the ratio as in Eq. 1 using the parameters obtained by stellar evolution calculations. Solid white lines are reference evolutionary tracks. The full drawn black line is the zero age main sequence. Overplotted as filled circles are photospheric microturbulent velocities ξ derived in a consistent way for hot massive stars by Trundle et al. (2007) and Hunter et al. (2008). Here, we use only data for stars with an apparent rotational velocity of $v_{\mathrm{rot}} \sin i < 80\,\mathrm{km\,s^{-1}}$. The uncertainty in the determination of ξ is typically $\pm 5\,\mathrm{km\,s^{-1}}$.

our grid of stellar models and the microturbulent velocity dataset of Trundle et al. (2007) and Hunter et al. (2008), these results have been confirmed also for stars in the SMC and the Galaxy.

The amount of energy that can be transported in the region of line formation by gravity waves depends on the dissipative processes in the radiative layer between the top of the convective region and the stellar surface. In this study we did not consider such a process. The propagation of waves in the region of interest needs to be investigated in a future work.

Clumping at the base of the stellar wind

Evidence has been accumulating that the winds of massive stars are inhomogeneous on different spatial scales. Indeed, evidence that O star winds are clumped is given by, among others, Puls et al. (2006). These authors investigate the clumping behavior of the inner wind (inside about two stellar radii) relative to the clumping in the outer wind (beyond tens of stellar radii) of a large sample of supergiant and giant stars. They find that in stars having strong winds, the inner wind is more strongly clumped than the outer wind, whereas those having weak winds have similar clumping properties in the inner and outer regions. The analysis only allows for such a relative statement. Interestingly, this type of radial behaviour is not consistent with hydrodynamic predictions of the intrinsic, *self-excited* line-driven instability (Runacres & Owocki 2002,2005). Such models predict a lower clumping in the inner wind than in the outer wind. If we compare the O stars investigated by Puls et al. (2006) with our models, the trend is such that stars with relatively strong clumping in the inner winds are in a regime where the subsurface convective velocity is larger. A correlation between clumping at the base of the wind and $\langle v_c \rangle$ may point to sub-photospheric convection as a possible excitation mechanism of clumping at the base of the wind. If mass loss is affected by this clumping, the subsurface convective region in massive stars may impact stellar evolution.

Acknowledgments. Matteo Cantiello acknowledges financial support from HELAS.

References

Aerts, C., et al. 2009 CoAst, 158, 66

Böhm-Vitense, E. 1958, Zeitschrift fur Astrophysik, 46, 108

Edmunds, M. G. 1978, A&A, 64, 103

Evans, C. J., Smartt, S. J., Lee, J.-K., et al. 2005, A&A, 437, 467

Goldreich, P., & Kumar, P. 1990, ApJ, 363, 694

Hunter, I., Lennon, D. J., Dufton, P. L., et al. 2008, A&A, 479, 541

Iglesias, C. A., & Rogers, F. J. 1996, ApJ, 464, 943

Iglesias, C. A., Rogers, F. J., & Wilson, B. G. 1992, ApJ, 397, 717

Lighthill, M. F. 1952, Royal Society of London Proceedings Series A, 211, 564

Puls, J., Markova, N., Scuderi, S., et al. 2006, A&A, 454, 625

Runacres, M. C., & Owocki, S. P. 2002, A&A, 381, 1015

Runacres, M. C., & Owocki, S. P. 2005, A&A, 429, 323

Stein, R. F. 1967, Sol. Phys., 2, 385

Stothers, R. F., & Chin, C.-W. 1993, ApJ, 408, L85

Trundle, C., Dufton, P. L., Hunter, I., et al. 2007, A&A, 471, 625

Yoon, S.-C., Langer, N., & Normann, C. 2006, A&A, 460, 199

DISCUSSION

Grevesse: How does one actually disentangle between micro- and macroturbulence?

Cantiello: Microturbulence changes the equivalent width of strong lines while macroturbulence does not. From an observational point of view this should allow to discern the two using curve of growth analysis.

Grevesse: How does fine structure affect so much the opacity in Fe?

Cantiello: The introduction of spin-orbit coupling into the atomic calculations for iron results into a huge increase in the number of spectral lines. The splitting of the LS (Russell-Saunders coupling) term into total angular momentum J-levels is also accompanied by *many inter combination transitions*, i.e; transitions that change the total spin, forbidden in the pure LS coupling. This results in a significative increase of the mean Rosseland opacity at $\log T \sim 5.3$. As a reference : Iglesias, Rogers & Wilson, 1992, in particular fig. 2 and fig. 8.

Dziembowski: Which of the convective zones (Z, HeII) plays the main role in your simulations?

Cantiello: We decided to focus only on Fe convective zones, since He convection is extremely inefficient in our models.

Comm. in Asteroseismology
Vol. 158, 2009, 38th LIAC/HELAS-ESTA/BAG, 2008
A. Noels, C. Aerts, J. Montalbán, A. Miglio and M. Briquet., eds.

On the origin of macroturbulence in hot stars

C. Aerts[1,2], J. Puls[3], M. Godart[4], M.-A. Dupret[5]

[1] Instituut voor Sterrenkunde, Celestijnenlaan 200D, B-3001 Leuven, Belgium
[2] Department of Astrophysics, IMAPP, Radboud University Nijmegen, PO Box 9010, 6500 GL, Nijmegen, the Netherlands
[3] Universitäts-Sternwarte, Scheinerstrasse 1, D-81679 München, Germany
[4] Institut d'Astrophysique et Géophysique, Université de Liège, allée du Six Août 17, B-4000 Liège, Belgium
[5] Observatoire de Paris, LESIA, 5 place Jules Janssen, 92195 Meudon Principal Cedex, France

Abstract

Since the use of high-resolution high signal-to-noise spectroscopy in the study of massive stars, it became clear that an ad-hoc velocity field at the stellar surface, termed macroturbulence, is needed to bring the observed shape of spectral lines into agreement with observations. We seek a physical explanation of this unknown broadening mechanism. We interprete the missing line broadening in terms of collective pulsational velocity broadening due to non-radial gravity-mode oscillations. We also point out that the rotational velocity can be seriously underestimated whenever the line profiles are fitted assuming a Gaussian macroturbulent velocity rather than an appropriate pulsational velocity expression.

Macroturbulence in hot massive stars

Velocity fields of very different scales occur in the atmospheres of hot massive stars. Apart from the rotational velocity, which is usually assumed to be uniform across the stellar disk and which can vary from zero speed up to the critical value (of several hundred $km\,s^{-1}$), line-prediction codes also include a certain amount of microturbulence (of order a few $km\,s^{-1}$) to bring the observed profiles into agreement with the data. Microturbulence is related to velocity fields with a scale that is shorter than the mean free path of the photons in the atmosphere (we refer to Cantiello et al. 2009, for a thorough explanation).

In recent years, the number of hot massive stars that have been studied with high-resolution spectroscopy with the goal to derive high-precision fundamental parameters has increased quite dramatically (e.g, Ryans et al. 2002, Lefever et al. 2007, Markova & Puls 2008 and references therein). This has led to the need to introduce an ad-hoc velocity field, termed macroturbulence, at the stellar surface in order to explain the high-quality data to an appropriate level. This need for macroturbulent broadening was, in fact, already emphasized by Howarth et al. (1997) from his study of massive stars from low-resolution UV spectroscopy from the space mission IUE. In contrast to microturbulence, macroturbulence refers to velocity fields with a scale longer than the mean free path of the photons. In practice, the studies listed above resulted in the requirement to introduce macroturbulence of the order of several tens of $km\,s^{-1}$, and quite often even supersonic velocity fields.

The abovementioned studies including macroturbulence rely on single snapshot spectra and did not take into account pulsational velocity fields so far, as it is done in time-resolved high-resolution spectroscopy of pulsating hot stars (e.g., Aerts & De Cat 2003). A natural step

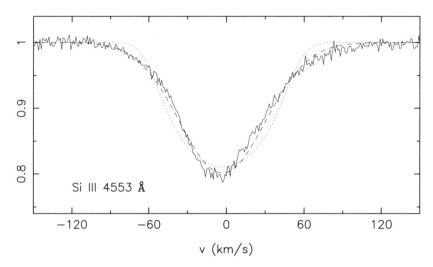

Figure 1: A simulated Si line in the spectrum of a massive hot star (full line) including, besides microturbulence and rotation, also the collective effect of numerous very low-amplitude gravity-mode oscillations which broaden the line wings. The input rotation was $v \sin i = 45 \, \mathrm{km \, s^{-1}}$. This profile is fitted with a model taking only rotational broadening and microturbulence into account (dotted line, estimated $v \sin i$ is $57 \, \mathrm{km \, s^{-1}}$) as well as with a model including microturbulence, rotation and Gaussian macroturbulence (dashed line, estimated $v \sin i$ and macroturbulence are, respectively, 5 and $32 \, \mathrm{km \, s^{-1}}$).

is to investigate whether the needed macroturbulence may simply be due to the omission of pulsational broadening in the line-prediction codes used for fundamental parameter estimation. In fact, for pulsating stars along the main sequence, one also needs to add some level of macroturbulence whenever one ignores (some of) the detected pulsations in line profile fitting of time-resolved or averaged spectra (e.g., Aerts & De Cat 2003; Morel et al. 2006). We investigate this hypothesis in the present paper.

Line-profile computations in the presence of stellar oscillations

We computed the excited non-radial oscillations with mode degree up to 10 for a stellar evolution model representative of the pulsating evolved B1Ib star HD 163899 with the Code Liégeois d'Évolution Stellaire (Scuflaire et al. 2008) and with the non-adiabatic pulsation code MAD (Dupret 2001). This model has $T_{\mathrm{eff}} = 18,200 \, \mathrm{K}$, $\log g = 3.05$, $R/R_{\odot} = 17.8$, $\log(L/L_{\odot}) = 4.5$, $M/M_{\odot} = 13$, $Z = 0.02$ and an age of 13 million years. The 241 excited $m = 0$ modes are all gravity modes with frequencies between 0.08 to 0.68 cycles per day and ratios of horizontal to vertical velocity displacement in the range 0.3 to 25. They give rise to 2965 rotationally split components. These were used to compute 504 sets of time-resolved spectroscopic line profiles of 50 profiles each, with peak-to-peak amplitudes for the radial velocity between 0.7 to $15 \, \mathrm{km \, s^{-1}}$ and for rotational velocities between 25 and $125 \, \mathrm{km \, s^{-1}}$, which is below 25% of the critical value, following the method of Aerts et al. (1992). Such radial-velocity amplitudes are well below those observed for hot B and A supergiants (e.g., Kaufer et al. 1997, Prinja et al. 2004) so that we can be sure not to have overestimated the effects of oscillations on the lines. These profiles were then fitted ignoring the oscillations but allowing for a Gaussian macroturbulent velocity.

Using a goodness-of-fit approach, we confirm the finding that the inclusion of an ad-

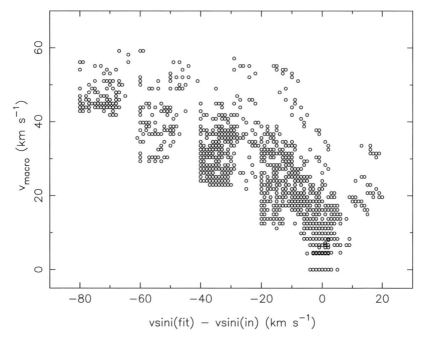

Figure 2: The macroturbulent velocity as a function of the rotational velocity estimate, derived from line profile fits ignoring the presence of pulsational broadening but allowing an ad-hoc Gaussian macroturbulence parameter, for the simulations described in the text.

hoc macroturbulence parameter leads to better fits than those obtained when only allowing rotational and microturbulent broadening (Fig. 1). The pulsational broadening ignored in the line profile fits is compensated by allowing a macroturbulent velocity. These ad-hoc velocities are sometimes in excess of the speed of sound to ensure a good fit (Fig. 2). At first sight, it might seem surprising to need such high macroturbulent speeds to explain the collective effect of oscillation modes that have by themselves only very low velocity amplitude. However, this is easy to understand if one realises that line widths depend on velocity squared, such that numerous small velocities add up to give a very significant effect in the overall line broadening when the collective effect of the modes is interpreted by a single ad-hoc parameter. This is particularly the case for gravity-mode oscillations which impact strongly on line wings. On the other hand, such low-amplitude modes do not alter seriously the observed quantities behaving linearly with velocity, such as the radial velocity variations of the star, because their collective effect tends to cancel out in this case.

As a side result of our line fits with macroturbulence, we report a risk to underestimate the projected rotational velocity appreciably when using the Fourier Transform (FT) method to estimate the projected rotational velocity from the first minimum of the FT (Simón-Díaz & Herrero 2007), as is illustrated in Fig. 3. While this method works well in general and is able to recover the correct input value of the rotational broadening for most of the cases, it is sometimes fooled when too asymmetric pulsational broadening is present and, in this case, one derives too low estimates for $v \sin i$. This is also the case for the results from a goodness-of-fit approach (see Fig. 1).

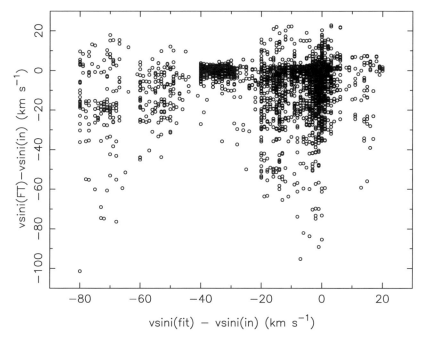

Figure 3: The excess of the rotational velocity estimated from the Fourier transform method (FT) versus the input value is plotted as a function of the excess of the estimated rotational velocity from line profile fits allowing for macroturbulence (fit) versus the input value.

Conclusions

We have shown that pulsational velocity broadening due to the collective effect of numerous low-amplitude gravity mode oscillations offers a natural and appropriate physical explanation for the occurrence of macroturbulence in hot massive stars. Our computations of course do not exclude other and/or additional physical interpretations of the macroturbulent velocities reported in the literature.

   We ignored rotational and non-adiabatic effects in the computations of the velocity eigenfunctions of the non-radial modes used for the line profile simulations. Codes to incorporate each of these effects separately are available (e.g., Townsend 1997, De Ridder et al. 2002) and their influence on the line profiles are well understood. In particular, they will not alter the line wings of the profiles dramatically as long as the rotational velocity remains below 25% of the critical velocity (Aerts & Waelkens 1993) and would thus not alter the main conclusions of our work, while implying a very serious increase in CPU time.

   From the observational side, high-precision multi-epoch observations of unblended metal lines are needed to evaluate appropriately the effect of pulsational broadening. Such type of data have not been used so far to estimate macroturbulence.

Acknowledgments. The computations for this research have been done on the VIC HPC supercomputer of the K.U.Leuven. CA is supported by the Research Council of K.U.Leuven under grant GOA/2008/04.

References

Aerts, C., & De Cat P. 2003, SSRv, 105, 453

Aerts, C., De Pauw, M., & Waelkens, C. 1992, A&A, 266, 294

Aerts, C., & Waelkens, C. 1993, A&A, 273, 135

Cantiello, M., Langer, N., Brott, I., et al. 2009, CoAst, 158, 61

De Ridder, J., Dupret, M.-A., Neuforge, C., & Aerts, C. 2002, A&A, 385, 572

Dupret, M.-A. 2001, A&A, 366, 166

Howarth, I. D., Siebert, K. W., Hussain, G. A. J., & Prinja, R. K. 1997, MNRAS, 284, 265

Kaufer, A., Stahl, O., Wolf, B., et al. 1997, A&A, 320, 273

Lefever, K., Puls, J., & Aerts, C. 2007, A&A, 463, 1093

Markova, N., & Puls, J. 2008, A&A, 478, 823

Morel, T., Butler, K., Aerts, C., et al. 2006, A&A, 457, 651

Prinja, R. K., Rivinius, Th., Stahl, O., et al. 2004, A&A, 418, 727

Ryans, R. S., Dufton, P. L., Rolleston, W. R. J., et al. 2002, MNRAS, 336, 577

Scuflaire, R., Théado, S., Montalbán, J., et al. 2008, ApSS, 316, 83

Simón-Díaz, S., & Herrero, A. 2007, A&A, 468, 1063

Townsend, R. H. D. 1997, MNRAS, 284, 839

Session 2
Physics and uncertainties and their effects on the
internal structure

Comm. in Asteroseismology
Vol. 158, 2009, 38th LIAC/HELAS-ESTA/BAG, 2008
A. Noels, C. Aerts, J. Montalbán, A. Miglio and M. Briquet., eds.

Modeling massive stars with rotation: the case of Nitrogen enrichments

A. Maeder, G. Meynet, S. Ekström, C. Georgy

Observatory of the Geneva University

Abstract

Recently, the concept of rotational mixing has been challenged by some authors (e.g. Hunter et al. 2008). We show that the excess N/H is a multivariate function $f(M, \quad age, \quad v \sin i, \quad multiplicity, \quad Z)$. To find a correlation of a multivariate function with some parameter, it is evidently necessary to limit the range of the other involved parameters as much as possible. When this is done, the concept of rotational mixing is supported by the observations. We also show that the sample data are not free from several biases. A fraction of $\sim 20\ \%$ of the stars may escape to the relation as a result of binary evolution.

Introduction

In the late 70's, the reality of mass loss in massive stars was debated as well as its effect on the evolution. When this eventually became accepted (Chiosi & Maeder 1986), the next question was whether the large differences in the populations of massive stars (for example the WR/O and WN/WC star numbers) were due to differences in mass loss with metallicity Z. Today, the debate concerns the reality of rotational mixing and its effects on massive star evolution. History never reproduces itself similarly, nevertheless there is a great parallelism in these debates, which are normal steps in the progress of knowledge and finally lead to a better understanding of stellar physics and evolution.

There is an impressive list of consequences of stellar rotation (Maeder & Meynet 2000), many of which are supported by observations: about the stellar shape, the temperature distribution at the surface, the mass loss and its asymmetries, on the size of the cores, the tracks in the HR diagram, the lifetimes, the surface composition, the chemical yields, the ratios of the different kinds of massive stars (blue, red supergiants, WR stars), the types of supernovae and the remnant masses, etc. In this Liege Colloquium, the reality of the rotational mixing was disputed by several authors on the basis of new observations, in particular the VLT-Flames survey (Hunter et al. 2008). Thus, we concentrate here on this problem, firstly by recalling some theoretical predictions concerning surface enrichments and secondly by carefully examining the observations.

Recall of theoretical predictions concerning rotational mixing

In the mass range of ~10 to 20 $M_\odot$ considered here, mass loss has a limited importance during the Main Sequence (MS) phase. The changes of abundances are expected to be mainly due to rotational mixing. The main effect producing element mixing is the diffusion by shear turbulence, which itself results from the internal Ω gradients built during evolution. To a smaller extent, meridional circulation makes some transport, however mainly of angular

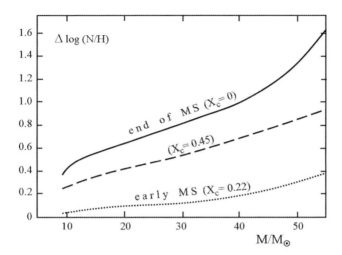

Figure 1: The differences in $\log(N/H)$ as a function of the initial masses at 3 stages during the MS phase for models with $Z = 0.02$ with average rotation velocities of 217, 209, 197, 183, 172, 168 km s^{-1} for respectively 12, 15, 20, 25 40 and 60 M$_\odot$. These 3 stages are indicated by the value of the central H–content X_c.

momentum. Mixing brings to the surface the products of CNO burning: mainly ^{14}N and ^{13}C enrichments, ^{12}C is depleted with limited ^{4}He enrichment and ^{16}O depletion.

Fig. 1 shows the predicted variations of $\log(N/H)$ during MS evolution as a function of the initial masses (Meynet & Maeder 2000). (N/H) is here the abundance ratio of N and H in numbers (the relative differences in mass and number are the same). Without rotational mixing, there would be no enrichment until the red supergiant stage. Rotation produces an increase (depending on velocity v) of N/H during the MS phase. The N excesses also depend on the ages t. The increase is modest during the first third of the MS phase, because the elements need some time to reach the surface, then it is more rapid. The N enrichments are larger for larger masses M. Thus we see that the N excesses are multivariate functions

$$\Delta \log(N/H) = f(M, t, v \sin i, \text{multiplicity}, Z) . \tag{1}$$

Models with lower initial metallicities Z have higher N enrichments for given M and v (Maeder & Meynet 2001; Meynet et al. 2006). The excesses become very strong at metallicities Z as low as 10^{-8}.

Fig. 2 shows, for given M and Z, the evolution of the N excesses with age for different initial velocities. This figure shows that there can be no single relation between $v \sin i$ and the N/H excesses for a mixture of ages. The same is true for a mixture of different masses.

Binarity may also affect the N and He enrichments due to tidal mixing and mass transfer. A binary star with low rotation may have a high N/H due to tidal mixing or due to the transfer of the enriched envelope of a red giant. At the opposite, a binary star may also have a high $v \sin i$ and no N/H excess, in the case of the accretion of an unevolved envelope bringing a lot of angular momentum. A nice illustration has been given (Martins, 2009).

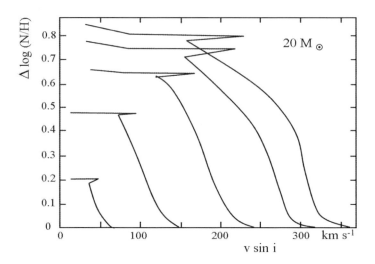

Figure 2: The evolution of the differences in log(N/H) during the MS phase as function of the actual rotation velocities (sin i being equal to 1 here) for models of 20 M$_\odot$ with $Z = 0.02$ and different initial velocities.

Study of the observed N/H excesses

The M, age and Z dependences

Let us start by examining the mass, age and Z dependences of the N/H excesses. The data for different groups of stars with different Z are summarized in Table 1. In the Galaxy ($Z \approx 0.02$), the main recent data sources (Herrero 2003; Venn & Przybilla 2003; Lyubimkov et al. 2004; Huang & Gies 2006; Trundle et al. 2007) support significant excesses of He or of (N/H). In the lowest mass range studied (6.6–8.2 M$_\odot$), small excesses of He/H are still present (Lyubimkov et al. 2004). In the LMC ($Z \approx 0.008$), the excesses are larger (Hunter et al. 2007; Trundle et al. 2007). In the SMC ($Z \approx 0.004$), still much larger N excesses are observed (Venn, Przybilla 2003; Heap & Lanz 2006; Trundle et al. 2007; Hunter et al. 2007).

Table 1: The largest $\Delta \log(N/H)$ values observed for different types of stars in the Galaxy, LMC and SMC (differences in dex with respect to the local values in the considered galaxy). The average is equal to about the half of the indicated values.

Types of stars	$\Delta \log(N/H)$ Galaxy	$\Delta \log(N/H)$ LMC	$\Delta \log(N/H)$ SMC
O stars	0.8 - 1.0	–	1.5 - 1.7
B–dwarfs $M < 20$ M$_\odot$	0.5	0.7 - 0.9	1.1
B giants, supg. $M < 20$ M$_\odot$	–	1.1 - 1.2	1.5
B giants, supg. $M > 20$ M$_\odot$	0.5 - 0.7	1.3	1.9

These data show the following facts, consistent with theoretical predictions:

- On the average, the N enrichments are larger for larger masses.

- The N enrichments are larger at lower Z.

- The He and N enrichments increase with the distance to the ZAMS (Huang & Geiss 2006). They are even larger in the giant and supergiant stages (Venn & Przybilla 2003). This property is also well observable for example in N11 (Fig. 34 of Hunter et al. 2007) and in NGC 2004 (Fig. 2 by Trundle et al. 2007).

The $v \sin i$ dependence of the N/H excesses

Several correlations of the N or He excesses with the observed $v \sin i$ have been performed. Huang and Gies (2006) and Lyubimkov et al. (2004) find a correlation of the He excesses with $v \sin i$ for B stars in the upper part of the MS band in agreement with model predictions.

In other comparisons (Hunter et al. 2008; Langer, this meeting; Brott, 2009), the authors conclude that "the observation ... challenges the concept of rotational mixing". They claim that "two groups of core hydrogen burning stars ... stand out as being in conflict with the evolutionary models". Group I contains rapid rotators with little chemical mixing, while Group II consists of low rotators with large N enrichments. We clearly disagree with the conclusions of Hunter et al. (2008), which mainly result from the fact that, instead of Eqn. (1), their analysis implicitly assumes that

$$\Delta \log(N/H) = f(v \sin i) . \tag{2}$$

We note the following points:

- Their sample contains a mixture of stars in the mass interval of 10 to 30 M$_\odot$. Fig. 1 shows that over this mass interval, the N/H excesses vary as much as by a factor of two for a given rotation velocity.

- The sample by Hunter et al. consists of stars in extended regions around the LMC clusters N11 and NGC 2004. As stated by the authors, their sample also contains field stars, which do not necessarily have the same age or degree of evolution as the cluster stars. Thus, *large differences of N/H are possible for given M and* $v \sin i$. Also, the two clusters do not have the same ages, N11 being younger than NGC 2004, so that the stars near the turnoff of N11 have a mass of about 20 M$_\odot$, while this is about 14 M$_\odot$ for NGC 2004 according to the HR diagrams by respectively Hunter et al. (2007) and Trundle et al. (2007).

- The completeness of the binary search is unknown.

To limit the severe effects of mass and age differences, we consider the two clusters separately. In N11, we limit the sample to the stars in the mass range 14 to 20 M$_\odot$ on the basis of the data provided by Hunter (2008) and in the formal MS band as given by Fig. 34 from Hunter et al. (2007). In NGC 2004, we take the mass interval 13 to 16 M$_\odot$ (same source) and in the formal MS band from Fig. 2 by Trundle et al. (2007). Figs. 3 and 4 show the results. Ideally these mass intervals should even be smaller.

For N11, we see that Group I (stars with high $v \sin i$ and low N/H) has essentially disappeared. There remain only 2 evolved binary stars, which is consistent with some scenarios of binary evolution. We suspect that Group I was also largely formed by stars of smaller ages and/or lower mass stars, where for a given $v \sin i$ the N excesses are smaller. A support to the latter possibility comes from the fact that the average mass of the stars in the region of Group I is 12.8 M$_\odot$ (for $v \sin i > 180$ km s^{-1} and $12 + \log(N/H) < 7.30$) and 17.1 M$_\odot$ for stars with $v \sin i > 180$ km s^{-1} and $12 + \log(N/H) > 7.30$. Group II is mainly made from evolved stars (open circles), which explains the low velocities and high N abundances. The two remaining stars in this group can easily be stars with a small $\sin i$, especially more than

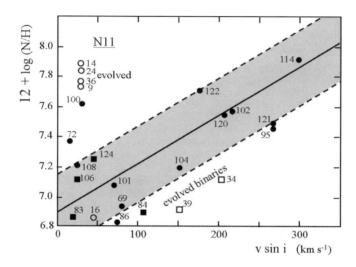

Figure 3: The N abundance (in a scale where log $H = 12.0$) as a function of $v \sin i$ for the MS stars (black dots) in N11 with masses between 14 and 20 $M_\odot$ according to Hunter (2008). The binaries are shown by a square. The evolved stars in a band of 0.1 dex in log T_{eff} beyond the end of the MS are shown with open symbols. The gray band indicates uncertainties of ± 0.25 dex.

the sample data by Hunter et al. (2008) is largely biased toward low rotators. The ratio of star numbers with $v \sin i \geq 250$ km s^{-1} to those with $v \sin i \leq 100$ km s^{-1} is 0.14, while in the clusters studied by Huang & Gies (2006), this ratio amounts to 0.40! Thus, the sample by Hunter et al. contains a large excess of slow rotators. We conclude that the bulk of stars in N11 shows a relation of the excess of N/H depending on $v \sin i$ (the mean square root of the data for the MS band stars is 0.23 dex from the data by Hunter (2008), the scatter in $v \sin i$ is not given). The amplitude of the (N/H) is about 0.6 dex for velocities of 200 km s^{-1}, slightly higher than at the corresponding mass in Fig. 1 for $Z = 0.02$.

For NGC 2004, the results are essentially similar. For most stars there is a relation between the excess of N/H and $v \sin i$. For Group I, there remains only star nb. 100 (which is not much for a group). This star is interesting. Its mass is 13 $M_\odot$, $v \sin i$ is 323 km s^{-1}, the highest of the whole sample. In reality, the velocity is still higher because the authors do not account for gravity darkening (Fig. 4 of Hunter al. (2008) tends to support this remark). This star might be a reaccelerated binary or simply a younger star in the field. Another possibility (which we favour) is that its parameters have been incorrectly appreciated due to the extreme rotation. In this respect, a log g vs. log T_{eff} diagram, e.g. Fig. 16 by Meynet & Maeder (2000), shows that a too high mass is assigned to a fast rotating star if in the log g vs. log T_{eff} diagram its mass is determined from nonrotating models. Thus, the mass of star nb. 100 could be lower than 13 $M_\odot$ which is the lower bound of our sample. To know whether this is what occurred for this star, the whole reduction process should be redone. For Group II, a large fraction consists as for N11 of evolved stars. Again for NGC 2004, the N enrichments increase with $v \sin i$ in agreement with theory.

We also note that the data used in the analysis by Hunter et al. (2008) are subject to several biases. Firstly, the sample contains no Be stars, while their number fraction is about 15 to 20 % in the LMC. This contributes to bias the sample toward low velocities. A second source of bias is that the $v \sin i$ determinations are based on models assuming that the stars

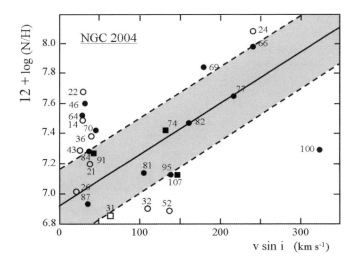

Figure 4: The N abundance as a function of $v \sin i$ for the MS stars in NGC 2004 with masses between 13 and 16 $M_\odot$. Same remarks as for Fig. 3.

are uniformly bright, with no account given to gravity darkening (now an observed effect). Thirdly, in the values of gravity used to estimate the masses, no account is given to the gravity change due to rotation. The effects of evolution and rotation should be disentangled before any mass is assigned.

Conclusions

To find a correlation for a multivariate function like N/H with some parameter like $v \sin i$, it is necessary to limit the range of the other involved parameters as much as possible. Otherwise, the conclusions may be erroneous. We note that data samples limited in mass and ages support an N enrichment depending on rotational velocities. Stars beyond the end of the MS phase do not obey to such a relation, because their velocities converge toward low values (see Fig. 12 by Meynet and Maeder 2000). A fraction, which we estimate to be ∼ 20 % of the stars, may escape from the relation as a result of binary evolution, either by tidal mixing or mass transfer.

Acknowledgments. We thank Dr. I. Hunter for having provided the observed data used in this study.

References

Brott, I., Hunter, I., de Koter, A., et al 2009, CoAst, 158, 55

Chiosi, C., & Maeder, A. 1986, Ann. Rev. Astron. Astrophys., 24, 329

Heap, S.R., Lanz, T., & Hubeny, I. 2006, ApJ, 638, 409

Herrero, A. 2003, in *CNO in the Universe*, Eds. C. Charbonnel, D. Schaerer, G. Meynet, ASP Conf. Ser., 304, 10

Huang, W., & Gies, D.R. 2006, ApJ, 648, 580 & 591

Hunter, I. 2008, private communication

Hunter, I., Dufton, P.L., Smartt, S., et al. 2007, A&A, 466, 277

Hunter, I., Brott, I., Lennon, D.J., et al. 2008, ApJ, 676, L29

Lyubimkov, L.S., Rostopochin, S.I., & Lambert, D.L. 2004, MNRAS, 351, 745

Maeder, A., & Meynet, G. 2000, Ann. Rev. Astron. Astrophys., 38, 143

Maeder, A., & Meynet, G. 2001, A&A, 373, 555

Martins, F. 2009, CoAst, 158, 106

Meynet, G., & Maeder, A. 2000, A&A, 361, 101

Meynet, G., Ekström, S., & Maeder, A. 2006, A&A, 447, 623

Venn, K.A., & Przybilla, N. 2003, in CNO in the Universe, eds. C. Charbonnel, D. Schaerer, G. Meynet, ASP Conf. Ser. 304, 20

DISCUSSION

Noels: Since the enrichment in CNO is strongly sensitive on the metallicity, I am wondering if it is correct to use the solar CNO/Fe ratio to compute models of "new born" massive stars in a low Z environment.

Maeder: At lower Z, the models composition accounts for the enhanced α-elements (C, O, Ne, Mg, Si,). When there is a primary N production for $Z < 10^{-3}$, the amount of new N produced is independent on the initial abundances.

Noels: In models computed with rotation, is there still an intermediate convective zone?

Maeder: This is an interesting point. Rotational mixing brings He in the region of the H-burning shell, which thus becomes weaker and this also weakens or makes the intermediate convective zone to disappear. The disappearance of the intermediate convective zone in turn allows the star to become a red supergiant. Thus, rotational mixing strongly affects (in a positive way) the blue to red supergiant ratio.

Cantiello: Your stressed the importance of the Gratton-Öpik term for the angular momentum transport in rotating massive stars. How reliable is the existence of Gratton-Öpik cell in a baroclinic star, i.e. a star where horizontal turbulence is strong (allowing for shellular approximation)? In particular, wouldn't the topology of the circulation be changed dramatically by such horizontal turbulence?

Maeder: The horizontal turbulence strongly affects the transport of chemical elements by meridional circulation. I refer you to the important study by Chaboyer & Zahn (1992) on this subject. The effect on the transport of angular momentum is likely small.

Aerts: Let us suppose that we have a seismic constraint from rotationally splitted detected frequency multiplets and we prove the rotation to be rigid and very slow everywhere in the star. If, in this case, the modeling requires an overshooting parameter $\alpha_{ov} \neq 0$, can we ascribe this to convective overshoot only, or are there still other effects that may cause an extent of the core?

Maeder: In case of solid body rotation, many instabilities (shears, baroclinic instabilities) are not present. However, the meridional circulation and its associated transport is still there, in addition to a certain amount of overshooting. This is interesting in the sense that it can allow to already possibly discriminate between certain effects.

Baglin: Coming back to the case raised by Conny Aerts, I would like to repeat that we will soon be able to measure the core rotation rate and the extent of the mixed region through seismology. Can we distinguish between overshooting and extension due to rotation?

Maeder: I think the answer will be yes, because the slope of the μ-gradient outside the core is different in the two cases. I expect a milder slope in the case of rotational mixing compared to overshooting.

Baglin: What is the present accuracy on the stellar structure of a massive star for which we assume that we have very precise fundamental parameters from observations?

Maeder: The basic structure of a massive star is simpler than the solar structure in some respects : simpler opacities, no large non-adiabatic convective layers. In other respects it is more complicated, in particular with mass loss and mixing. However, for a given physics, the models by the various authors agree well.

Comm. in Asteroseismology
Vol. 158, 2009, 38th LIAC/HELAS-ESTA/BAG, 2008
A. Noels, C. Aerts, J. Montalbán, A. Miglio and M. Briquet., eds.

Modelling convection and semiconvection in massive stars

C. Chiosi

Department of Astronomy, University of Padova, Vicolo dell'Osservatorio 2, 35122 Padova, Italy

Abstract

Massive stars the ideal laboratory to investigate three important physical phenomena: mass loss by stellar wind, convective instability and associated mixing, rotation, and their interplay. In this note, we will focus on the convective instability, summarizing the results obtained over the past three decades and highlighting some points that are still controversial.

Mass loss, Convection, and Mixing

Mass loss by stellar wind dominates the entire evolution of a massive star. The situation has been reviewed several times (Chiosi & Maeder 1986, Chiosi et al 1992, Maeder & Conti 1994, Chiosi 1998). After the first pioneering studies, e.g. Chiosi et al (1978) where the relationship between mass loss in blue and mass loss in the red was pointed out, it became soon evident that mass loss alone was not sufficient to explain the properties of massive stars and that other changes to the standard theory of stellar structure were needed. Indeed even today our knowledge of the extension of convective regions (cores, shells and envelopes) and associated mixing is rather poor. The local Schwarzschild (1958) criterion $\nabla_R = \nabla_A$ (SC) provides the simplest evaluation of the extension of convective regions and the Mixing Length Theory (MLT) simplifies the complicated pattern of motions therein by saying that full and instantaneous mixing takes place. *The reality is, however, more complicated than this simple picture.* First of all, inconsistencies are known to develop at the border of convective regions which give rise to the so-called *H-semiconvection* during the core H-burning of high mass stars (HMS), *He-semiconvection* and *breathing convection* in the core He-burning phase of low (LMS) and intermediate mass stars (IMS). Secondly, convection and mixing are non-local phenomena that in principle can also affect (*overshoot*) into regions that are stable according to the SC. Finally, mixing requires a certain amount of time to occur, and may not be complete. Unfortunately as soon as the classical scheme is abandoned, things lose simplicity and a unique solution does not exist, thus making uncertain the effects of convection and mixing. Nevertheless, it is worth recalling that (i) the size of the central convective core affects the luminosity, effective temperature, and lifetime of the corresponding evolutionary phase; (ii) the inward extension of the external convection may affect the surface abundances; (iii) the size of the intermediate convective shells has less straightforward effects: in HMS it may affect the extension of the blue loops, in IMS and LMS it may alter the surface abundances during the late evolutionary stages (RGB and AGB).

Hydrogen semiconvection. During the core H-burning phase of HMS, radiation pressure and electron scattering opacity give rise to a large convective core surrounded by an H-rich region, which is potentially unstable to convection if the original gradient in chemical abundance is maintained, but stable if suitable mixing is allowed to take place so that neutrality is restored. Negligible energy flux is carried by the process. The chemical gradient depends on which neutrality condition is used, either Schwarzschild (1958) or Ledoux (1947), $\nabla_R = \nabla_A + (\phi/\delta)\nabla_\mu$ (LC). The former gives smoother chemical profiles and in some cases leads to the onset of an intermediate fully convective layer. The Ledoux criterion is a stronger condition favouring stability with respect to the SC. Similar instability also occurs during the early shell H-burning stages. The effects of H-semiconvection and associated stability condition on the evolution of HMS have been first studied by Chiosi & Summa (1970).

Helium Semiconvection. As He-burning proceeds in the convective core of stars of any mass, the C-rich mixture inside the core becomes more opaque than the C-poor material outside; therefore the radiative temperature gradient increases within the core. The resulting super-adiabaticity at the edge of the core leads to a progressive increase (local convective over-shoot) in the size of the convective core during the early stages of He-burning (Schwarzschild 1970; Castellani et al 1971a,b). Once the convective core exceeds a certain size, overshoot is no longer able to restore the neutrality condition at the border due to a characteristic turn-up of the radiative gradient. The core splits into an inner convective core and an outer convective shell. As further helium is captured by the convective shell, this latter tends to become stable, leaving behind a region of varying composition in which $\nabla_R = \nabla_A$. This type of mixing is called He-semiconvection, whose extension varies with the star mass, being important in LMS and IMS up to about 5 $M_\odot$, and negligible in more massive stars.

Helium Breathing Convection. In all computed models, when $Y_c \leq 0.1$, the convective core may undergo recurrent episodes of rapid increase followed by an equally rapid decrease until it engulfs the whole semiconvective region. Castellani et al (1985) have named this phase "breathing pulses of convection". Semiconvection increases the core He-burning lifetime (by approximately a factor of two), whereas breathing convection increases the mass of the C-O core left over at the end of He-burning phase. This fact will greatly shorten the early AGB phase. Models with semiconvection and models with semiconvection plus breathing convection have different predictable effects on the expected ratio of the number of AGB stars to the number of HB stars in well-studied globular clusters. Chiosi (1986) argued that breathing convection is most likely an artifact of the idealized algorithms used in describing mixing. Given that breathing convection is a consequence of the time-independent treatment of semiconvection, and that both are based on local descriptions of mixing, the question arises whether non-local, e.g. full convective overshoot (see below), and/or time-dependent mixing may overcome the above difficulties.

Detailed reviews of all these subjects are in Chiosi & Maeder (1986), Chiosi et al (1992), and Chiosi (1998).

Does the classical theory fully agree with the observations?

Stellar models calculated with the above prescriptions are often named *the classical models*. They are the body of theoretical predictions to which observational data are compared. In the following we will shortly report on some observational problems encountered with the interpretation of *Blue and Red supergiant and Wolf-Rayet Stars* that likely suggest larger convective regions in stellar interiors.

Super-Giant Stars. The HRD for super-giant stars in the solar vicinity (Blaha & Humphreys 1989), and in the MCs (Massey et al 1995) together with the position and relative frequency of Wolf-Rayet (WR) stars have so far eluded precise quantitative explanations. The discussion below will be limited to the Milky Way and LMC and to stars brighter than $M_{bol} = -7.5$, to minimize photometric incompleteness. Key features to consider are:

The missing blue gap. There is no evidence of the so-called blue Hertzsprung gap (BHG) between the core H- and He-burning phase. The gap is the observational counterpart of the very rapid evolution across the HRD following core H-exhaustion and prior to stationary core He-burning. In contrast, we observe a continuous distribution of stars all across this region.

Many of the stars in the BHG show evidence of CNO processed and He-rich material at the surface (see Kudritziski (1998) and references) suggesting an advanced stage of evolution. *The relative frequencies of stars across the HRD.* Star counts in the HRD of supergiant stars in the Milky Way and LMC performed in the luminosity interval $-7.5 \geq M_{bol} \geq -9$ (roughly corresponding to main sequence stars in the mass interval 25 to 40 $M_\odot$), and the comparison with the evolutionary models calculated with mass loss, semiconvection, and overshoot (and modern input physics) reveals that there is a clear shortage of main sequence stars, which is more pronounced for the LMC sample. Indeed, recalling that the ratio N_{MS}/N_{PMS} cannot be too different from 10, it means that the group of stars supposedly in the main sequence phase should amount to about 80-90% of the total. The extension of the MS band towards the red depends on mass loss and mixing: it reaches the spectral type O9.5 in constant mass models with semiconvection and the spectral type B0.5 in models with mass loss and moderate convective overshoot. In this latter case, the percentage of stars falling in the MS band amount to about 80%. In reality the percentage is subjected to decrease if one takes WR into account for which post-MS stages are more appropriate (see Chiosi 1998). If the above shortage is not due to strong photometric incompleteness, it seems as if MS band ought to extend to even lower temperature. Possible mechanisms making the MS band wider have already been discussed by Chiosi (1998) and will not be repeated here.

Wolf-Rayet Stars. WR stars are commonly understood as the descendants of early type stars initially more massive than about $40 M_\odot$ that owing to severe mass loss do not evolve to the red side of the HRD, but soon after central H-exhaustion or even in the middle of this phase reverse their path in the HRD towards higher effective temperatures, first at constant luminosity and later at decreasing luminosity, see Maeder & Conti (1994) and Chiosi (1998). During all these evolutionary stages, the stellar models (no matter whether with semiconvection, standard overshoot or diffusive overshoot) have the pattern of surface chemical abundances typical of the WR objects. However, while the surface abundances and the decrease in the luminosity passing from WNL to WNE and WC stars indicated by the data (Hamann et al 1993) are matched by the models, their effective temperatures are much hotter than those possessed by real WR stars. It is often argued that the discrepancy in the effective temperature can be cured by applying the well-known correction for an expanding atmosphere, see Bertelli et al (1984), Maeder & Meynet (1987) and Hamann et al (1993). However, among the galactic WN stars studied by Hamann et al (1993), only WNE stars with strong emission lines show photospheric radii larger than the hydrostatic ones. So that the above correction does not apply to the majority of WR stars. Finally, since the minimum initial mass of the models able to reach the so-called WR configuration is greater than about $40 M_\odot$, there is the additional difficulty that the relative number of WR stars with respect to the progenitor OB type stars exceeds the expectation based on the possible duration of the WR phase and initial mass function, see Chiosi et al (1992), Deng et al (1996a,b) and Chiosi (1998) for details. Finally, possible evolutionary schemes linking blue and red supergiants, Wolf Rayet Stars, and progenitors of type II Supernovæ have been presented by Chiosi & Maeder (1986) and ever since adopted.

Convective overshoot and mixing in stellar models

H-semiconvection, He-semiconvection, breathing convection, and local overshoot are different attempts to cure the inconsistencies implicit in the classical scheme: simple Schwarzschild criterion and MLT. In a way or another they seek to answer the following questions:
(1) What determines the extension of the convectively unstable regions (either core or envelope or both) together with the extension of the surrounding regions formally stable but that, in a way or another, are affected by mixing? In other words: how far can convective elements penetrate into formally stable regions?
(2) What is the thermodynamic structure of the unstable and potentially unstable regions?

(3) What is the time-scale of mixing? Is it instantaneous or over a finite (long) period of time? What is the mechanism securing either full or partial homogenization of the unstable regions?

Attempts to answer the above questions have generated more complex schemes aimed at providing a sound model for overshoot and associated mixing. If the physical ground of convective overshoot is simple, its formulation and efficiency are much more uncertain giving rise to a variety of solutions and evolutionary models. This uncertainty is reflected in the variety of solutions both in local and non-local formulations and associated evolutionary models that have been proposed over the years. The situation has been reviewed by Chiosi (2007 and references). In addition to the studies explicitly mentioned below we recall here Xiong (1989, 1990), Canuto & Dubovikov (1998), Canuto (1998a,b), Maeder & Meynet (1991) and references.

The vast majority of stellar models in literature (see the Padova and Geneva libraries) stand on the ballistic scheme proposed long ago by Maeder (1975) and Bressan et al (1981, 1986), Bertelli et al (1985) to evaluate the mass extension of the overshoot region. The scheme is non-local but makes use of the MLT and hence contains the ML parameter. It may differ in the assumption made for the extension and thermal structure of the overshoot region. Mixing is always assumed to be instantaneous. The overshoot distance at the edge of the convective core has been proposed between nearly zero and about $2 \times H_P$. As many evolutionary results depend on the extension of the convective regions, this uncertainty is most critical. Because a generally accepted theory for overshoot is not yet available, most of those studies sought to constrain the efficiency of overshoot by comparing theoretical predictions with observations.

In addition to the convective core, overshoot may occur at the bottom of the convective envelope during the various phases in which this develops, such as on the RGB. The effect of envelope overshoot on stellar models of LMS and IMS has been studied by Alongi et al (1991), whereas that for HMS by Chiosi et al (1991).

Finally, a much debated question is whether steep gradients in molecular weight may constitute almost insuperable barriers against the penetration of convective elements into stable radiative regions. According to Maeder (1975) and Deng et al (1996a) the gradients in molecular weight can be eroded by overshoot; the opposite according to Canuto (1998b).

Diffusive overshoot and mixing. Special mention deserve the few cases in literature in which convective overshoot and associated mixing have been treated as a diffusive process taking into account the existence of many scale lengths, i.e. Xiong (1989, 1990), Deng et al (1996a,b), Salasnich et al (1999), Ventura et al (1998), Deng & Xiong (2008). The studies share several common aspects but differ in other specific points and also the mass range in which they have been applied.

The Deng et al (1996) model. The assumption of straight mixing is abandoned, the contribution of all possible scales from very small to the size of the whole mixed region is included by means of the concept of the characteristic scale most effective for mixing, and important properties of turbulence that are usually neglected, i.e. *intermittency* and *stirring* are taken into account. However, it still stands on the MLT as far as the derivation of the maximum extension L_{OV} of overshoot region and the velocities in the unstable regions are concerned. An estimate of the maximum overshoot distance L_{OV} is given by the sum of all possible scale lengths, i.e. $LOV \simeq 1/(1 - f) \times l_0$ where l_0 is the maximum scale typical of the MLT or equivalently λH_P with λ the mixing length parameter, and f is the breaking factor between two successive generations of turbulent elements whose dimensions decrease by a factor of two ($f \simeq 0.5$). Furthermore the *energy transport* across the unstable region (full convection and overshoot) is assumed to be adiabatic. Finally, to describe both motions and mixing, at all scales the diffusion equation is used $\frac{dX}{dt} = \left(\frac{\partial X}{\partial t} \right)_{nucl} + \frac{\partial}{\partial m_r} \left[(4\pi r^2 \rho)^2 D \frac{\partial X}{\partial m_r} \right]$ where X is the chemical abundance of the generic element consideration and D is the diffusion coefficient which expressed as $D = (1/3)F_i F_s v_d L$ where, v_d is the velocity of a suitable mean effective scale driving mixing, and L is the dimension of the region interested by diffusion. The characteristic velocity v_d is not known a priori. It must be derived from the physical conditions under which mixing occurs. The factor F_i accounts for intermittency and the factor F_s accounts for the *stirring efficiency* at the largest scale.

Intermittency. In brief, the small scale elements become less and less able to fill the whole volume or in other words mixing becomes less efficient. Using the so-called β-model of intermittency (Frisch 1977), they derive $F_i = (l_d/l_0)^{3/2}$.

Stirring. The largest eddy in a turbulence field works as a rigid stick stirring the material in a mixer and inducing smaller scale motions. However, if the stick (largest eddy) is comparable in size with the dimension of the total region to be mixed, the net mixing efficiency turns out to be much less. To take it into account we correct the diffusion coefficient by the factor $F_s = (L/l_0 - 1)^3$ for $l_0 \leq L$. This correction turns out to be important only in the convective envelope because it extends over several pressure scale heights.

Diffusion Coefficient and the Most Effective Scale l_d. The diffusion coefficient may change with the convective region under consideration, i.e core, intermediate shell (not always present) and external envelope each of which in turn is surrounded by its overshoot zone. Suitable expressions are proposed by Deng et al (1996a) for each region. In a turbulent medium, all scales from the dimension of the unstable zone itself down to that of the dissipative processes, are possible, however there will be a typical scale l_d at which mixing is most efficient. The minimum scale is the Kolmogorov micro-scale $l_K = (\nu^3/\bar{\epsilon})^{1/4}$ where ν is the kinematic viscosity and $\bar{\epsilon}$ is the kinetic energy flux injected into the turbulence field. If $l_d = l_K$ no mixing would occur. In contrast, if $l_d = l_0$ almost instantaneous mixing takes place. Numerical experiments yield $l_d = P_{dif} \times 10^{-5} l_0$ where l_0 is in units of H_p and P_{dif} is a fine tuning parameter of the order of unity. By slightly changing P_{dif} all existing evolutionary schemes are recovered, going from the semiconvective type of models (Langer 1989a,b) to the fully homogenized overshoot models (Bressan et al. 1981; Bertelli et al. 1985; Alongi et al. 1993).

The Ventura et al (1998) model. The Deng et al (1996) model has been further refined by Ventura et al (1998) who considered (i) Full Spectrum of Turbulence (FST, billions of eddy scales are considered) with the appropriate distribution of convective fluxes taken from Canuto et al (1996); (ii) Full coupling of nuclear evolution and turbulent transport by means of a diffusive scheme; (iii) Convective overshoot described by assuming that the turbulent velocity exponentially vanishes outside the formally convective region according to an e-folding free parameter, tuned to fit observations. They also provide a more physically consistent definition of the scale length ζ at the boundary of a convective region (unknown quantity). In the MLT this is simply approximated to $\Lambda = \alpha H_P$, where α is a free parameter. Depending on the micro-physics $1.5 < \alpha < 2.2$. In the FST the definition is as follows. Far from the boundaries, it has to approach the hydrostatic scale length H_P. At each layer of a convective region ζ_{up} and ζ_{low} are derived and their harmonic mean $< \zeta >$ is used. Close to the boundaries $\zeta \to \zeta_{up}$ or ζ_{low}. In deep layers, recalling that a convective structure has a polytropic index $n = 1.5$ and $\zeta = (1+n)H_P$, one gets $\zeta = 2.5 \times H_P$ and $< \zeta > = z_{up}/2 = \zeta_{low}/2 = 1.25 \times H_P$. In the FST, Λ should also include overshoot ($\Lambda = \zeta + L_{OV}$). Therefore it is assumed that $L_{OV} = \beta H_{P,top}$ (or $L_{OV} = \beta H_{P,bot}$), where β is a fine tuning parameter. With the fluxes of Canuto et al (1996) the solar fit requires $\beta = 0.1$. Finally, the diffusion coefficient is $D = 16\pi^2 R^4 \rho^2 \tau^{-1}$, where the diffusion time scale τ is given by the one-point density -radial velocity relationship $< \rho' u' > = -\tau \partial \rho / \partial r$ which unfortunately is not known. The diffusion coefficient is approximated to $D = (1/3)u l_d$, where u is the average turbulent velocity, and $l_d = \Lambda$. The velocity is given by $u = u_b \exp[\pm(1/\chi f_{thick}) \ln(P/P_b)]$ where u_b and P_b are the velocity and pressure at the boundary, χ is a parameter and f_{thick} is the thickness of the convective layer in units of H_ρ. In many respects this scheme resembles the one proposed by Salasnich et al (1999). Both contain adjustable parameters.

Stellar Models with Overshoot (and mass loss)

From the body of literature on this subject the following common results can be singled out:
(1) The core H-burning phase of all stars more massive than $M \geq M_L$ runs at higher luminosities, stretches to lower T_{eff}, and lasts longer as compared to classical models.
(2) In the mass range $M_L \leq M \leq 2M_\odot$, the onset of the convective core and associated overshoot takes place gradually as required by the turn-off morphology of old clusters.
(3) The over-luminosity caused by overshoot during the core H-burning phase still remains during the shell H- and core He-burning phases. As a consequence of it, the lifetime of the He-burning phase (t_{He}) of IMS and HMS gets shorter in spite of the larger mass of the convective core. Therefore the ratio t_{He}/t_H gets a factor of 2 lower than in classical models.
(4) As the helium cores of all LMS have nearly the same mass, the inclusion of convective overshoot leads to results similar to those obtained with the classical semiconvective scheme (Bressan et al. 1986).
(5) Due to the larger masses of the He and C-O cores left over at the end of core H- and He-burning phases, respectively, the critical masses M_{up} and M_{HeF} are about 30% smaller than in classical models.
(6) In models with straight overshoot (no diffusion) the loops of IMS and HMS tend, however, to be less extended than in classical models.
(7) The evolution of HMS ($M \geq 30M_\odot$) is complicated by the presence of mass loss all over their evolutionary history so that well-behaved loops are destroyed. Nevertheless, models may evolve according the general scheme blue - red - blue and may appear as WR stars of different morphological type.
(8) Models with diffusive overshoot share the same properties of models with straight overshoot. They differ only in some quantitative details, for instance the wider loops of IMS stars and the higher τ_{He}/τ_H ratio Deng et al (1996a,b).
(9) Models of HMS with straight and diffusive overshoot and standard mass loss rates are still unable to fill the BHG and solve the problems with the formation of WR stars, see the discussion in Deng et al (1996b).
(10) The situation is much better with the new models by Salasnich et al (1999), in which the ballistic model is abandoned and the decay model of overshoot by Grossman (1996) is applied, and perhaps more important the mass-loss rates during the red super-giant phase are deeply revised. These are given by the relation $log(\dot{M}) = 2.1 \times log(L/L_\odot) - 14.5$ which yields values that are significantly higher than those by de Jager et al (1988). See Salasnich (1999) for all details.
(11) The new models possess very extended blue loops which may reach the main sequence region. Since in this phase the models spend $\sim 50\%$ of the He-burning lifetime, *they are able to solve for the first time the mystery of the missing BHG.*
(12) Finally, they offer a new channel for the formation of faint WR. The progenitors of these stars are not the massive objects ($M \geq 60M_\odot$) evolving *vertically* in the HRD, but the less massive ones (in the range $15 - 20M_\odot$) evolving *horizontally* in HRD and undergoing substantial mass loss as RSG.

References

Alongi, M., Bertelli, G., Bressan, A., & Chiosi, C. 1991, A&A, 224, 95

Alongi, M., Bertelli, G., Bressan, A., et al. 1993, A&AS, 87, 85

Bertelli, G., Bressan, A., & Chiosi, C. 1984, A&A, 130, 279

Bertelli, G., Bressan, A., & Chiosi, C. 1985, A&A, 150, 33

Blaha, C., & Humphreys, R. M. 1989, AJ, 98, 1598

Bressan, A., Bertelli, G., & Chiosi, C. 1981, A&A, 102, 25

Bressan, A., Bertelli, G., & Chiosi, C. 1986, MemSAIt, 57, 411

Canuto, V. 1998a, ApJ, 505, L47

Canuto, V. 1998b, ApJ, 508, L103

Canuto, V., & Dubovikov, M. 1998, ApJ, 493, 834

Canuto, V., Goldman, I., & Mazzitelli, I. 1996, ApJ, 473, 550

Canuto, V., & Mazzitelli, I. 1991, ApJ, 370, 295

Canuto, V., & Mazzitelli, I. 1992, ApJ, 389, 724

Castellani, V., Chieffi, A.. Pulone, L., & Tornambé, A. 1985, ApJ, 296, 204

Castellani, V., Giannone, P., & Renzini, A. 1971, Astrophys. Space Sci., 10, 340

Castellani, V., Giannone, P., & Renzini, A. 1971, Astrophys. Space Sci., 10, 355

Chiosi, C. 1986, In *Nucleosynthesis and Stellar Evolution, 16th Saas-Fee Course, ed. B. Hauck, A. Maeder, G. Meynet*, p. 199. Geneva: Geneva Obs.

Chiosi, C. 1998, In *Stellar Astrophysics for the Local Group*, eds. A. Aparicio, A. Herrero, F. Sanchez, (Cambridge: Cambridge Univ), p. 1

Chiosi, C. 2007, In *IAU Symposium 239*, eds. F. Kupka, I. Roxburgh, K. Chan, p. 235

Chiosi, C., Bertelli, G., & Bressan, A. 1991. In *Instabilities in Evolved Super and Hypergiants*, eds. C. de Jager, H. Nieuwenhuijzen, Amsterdam: North Holland

Chiosi, C., Bertelli, G., & Bressan, A. 1992, ARA&A, 30, 235

Chiosi, C., & Maeder, A. 1986, ARA&A, 24, 329

Chiosi, C., Nasi, E., & Sreenivasan, S. R. 1978, A&A, 63, 103

Chiosi, C., & Summa, C. 1970, Ap&SS, 8, 478

Deng, L., Bressan, A., & Chiosi, C. 1996a, A&A, 313, 145

Deng, L., Bressan, A., & Chiosi, C. 1996b, A&A, 313, 159

Deng, L., & Xiong, D. R. 2008, MNRAS, 386, 1979

de Jager, C., Nieuwenhuijzen, H., & van der Hucht, K. A. 1988, A&AS, 72, 295

Frisch, U. 1977, Lect. Not. Phys., 71, 325.

Grossman, S. A. 1996, MNRAS, 279, 305

Hamann, W. R., Koesterke, L., & Wessolowski, U. 1993, A&A, 267, 410

Kudritziski R. 1998, In *Stellar Astrophysics for the Local Group*, eds. A. Aparicio, A. Herrero, F. Sanchez, (Cambridge: Cambridge Univ), p. 149

Langer, N. 1989a, A&A, 210, 93

Langer, N. 1989b, A&A, 220, 135

Ledoux, P. 1947, ApJ, 94, 537

Maeder, A. 1975, A&A, 40, 303

Maeder, A. 1990, A&AS, 84, 139

Maeder, A., & Conti, P. S. 1994, ARA&A, 32, 327

Maeder, A., & Meynet, G. 1991, A&AS, 89, 451

Massey, P., Lang, C., DeGioia-Eastwood, K, & Garmany, C. 1995, ApJ, 438, 188

Salasnich, B., Bressan, A., & Chiosi, C. 1999, A&A, 342, 131

Schwarzschild, M. 1958, *Structure and Evolution of the Stars* Princeton: Princeton Univ. Press

Schwarzschild, M. 1970, QJRAS, 11, 12

Ventura, P., Zeppieri, A., Mazzitelli, I., & D'Antona, F. 1998, A&A, 334, 953

Xiong, D. R. 1989, A&A, 213, 176

Xiong, D. R. 1990, A&A, 323, 31

DISCUSSION

Cox: What is the cepheid mass discrepancy ?

Chiosi: The evolutionary mass turns out to be larger than the pulsation mass. As stellar model computed with convective overshoot at given mass are brighter, the discrepancy can be removed. At given luminosity the evolutionary mass is then smaller and in agreement with the pulsation mass. The bump cepheids show this very clearly.

Noels: Do you have an explanation for the very different behaviour of the loops during He-burning when the Schwarzschild's or the Ledoux's criterion is used ?

Chiosi: Stellar models computed with semiconvection according to the Schwarzschild's criterion possess a thin convective layer inside the semiconvective region. The chemical profile presents a steep slope against which the H-burning shell stays. Such models soon reach thermal equilibrium and, whithout going to the red to ignite core He-burning, do so in the blue and slowly move towards the red. Models computed with the Ledoux's criterion do not develop a thin convective shell. The H-burning shell can thus easily move across the chemical profile and thermal equilibrium is not reached before going to the red. A loop in the HR diagram is made during core He-burning that starts in the red and goes to near completion at high effective temperature.

Puls: Just one short comment. The so-called observed parameters (e.g. stellar parameters, abundances etc) depend on underlying models, and might be thus affected by errors.

Comm. in Asteroseismology
Vol. 158, 2009, 38th LIAC/HELAS-ESTA/BAG, 2008
A. Noels, C. Aerts, J. Montalbán, A. Miglio and M. Briquet., eds.

Modelling massive stars with mass loss

P. Eggenberger[1], G. Meynet[2], A. Maeder[2]

[1] Institut d'Astrophysique et de Géophysique
Université de Liège, Allée du 6 Août 17 - B 4000 Liège - Belgique
[2] Observatoire de Genève
Université de Genève, Ch. des Maillettes 51 - CH 1290 Sauverny - Suisse

Abstract

Mass loss plays a major role in the evolution of massive stars. Its effects on the modelling of massive stars (in particular internal structure, evolutionary tracks in the HR diagram, lifetimes, and surface abundances) will first be presented in detail. The modelling of Wolf-Rayet stars will then be examined. The interaction between mass loss and rotation, as well as between mass loss and pulsation, will finally be briefly discussed.

Introduction

Massive stars are stars with an initial mass larger than about 9 M$_\odot$. These stars are especially interesting since they strongly influence the spectral and chemical evolution of galaxies. They are indeed the main nuclear reactors forming the heavy elements, as well as the main source of UV radiation. These stars are characterized by a high temperature to density ratio leading to a high radiation to gas pressure ratio. This of course favours mass loss by stellar winds, which plays a major role in the evolution of massive stars.

To be able to include the effects of mass loss in a stellar evolution code, one needs some relation between the mass loss rates and the global stellar properties. A simple parametrization that has been very widely used in stellar evolution computations was given by de Jager et al. (1988). The mass loss rates increase with the luminosity and, at a given luminosity, increase when the effective temperature decreases. During the evolution of a massive star from the blue to the red part of the HR diagram, the mass loss rates will therefore increase and the effects of mass loss will become more and more important. Another important feature of the mass loss rates is the variation with the metallicity Z. The mass loss rates are indeed found to increase with the metallicity; this dependency can be simply parametrized as $\dot{M} \sim (Z/Z_\odot)^\alpha$ with $Z_\odot$ being the solar metallicity and α being between about 0.5 and 0.8 according to stellar wind models (Kudritzki & Puls 2000, Vink et al. 2001).

Evolution in the HR diagram

As a starting point to discuss the effects of mass loss in the HR diagram, we first consider the evolution of massive stars at constant mass. Evolutionary tracks at constant mass are shown for stars between 9 and 120 M$_\odot$ in Fig. 1 (left). For these models without mass loss, the main sequence band becomes wider when the mass increases, as a result of the larger convective cores of the more massive stars. The He-burning phase is found to form a kind of 'horn' in the HR diagram. Figure 1 (left) shows that most of this phase is spent in the blue part of the HR

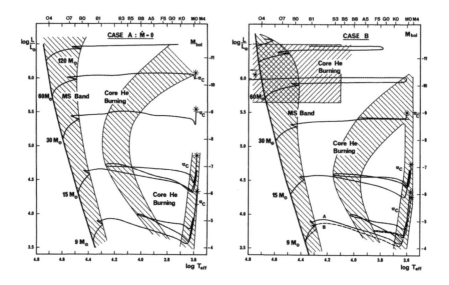

Figure 1: Evolutionary tracks in the HR diagram for stellar models computed without mass loss *(Left)* and with mass loss *(Right)*. Hatched areas correspond to the main phases of H and He burning. From Maeder (1981).

diagram (except for the most massive stars). This is due to the presence of an intermediate convective zone on top of an H-burning shell, which prevents the star from quickly evolving to the red part of the HR diagram (see e.g. Chiosi & Maeder 1986). Consequently, stellar models computed without mass loss predict almost no red supergiants for masses between about 20 and 50 $M_\odot$. This is of course in contradiction with observations, since many red supergiant stars are found in this mass range. Moreover, models at constant mass predict many red supergiants at very high luminosities, in contradiction with the observed upper limit distribution of luminous stars in the HR diagram known as the Humphreys-Davidson limit (Humphreys & Davidson 1979).

When the effects of mass loss are included in the computation, the luminosity of the star is reduced, as a result of the mass decrease. The star is nevertheless overluminous for its actual mass. Models including mass loss also exhibit larger core mass fraction, leading to a wider main sequence. The main sequence lifetime is found to slightly increase (by about 5 to 10%) when the effects of mass loss are taken into account. Figure 1 (right) shows the evolutionary tracks for massive stars computed with mass loss. Comparing HR diagrams computed with and without mass loss, one notes that the larger core mass fraction of models including mass loss lead to a wider main sequence. However, the main sequence becomes narrower for the most massive stars. By removing the surface layers of the star, mass loss leads indeed to a decrease of the envelope opacity. The stellar radius is therefore reduced and a blueward evolution is favoured. Concerning the He-burning phase, Fig. 1 (right) shows that there is a shift of the 'horn' to the red part of the HR diagram. This is due to the decrease or even the absence of an intermediate convective zone during the post main-sequence phase, resulting from the envelope reduction due to mass loss. Models with mass loss predict more red supergiants than without mass loss, in better agreement with the observations. For stars more massive than about 60 $M_\odot$, the effects of mass loss are very important and are able to remove the entire stellar envelope leaving a bare core: the star evolves to the Wolf-Rayet phase.

Wolf-Rayet stars

Wolf-Rayet (WR) stars are nice illustrations of mass loss effects on massive stars. These stars are characterized by high mass loss rates and strong emission lines with highly non-solar chemical abundances (see the recent review on the properties of WR stars by Crowther 2007). The WR spectra can be divided in two groups: the WN and WC stars. WN stars show products of the H-burning phase at their surface. For these stars, the HeII and N lines dominate the spectrum. WN stars can be subdivided in two groups: the late type WNL and the early type WNE stars. WNL are generally more luminous than WNE stars and contain some H contrary to WNE stars, which have no H left. WC stars show products of the He-burning phase at the surface. HeII, C and O lines dominate their spectra. They are also divided in late type WCL and early type WCE stars, with WO stars corresponding to extreme cases of WCE stars with higher O/C ratios.

In order to explain the abundance anomalies observed at the surface of WR stars, one has to find a mechanism able to efficiently remove the surface layers of the star. There are two main ways to form a WR star: through Roche lobe overflow in binary systems or through stellar wind losses in single stars. Figure 2 shows the typical evolution of the internal structure for a single massive star of $60\,M_\odot$. The removal of the external layers by stellar winds progressively reveals the internal layers. Near the end of the main sequence, the surface reaches layers that were initially in the stellar core. Products of H-burning can be observed at the surface and the star therefore becomes a WR star. During the post-main sequence phase of evolution, a H-burning shell is active. It then becomes inactive when its temperature decreases due to mass loss. The surface then reaches these layers resulting in a rapid decrease of the H abundance. The star evolves from WNL to WNE. As evolution continues, mass loss reveals layers that were in the He-burning core and enhancements of C and O are observed at the surface: the star becomes a WC star. The evolution of the surface chemical abundances of a massive star with mass loss can thus be briefly summarized as follows: first the initial abundances are observed at the surface. Then, intermediate abundances due to partial CNO processing with possible dilution effects are observed, before a phase of CNO equilibrium with the presence of H is reached. This phase is then followed by abundances of CNO elements at equilibrium but without H. Finally, products of He-burning are seen at the surface.

An important observational property of WR stars concerns the number frequency of WC stars with respect to WN stars, and in particular the variation of this number ratio with metallicity. Eldridge & Vink (2006) found that the variation of the WC/WN ratio with metallicity is very sensitive to the adopted mass loss rates during the WR phase. They showed that the observed relative population of WC to WN stars at different metallicities can be closely reproduced by using mass loss rates that scale with the initial metallicity during the WR phase. This scaling of the mass loss rates was first suggested by Crowther et al. (2002) from observations of WC stars, while Vink & de Koter (2005) showed how the mass loss rate is predicted to vary with initial metallicity for late type WC and WN stars. We thus see that standard models of massive stars with mass loss are able to correctly reproduce the variation of the WC/WN ratio. In the context of the transition between WN and WC stars, there is however another observational constraint that standard models are not able to reproduce: the number of WN/C stars. WN/C stars are transition stars between the WN and WC phase that are characterized by the simultaneous presence of both H- and He-burning products at the surface. While observations reveal that WN/C stars represent about 4% of the WR stars (van der Hucht 2001), standard models are not able to explain the existence of these stars because of the strong chemical discontinuity at the edge of the convective core in the He-burning phase. A smooth chemical transition is needed to produce the WN/C stars (Langer 1991).

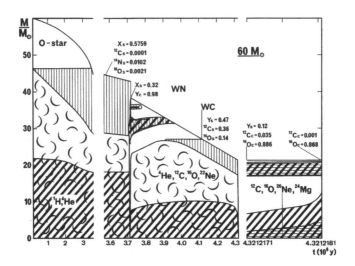

Figure 2: Evolution of the internal structure of a 60 $M_\odot$ star up to central C-exhaustion. Cloudy regions represent convective zones, heavy diagonals correspond to layers where the nuclear energy rates are larger than $10^3 \, \mathrm{erg \, g^{-1} \, s^{-1}}$. Vertically hatched areas indicate zones of variable H and He contents, while horizontally hatched regions correspond to zones of variable ^{12}C, ^{16}O, and ^{20}Ne contents. From Maeder & Meynet (1987).

Another important observational feature of massive stars is the number ratio of WR to O-type stars at different metallicities. Standard models with current mass loss rates are not able to reproduce these observations, since they severely underestimate the fraction of WR to O-type stars. A better agreement between theoretical predictions and observations can be obtained by increasing the value of the mass loss rates included in the computation (Meynet & Maeder 1994). This is due to the fact that higher mass loss rates increase the total lifetime in the WR phase and lower the value of the minimum initial mass required to form a WR star. These high mass loss rates are however in disagreement with recent determinations accounting for the effects of clumping in the winds of massive stars that suggest significantly lower values. We thus see that models including only mass loss do not predict a sufficient number of WR stars. To improve the situation, a first possibility is to account for the formation of Wolf-Rayet stars in binary systems. Models of massive star populations including mass transfer in binary stars are found to better reproduce the observed WR/O ratios by increasing the predicted number of WR stars (e.g. Eldridge et al. 2008). However, observational studies of WR populations in the Galaxy (van der Hucht 2001, 2006) and in the Magellanic Clouds (Foellmi et al. 2003ab) set some limit to the role of binaries in the formation of WR stars. These observations indeed reveal a fraction of WR stars in binaries of about 24%, 15% and 45% for the Galaxy, the Large Magellanic Cloud and the Small Magellanic Cloud, respectively. We thus see that binary mass transfer is an important channel for the formation of WR stars, but that binary evolution alone cannot explain all the observations. The effects of rotation and of enhanced mass loss due to Luminous Blue Variable (LBV)-like eruptions on the evolution of single massive stars have to be considered.

Interaction of mass loss with rotation and pulsation

Rotation is one of the key processes that have a strong impact on the physics and evolution of massive stars (see the review by Maeder & Meynet 2000). Mass loss and rotation interact in many ways. First, the evolution of rotational velocities of massive stars is sensitive to mass loss. The decrease of the rotation velocity of a massive star during its evolution is indeed found to be more efficient when the mass loss increases, due to the larger loss of angular momentum at the stellar surface. Secondly, rotation is found to change the mass loss rates. The line driven mass loss rates are indeed enhanced by rotation. Moreover, the mass loss rate per surface unit varies as a function of the colatitude leading to wind anisotropies induced by rotation. Mass loss of rotating stars can also be increased by reaching the critical limit or by surface chemical enrichments induced by rotational mixing. The latter process is especially important at low metallicity (see Maeder et al 2009).

In the context of WR stars, the inclusion of rotation in the models leads to many interesting results. Globally, rotation acts in two different ways. First, it increases the mass loss rates and favours the removal of the external layers. Secondly, rotational mixing brings He to the stellar surface, thereby favouring the appearance of the He bare core. Inclusion of rotation therefore results in an earlier entry into the WR phase. The WR lifetime, as well as the duration of the WNL phase, are then increased. Moreover, the threshold mass required to form a WR star is lowered by rotation. As a result, rotating models of WR stars are able to correctly reproduce the variation of the number ratio of WR to O-type stars with metallicity, as well as of the fraction of supernovae of type I_b/I_c to type II supernovae. Models including rotation also predict the existence of transition WN/C stars for initial masses between about 30 and 50 $M_\odot$, in good agreement with observations. However, these models are found to produce too many WN stars at solar and higher metallicities. Consequently, the observed WC/WN ratios can be correctly reproduced in the metal-poor region but not at solar and higher metallicities (Meynet & Maeder 2005). As discussed below, the evolution through the LBV phase and the associated strong enhancement of the mass loss rates may play an important role in shaping the WC/WN ratio.

Stars with initial masses above about 30 $M_\odot$ at solar metallicity may evolve into a LBV phase (see Cox & Guzik 2009). Observations of LBV stars show that during outburst phases very high mass loss rates can be reached. These outbursts, which are shell ejections rather than steady stellar winds, involve other processes in addition to the effects of the radiation pressure (Smith & Owocki 2006). In this context, pulsation and in particular strange mode instabilities may play a role (for more details about strange modes and instabilities, see Glatzel 2009 and Saio 2009). The LBV phase may also play a key role for populations of WR stars. As mentioned above, rotating models computed by Meynet & Maeder (2005) predict too low values for the WC/WN ratio at solar and higher metallicities. In these computations, it was assumed that a star entering the WR stage during the main sequence phase avoids the LBV phase. As discussed by Meynet et al. (2008), this hypothesis is probably not correct. A more realistic solution is to consider that a star becoming a WR star during the main sequence enters a LBV phase after the core H-burning phase, before evolving back into the WR regime. When this solution is applied to rotating models, reasonable values for both the WR/O and the WC/WN ratios are obtained at solar metallicity. Both ratios are not reproduced by non-rotating models computed with the same hypothesis. Only the case at solar metallicity has been currently computed, but such a scenario is expected to lead also to results in good agreement with observations at other metallicities.

As seen before, the inclusion of mass loss results in a change of the global properties and internal structure of a stellar model and thereby modifies the pulsational properties of a star. This is nicely illustrated in the case of B supergiants for which oscillation modes are detected. For these stars, the presence of an intermediate convective zone seems to be required in order to correctly reproduce the observed oscillation frequencies (Saio et al. 2006). Interestingly, the structure of this intermediate convective zone is very sensitive to many physical processes and in particular to the adopted mass loss rates (see Dupret et al. 2009 and Godart et al. 2009). Finally, we note that the observation of oscillations for WR stars are also very promising in order to improve our understanding of the modelling of massive stars. This has been recently illustrated by the detection of a pulsation period of 9.8 h in the star WR 123 (Lefèvre et al. 2005) and by the following theoretical studies aiming at correctly reproducing this observation (see Glatzel 2009).

Acknowledgments. PE is thankful to the Swiss National Science Foundation for support.

References

Chiosi, C., & Maeder, A. 1986, ARA&A, 24, 329

Cox, A. & Guzik, J. 2009, CoAst, 158, 259

Crowther, P.A. 2007, ARA&A, 45, 177

Crowther, P.A., Dessart, L., Hillier, D.J., et al. 2002, A&A, 392, 653

de Jager, C., Nieuwenhuijzen, H., & van der Hucht, K.A. 1988, A&AS, 72, 259

Dupret, M.-A., Godart, M., Noels, A., & Lebreton, Y. 2009, CoAst, 158, 239

Eldridge, J.J., & Vink, J. 2006, A&A, 452, 295

Eldridge, J.J., Izzard, R.G., & Tout, C.A. 2008, MNRAS, 384, 1109

Foellmi, C., Moffat, A.F.J., & Guerrero, M. 2003a, MNRAS, 338, 360

Foellmi, C., Moffat, A.F.J., & Guerrero, M. 2003b, MNRAS, 338, 1025

Glatzel, W. 2009, CoAst, 158, 252

Godart, M., Dupret, M.-A., & Noels, A. 2009, CoAst, 158, 308

Humphreys, R., & Davidson, K. 1979, ApJ, 232, 409

Kudritzki, R.-P., & Puls, J. 2000, ARA&A, 38, 613

Langer, N. 1991, A&A, 248, 531

Lefèvre, L., Marchenko, S.V., Moffat, A.F.J., et al. 2005, ApJ, 634, L109

Maeder, A. 1981, A&A, 102, 401

Maeder, A., & Meynet, G. 1987, A&A, 182, 243

Maeder, A., & Meynet, G. 2000, ARA&A, 38, 143

Maeder, A., Meynet, G., Ekstrom, S., et al. 2009, CoAst, 158, 72

Meynet, G., & Maeder, A. 1994, A&A, 287, 803

Meynet, G., & Maeder, A. 2005, A&A, 429, 581

Meynet, G., Ekström, S., Maeder, A., et al. 2008, in Proceedings of the IAU Symposium 250, 147

Saio, H., Kuschnig, R., Gautschy, A., et al. 2006, ApJ, 650, 1111

Saio, H. 2009, CoAst, 158, 245

Smith, N., & Owocki, S.P. 2006, ApJ, 645, L45

van der Hucht, K.A. 2001, NewAR, 45, 135

van der Hucht, K.A. 2006, A&A, 458, 453

Vink, J.S., & de Koter, A. 2005, A&A, 442, 587

Vink, J.S., de Koter, A., & Lamers, H.J.G.L.M. 2001, A&A, 369, 574

DISCUSSION

Baglin: The observational mass loss as a function of luminosity and temperature used in stellar modeling includes rotating stars and is based on average values. How do you take this into account in stellar modeling?

Eggenberger: We have indeed to account for the fact that the empirical values for the mass loss rates used for non-rotating stars are based on stars covering the whole range of rotational velocities. This is done by the convolution of the rotation effects on the mass loss rates over the observed distribution of rotational velocities, taking also into account that the orientation axes are randomly distributed. This leads to an estimated correction factor of about 0.8, which has to be applied to the mass loss rates for the main sequence OB stars.

Puls: (i) Regarding mass loss in the LBV phase, there is not only the possibility of inducing mass loss by strange mode-oscillations but there is also the possibility of very strong continuum driven mass loss which might be responsible for the giant outbursts (see the work by Owocki & Shaviv). (ii) You mentioned that the mass loss rates of O and WR stars in the evolutionary codes are described by certain scaling relations. One should point out, however, that there is a significant difference between both. For O stars, the mass loss seems to be fairly understood, and theoretical and observed scaling relations agree quite well. For WR stars, there are "only" observed relations, with a significant scatter if compared to individual objects. Even worse, there are almost no theoretical "self-consistent" models of WR winds (except for one model by Gräfener et al. which is not very representative). Thus, in particular the Z-dependence of WR winds cannot be considered as really understood.

Noels: If the mass loss rate is decreased by a factor 3 to 10, is it still possible to form a WR star from the evolution of a single massive star, from a theoretical point of view?

Meynet: Mass loss during the O-type star phase is a key parameter for deciding when a star is into the WR phase and therefore its duration. Lower mass loss will delay or even prevent the star from becoming a WR star. In case very low mass rates for O-type stars would be confirmed then, one will be obliged to consider other possibilities: WR formation through RLOF in close binary systems, or fast rotation, or heavy mass loss in other phases of evolution. However it has to be checked for all these possibilities if they can account for the observed variations with the metallicity of the WR/O, WN/WC number ratios.

Comm. in Asteroseismology
Vol. 158, 2009, 38th LIAC/HELAS-ESTA/BAG, 2008
A. Noels, C. Aerts, J. Montalbán, A. Miglio and M. Briquet., eds.

Rotational mixing in tidally locked massive main sequence binaries

S.E. de Mink, M. Cantiello, N. Langer, O.R. Pols

Astronomical Institute Utrecht, Princetonplein 5, 3584 CC Utrecht, The Netherlands, S.E.deMink@uu.nl

Abstract

One of the main uncertainties in evolutionary calculations of massive stars is the efficiency of internal mixing. It changes the chemical profile inside the star and can therefore affect the structure and further evolution.

We demonstrate that eclipsing binaries, in which the tides synchronize the rotation period of the stars and the orbital period, constitute a potentially strong test for the efficiency of rotational mixing. We present detailed stellar evolutionary models of massive binaries assuming the composition of the Small Magellanic Cloud. In these models we find enhancements in the surface nitrogen abundance of up to 0.6 dex.

Introduction

The inclusion of rotation into stellar evolution models has been shown to be very successful in explaining various observed characteristics of stars (see Maeder & Meynet 2000, for a review). It can have a large effect on the internal distribution of elements, as it leads to instabilities in the star, resulting in internal mixing.

The two most important mixing processes induced by rotation are *Eddington-Sweet circulations*, which consist of large scale meridional currents originating from a thermal imbalance between pole and equator in rotating stars (von Zeipel 1924; Eddington 1925, 1926; Vogt 1925) and *shear mixing*, which results from eddies formed between two layers of the star rotating at different angular velocities.

Near the center of a massive main sequence star hydrogen is converted into helium, and carbon and oxygen into nitrogen. Rotational mixing can bring this processed material to the surface, where it can be observed in the stellar spectra. Therefore rotation has been proposed as an explanation for the enhanced nitrogen abundances observed in a fraction of massive early type stars (e.g. Walborn 1976; Maeder & Meynet 2000; Heger & Langer 2000).

Although the effects of rotation on stellar evolution have been studied by various authors, we are still left with many questions. An essential question concerns the efficiency of rotational mixing. Attempts to constrain it have remained inconclusive due to limited sample sizes and/or a strong bias towards stars with small projected rotational velocities (e.g. Gies & Lambert 1992; Fliegner et al. 1996; Daflon et al. 2001; Venn et al. 2002; Korn et al. 2002; Huang & Gies 2006; Mendel et al. 2006).

The recent VLT-flames survey of massive stars (Evans et al. 2005) provided for the first time a large sample of massive stars covering a wide range of projected rotational velocities with accurate abundance determinations (Hunter et al. 2008). Brott et al (2008, this Vol.) demonstrated that the properties of the VLT-flames sample cannot be reproduced by simulations of a population of rotating single stars. This raises the question whether other processes, besides rotational mixing, play an important role in explaining helium and nitrogen enhancements of massive main sequence stars. For example, the observed enhancements could also be explained by binary interactions (Langer et al. 2008). In this case a downward revision of efficiency of the rotational mixing in single stars might be required.

Clearly, a strong and conclusive observational test for the efficiency of rotational mixing is needed. In this contribution we propose to use detached eclipsing binaries for this purpose.

Eclipsing binaries as laboratories for rotational mixing

Eclipsing binaries have frequently been used to test stellar evolution models as they provided the only method (until the development of asteroseismological techniques) for accurate determinations of stellar masses, radii and effective temperatures. Even beyond our own Galaxy, in the Magellanic Clouds, masses of O and early B stars have been determined with accuracies of 10% (Hilditch et al. 2005). Rotational mixing is more important in more massive stars (e.g. Heger et al. 2000). Therefore, it is very useful to know the stellar mass for quantitative testing of the efficiency of rotational mixing: it enables a direct comparison with a corresponding stellar evolution model with the proper mass.

In close binaries, with orbital periods (P_{orbit}) less than a few days, the tides are so strong that the stars rotate synchronously with the orbital period: $P_{spin} = P_{orbit}$. This enables us to determine the rotation rate directly from the orbital period. This is an important advantage of using binaries for testing rotational mixing with respect to single stars, for which fitting of spectral lines allows only for the determination of $v \sin i$, where v is the rotational velocity at the equator. The inclination i of the rotation axis is generally not known.

Here, we propose to use detached eclipsing binaries consisting of two main sequence stars residing within their Roche lobes. Detailed calculations of binary evolution show that if one of the stars fills its Roche lobe during the main sequence, it does not detach again before hydrogen is exhausted in the core, except maybe for a very short thermal timescale (Wellstein et al. 2001; De Mink et al. 2007). If we turn the argument around we find that, in a binary with two detached main sequence stars, we can safely exclude the occurrence of mass transfer since the onset of core H burning. The stars have lived their lives similar to rotating single stars. This is a third major advantage of using eclipsing binaries with respect to single stars. A fast rotating single star may in contrast be the result of a merger of two stars in a former binary. Moreover an apparently single star may have been affected by mass transfer, while its companion may be very hard to detect, being a faint low mass star in a wide orbit.

To test rotational mixing we need determinations of the surface abundances. If the spectra of a binary are of high quality, one can determine the surface abundances, as is done for single stars, after disentangling the composite spectra (e.g. Leushin 1988; Pavlovski & Hensberge 2005; Rauw et al. 2005). In the remainder of this paper we discuss what type of binaries are suitable for testing rotational mixing.

Stellar evolution code

To investigate to what extent the surface abundances of close detached binaries are affected by rotational mixing we model their evolution using a detailed 1D stellar evolution code, described by Yoon et al. (2006), which includes the effects of rotation on the stellar structure, the transport of angular momentum and chemical species via rotationally induced hydrodynamic instabilities (Heger et al. 2000) and magnetic torques (Spruit 2002; Heger et al. 2005). Brott et al. (2008, this Vol.) calibrated the efficiency of mixing processes in single star models using the data from the VLT-flames survey (Hunter et al. 2008).

Tidal interaction is implemented as described in Detmers et al. (2008) using the timescale for synchronization given by Zahn (1977, eq. 6.1). The tides act on the outer layers of the star. Angular momentum is redistributed in the stellar interior by magnetic coupling and

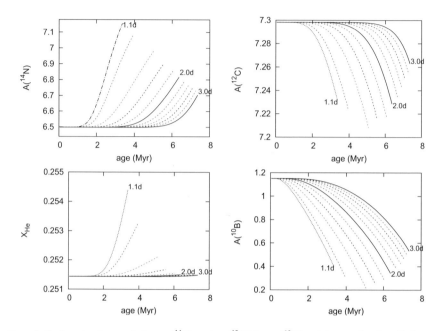

Figure 1: Surface abundances of nitrogen (^{14}N), carbon (^{12}C), boron (^{10}B) and the mass fraction of helium at the surface versus time for a 20 $M_\odot$ star with a 15 $M_\odot$ close companion. Note the different vertical scales. The abundance of an element X is given in the conventional units: $A(X) = \log_{10}(n_X/n_H) + 12$, where n_X and n_H refer to the number fractions. The different lines show the evolution assuming different initial orbital period, varying between 1.1 and 3 days. The tracks are plotted from the onset of central H burning until the onset of Roche lobe overflow.

rotational instabilities. We note that the binary systems modeled here are so tight that the stars are synchronized throughout their main sequence evolution.

Results

With the Small Magellanic Cloud (SMC) sample of Hilditch et al. (2005) in mind, which contains 21 detached systems[1] with orbital periods of a few days and masses of the primary component up to 20 $M_\odot$, we chose to model the following binary systems. For the mass of the primary component we adopt 20 $M_\odot$, for the secondary component 15 $M_\odot$ and we adopt initial orbital periods of up to three days. We assume a composition representative of the small Magellanic cloud, which is relatively metal-poor and has a high carbon to nitrogen ratio. The evolution is followed from the onset of central hydrogen burning at zero age until the primary star fills its Roche lobe.

In all computed models the tides are efficient enough to keep both stars in synchronous rotation with the orbit. The shorter the orbital period, the faster the rotation of the stars, the more efficient rotational mixing, the faster the surface abundances change with time. On the other hand, the systems with short orbital periods are tighter and therefore the stars will

---

[1] Possibly only 20 systems are detached. For two of the systems an alternative semi-detached solution exists. For one of these systems a comparison to binary evolution models including the effects of mass transfer showed that the semi-detached solution was more consistent than the detached solution (De Mink et al. 2007).

fill their Roche lobe at an earlier stage, leaving less time to modify their surface abundances. These effects are illustrated in Fig. 1.

Nitrogen is produced in the core and in the layers just above, as carbon and oxygen are consumed. Due to rotational mixing the nitrogen surface abundance increases and the carbon abundance decreases accordingly (see Fig. 1). Helium is produced deeper inside the star on a much longer timescale (the nuclear timescale). Some helium can be mixed up, but the helium surface enhancements achieved in our models are very small (less than 1%). Another element that acts as a tracer of rotational mixing is boron. This element can only survive in the coolest outermost layers of the star. Rotational mixing will bring it to hotter layers where it is destroyed. It is one of the elements most sensitive to rotational mixing. It is, however, hard to observe due to its low abundance, especially in the metal poor SMC. The increase in the nitrogen abundance in our models is up to three times larger than the typical error bar for surface abundance measurements in the VLT-flames survey (0.2 dex). Nitrogen may therefore be the most suitable element to test rotational mixing.

Conclusion

We have argued that eclipsing binaries can provide a potentially stringent test for the efficiency of rotational mixing in massive stars. The stellar parameters and rotation rate can be accurately determined enabling direct comparison to stellar evolution models. Therefore even one well-determined system could be used as a test case. By performing detailed evolutionary calculations of close massive binaries, we show that (with currently assumed rotational mixing efficiencies) we expect nitrogen enhancements of up to 0.6 dex for binaries such as those in the sample of Hilditch et al. (2005).

At present, it is not clear whether the presence of a binary companion can lead to extra mixing, on top of the rotationally induced mixing. If so, our proposed test would still constrain the efficiency of rotational mixing in single stars by providing an upper limit to this quantity. This will be discussed in a forthcoming paper.

References

Daflon, S., Cunha, K., Butler, K., & Smith, V. V. 2001, ApJ, 563, 325

De Mink, S. E., Pols, O. R., & Hilditch, R. W. 2007, A&A, 467, 1181

Detmers, R. G., Langer, N., Podsiadlowski, P., & Izzard, R. G. 2008, A&A, 484, 831

Eddington, A. S. 1925, The Observatory, 48, 73

Eddington, A. S. 1926, The Internal Constitution of the Stars (Cambridge: Cambridge University Press, 1926)

Evans, C. J., Smartt, S. J., Lee, J.-K., et al. 2005, A&A, 437, 467

Fliegner, J., Langer, N., & Venn, K. A. 1996, A&A, 308, L13

Gies, D. R. & Lambert, D. L. 1992, ApJ, 387, 673

Heger, A. & Langer, N. 2000, ApJ, 544, 1016

Heger, A., Langer, N., & Woosley, S. E. 2000, ApJ, 528, 368

Heger, A., Woosley, S. E., & Spruit, H. C. 2005, ApJ, 626, 350

Hilditch, R. W., Howarth, I. D., & Harries, T. J. 2005, MNRAS, 357, 304

Huang, W. & Gies, D. R. 2006, ApJ, 648, 591

Hunter, I., Brott, I., Lennon, D. J., et al. 2008, ApJL, 676, L29

Korn, A. J., Keller, S. C., Kaufer, A., et al. 2002, A&A, 385, 143

Langer, N., Cantiello, M., Yoon, S.-C., et al. 2008, in IAU Symposium, Vol. 250, IAU Symposium, 167–178

Leushin, V. V. 1988, Soviet Astronomy, 32, 430

Maeder, A. & Meynet, G. 2000, ARAA, 38, 143

Mendel, J. T., Venn, K. A., Proffitt, C. R., Brooks, A. M., & Lambert, D. L. 2006, ApJ, 640, 1039

Pavlovski, K. & Hensberge, H. 2005, A&A, 439, 309

Rauw, G., Crowther, P. A., De Becker, M., et al. 2005, A&A, 432, 985

Spruit, H. C. 2002, A&A, 381, 923

Venn, K. A., Brooks, A. M., Lambert, D. L., et al. 2002, ApJ, 565, 571

Vogt, H. 1925, Astronomische Nachrichten, 223, 229

von Zeipel, H. 1924, MNRAS, 84, 665

Walborn, N. R. 1976, ApJ, 205, 419

Wellstein, S., Langer, N., & Braun, H. 2001, A&A, 369, 939

Yoon, S.-C., Langer, N., & Norman, C. 2006, A&A, 460, 199

Zahn, J.-P. 1977, A&A, 57, 383

DISCUSSION

Zahn: Do your calculations of binary evolution take into account the angular momentum loss through the wind?

de Mink: Yes, we assume isotropic wind mass loss and take into account the associated angular momentum loss. However, the effect is not very important for these relatively unevolved binaries, especially not at the metallicity of the LMC and SMC where the stellar winds are weak.

Cox: Can you give us the names of the variable stars?

de Mink: Hilditch et al. (2005) published the stellar parameters of 21 massive close eclipsing binaries in the SMC. These systems would be suitable for testing rotational mixing, especially the more massive systems with shorter orbital periods; To test whether "quasi-chemically homogeneous evolution" occurs in tidally locked detached binaries, I would suggest to look at the very massive binaries in the LMC : R136-38 (Massey et al. 2002) or [L72] L4 54-425 (Williams et al. 2008) but there are certainly more suitable systems out there.

Comm. in Asteroseismology
Vol. 158, 2009, 38th LIAC/HELAS-ESTA/BAG, 2008
A. Noels, C. Aerts, J. Montalbán, A. Miglio and M. Briquet., eds.

On the dynamics of a radiative rapidly rotating star

M. Rieutord, F. Espinosa Lara[1]

Laboratoire d'Astrophysique de Toulouse-Tarbes, CNRS et Université de Toulouse, 14 avenue E. Belin,
31400 Toulouse, France

Abstract

The envelope of massive rapidly rotating stars is pervaded by baroclinic fluid flows, namely a differential rotation with a meridional circulation and likely a small-scale turbulence. We present here some of the first results of the ESTER project (Evolution STEllaire en Rotation). More specifically, using for the first time the spheroidal geometry, we give the shape of the differential rotation and show that the baroclinic torque imposes a fast rotating core and a slow envelope, together with a slow pole and a fast equator. The angular velocity of the core is 50% larger than that of the envelope.

Introduction

Rotation is one of the physical ingredients of stellar models of which the consequences are still not fully understood, either in the domains of stellar evolution, stellar structure or stellar oscillations. New data coming from interferometers (like the VLTI) or space missions like CoRoT, which give a detailed view of rotating stars, now call for better models. Even the chemical evolution of galaxies requires a good knowledge of the effects of stellar rotation since first stars, lacking metals, are thought to have been fast rotators (e.g. Chiappini et al. 2008).

The effects of rotation are in the first place to break the spherical symmetry of the stars. The centrifugal flattening induces new baroclinic flows which pervade the radiative regions. Hence, a differential rotation and meridional circulation arise and generate the now famous rotational mixing (Zahn 1992). On the side of asteroseismology, rotation strongly modifies the low-frequency part of the spectrum thanks to the effects of the Coriolis acceleration (e.g. Dintrans & Rieutord 2000). On the high frequency range, the centrifugal distortion is also very important as it changes the size and shape of the resonant cavity (Reese et al. 2006).

The case of rotation is therefore well motivated by the recent developments in stellar physics; unfortunately, it is not easy to handle. Rotation brings a new dimension into the models, enlarging a lot the parameter space of the models. Moreover, angular momentum is not conserved during the life of a star and this evolution couples to an another difficult question: that of mass loss.

In order to progress in this challenging problem, we have started the construction of two-dimensional models of isolated rotating stars. The ambition of this project, called ESTER for Evolution STellaire en Rotation, is to devise models, which are physically self-consistent, including the necessary baroclinic flows.

Presently, rotation is included through its mean effects in 1D models using astute averaging of the fluid flows (Zahn 1992, Maeder & Meynet 2000). Such models aim at studying the

---

[1]Present address: GEPI, bat. Copernic A, Observatoire de Meudon, 5 pl. J. Janssen 92195 Meudon Cedex, France

evolutionary consequences of rotational mixing. These models are however limited to slowly rotating stars. When centrifugal distortion is significant, 2D barotropic models have been proposed (e.g. Roxburgh 2004, Jackson et al. 2004). However, such models are in our opinion rather pseudo-2D models since only the Poisson equation is solved in two dimensions. The barotropic equation of state indeed implies that thermodynamical variables depend only on the total potential (centrifugal plus gravitational), and are thus solutions of ordinary differential equations. This implies no fluid flows and differential rotation needs to be prescribed (e.g. Jackson et al. 2004).

A true two-dimensional model is therefore quite challenging as one actually needs to solve the equations of motions of a self-gravitating compressible fluid, with nuclear reactions, radiative heat transfer, etc. in spheroidal geometry.

In this contribution, we present the way this problem can be solved and some of the first results, which have been obtained following the work of Espinosa Lara & Rieutord 2007. Conclusions follow.

The model

Let us first recall the equations that need to be solved. These are

$$
\begin{cases}
\Delta\phi = 4\pi G\rho \\
\rho T \mathbf{v} \cdot \nabla s = -\nabla \cdot \mathbf{F} + \varepsilon \\
\rho\left(2\boldsymbol{\Omega} \times \mathbf{v} + \mathbf{v} \cdot \nabla\mathbf{v}\right) = -\nabla p - \rho\nabla\left(\phi - \tfrac{1}{2}\Omega^2 r^2 \sin^2\theta\right) + \mathbf{F}_v \\
\nabla \cdot (\rho\mathbf{v}) = 0
\end{cases}
$$

which are respectively Poisson equation, the equation for entropy s, the momentum equation and the equation of continuity for a steady state configuration. Standard notations have been used, but note that ρ is the density and Ω the angular velocity of the frame; we also introduced

- the viscous force: $\mathbf{F}_v = \mu\left(\Delta\mathbf{v} + \dfrac{1}{3}\nabla\left(\nabla \cdot \mathbf{v}\right)\right)$

- the energy flux (radiative): $\mathbf{F} = -\chi\nabla T$

The microphysics need to specify the opacities, the nuclear heating and the equation of state. For the sake of simplicity we use analytical expressions of these quantities, namely

- Nuclear reactions (pp-chain) $\varepsilon = 8.37\,10^{10}X^2\rho^2\,T^{-2/3}e^{-bT^{-1/3}}$ (cgs) and $b = 3600$ (values from the CESAM code, Morel, 1997).

- Opacity (Kramer's type law)

$$
\kappa = \kappa_0\,T^{-\beta}\rho^\eta \quad \Longrightarrow \quad \chi = \frac{16\sigma\,T^3}{3\kappa\rho} = \chi_0\,T^{\beta+3}\rho^{-\eta-1}
$$

with $\kappa_0 = 7.155\,10^{13}$ (cgs), $\beta = 1.98$ and $\eta = 0.14$ (Christensen-Dalsgaard & Reiter, 1995).

- Equation of state: $p = \mathcal{R}_M\rho T + \dfrac{a}{3}T^4$

The partial differential equations need also to be completed by boundary conditions. These conditions imply

- Matching of the gravitational potential with the vacuum solution.

- Setting the pressure at the pole.

- Using stress-free conditions for the velocity field.

$$\mathbf{v} \cdot \mathbf{n} = 0 \qquad \mathbf{n} \times ([\sigma]\mathbf{n}) = \mathbf{0}$$

where $[\sigma]$ is the viscous stress.

- Imposing that the star locally radiates like a black body.

$$\mathbf{n} \cdot \nabla T + \sigma_T T = 0$$

where $\sigma_T = 3\rho\kappa/16$ and κ is the opacity. We use a constant value of σ_T for the entire surface.

When the solutions are determined, we may find regions of the star which are convectively unstable. In this case we cannot apply the mixing-length prescription, which is essentially one-dimensional. Using the concept of turbulent diffusion, we assume that convection zones diffuse entropy and therefore we solve a diffusion equation for this quantity; thus the heat flux is

$$\mathbf{F} = -\chi_r \nabla T - \frac{\chi_{\text{turb}} T}{c_p} \nabla s$$

The numerical resolution of the equations needs a full account of the spheroidal geometry. For this, we adopt the technique of Bonazzola et al. (1998), which uses a mapping of the spheroidal coordinates towards the spherical ones. These spheroidal coordinates are not orthogonal. Then, the discretization is done through a spectral decomposition of the solutions using the spherical harmonics for the horizontal part and the Chebyshev polynomials for the radial one. The use of a spectral method is important as it minimizes the grid size for a given precision.

Results

We have solved these equations using microphysics similar to that of the Sun's interior; hence, we neglect radiation pressure and impose a central temperature of $T_c = 1.33\,10^7$ K. Because of numerical difficulties, the surface pressure is imposed to be 10^{-5} times the central pressure. In non-rotating models the surface pressure is around 10^{-12} or less. Hence, our model does not compute (yet!) the outer 10% (in radius) of the star. We also impose a rotation rate which is $\sim 82\%$ of the critical velocity. Effects of viscosity are measured, with respect to the Coriolis force, by the Ekman number, which we set as small as possible; we take $E = 10^{-8}$. Since Prandtl number is also very small we set it to zero; this is a first step which is equivalent to the neglect of entropy advection by meridional currents (see Rieutord 2006 for a discussion of the effects of non-zero Prandlt number). With this setup, the model yields a "star" which has the following characteristics:

$$M = 1.105 M_\odot, R = 0.9674 R_\odot, L = 0.829 L_\odot, \rho_c = 85.6\,\text{g/cm}^3, V_{\text{eq}} = 400\,\text{km/s}$$

However, the most interesting output of the model are the baroclinic flows generated by the combination of the centrifugal distortion and the separate solving of the temperature and pressure fields. In Fig. 1 we show the differential rotation (the $v_\varphi/r\sin\theta$-component of the baroclinic flow). Comparing this result to the one obtained by Espinosa Lara & Rieutord (2007), we see that the boundary conditions do not much influence this flow. This results thus

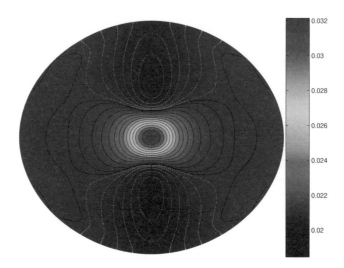

Figure 1: The differential rotation in a fully radiative star. Dotted contours show negative values with respect to the mean rotation rate.

confirm that in the radiative envelope the baroclinic torque generates a differential rotation which has a slow pole and a fast equator, as well as a slow envelope and a fast core. In our setup, the core is rotating approximately 1.5 faster than the envelope. This result may be compared to the fast rotating core inferred by Zorec et al. 2007 for the Be star Achernar. In Fig. 2 we show the associated meridional circulation. Again, the spheroidal solutions confirm the ones obtained with a spherical container.

Conclusions

In this contribution we have shown the first self-consistent model of the radiative core of a rapidly rotating star using spheroidal geometry. The results confirm and extend those obtained by Espinosa Lara & Rieutord (2007) who confined a rotating star in a spherical container so as to be able to use spherical coordinates. The comparison of the flows shows that the outer boundary conditions fortunately do not influence the interior solution qualitatively. However, it may well be that the ratio of surface angular velocity to the central one increases as the surface pressure is reduced.

The models will now be extended to allow much lower surface pressure so that the full star can be computed. They will also include convection zones whose structure will be computed by solving the diffusion equation on the entropy. Further steps will then replace the analytical formulae of microphysics with the most recent tables of opacities, equation of states and nuclear reactions networks, before the jump to a full-time evolution.

Acknowledgments.   We are very grateful to the CNRS for the continuous support of the ESTER project through the Programme National de Physique Stellaire.

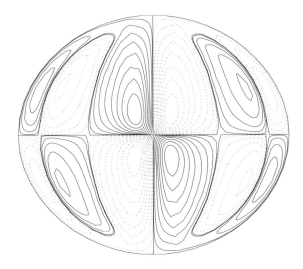

Figure 2: The associated meridional circulation. The amplitude of the flow is of the order of $0.1 E V_{eq}$.

References

Bonazzola, S., Gourgoulhon, E., & Marck, J.-A. 1998, Phys. Rev. D, 58, 104020

Chiappini, C., Ekström, S., Meynet, G. et al. 2008, A&A, 479, L9

Christensen-Dalsgaard, J., & Reiter, J. 1995, ASP Conf. Ser., 76, 136

Dintrans, B., & Rieutord, M. 2000, A&A, 354, 86

Espinosa Lara, F., & Rieutord, M. 2007, A&A, 470, 1013

Jackson, S., MacGregor, K. B., & Skumanich, A. 2004, ApJ, 606, 1196

Maeder, A., & Meynet, G. 2000, ARA&A, 38, 143

Morel, P. 1997, A&AS, 124, 597

Reese, D., Lignières, F., & Rieutord, M. 2006, A&A, 455, 621

Rieutord, M. 2006, A&A, 451, 1025

Roxburgh, I. 2004, A&A, 428, 171

Zahn, J.-P. 1992, A&A, 265, 115

Zorec, J., Frémat, Y., & Domiciano de Souza, A. 2007, ASPC, 361, 542

Session 3
Atmosphere, mass loss and stellar winds

Comm. in Asteroseismology
Vol. 158, 2009, 38th LIAC/HELAS-ESTA/BAG, 2008
A. Noels, C. Aerts, J. Montalbán, A. Miglio and M. Briquet., eds.

Stellar and wind properties of massive stars

F. Martins[1]

[1] GRAAL-CNRS, Université Montpellier II, Place Eugene Bataillon, F-34095, Montpellier - France

Abstract

Various recent studies have changed our understanding of the stellar and wind properties of massive stars. Here, we present an overview of such results in the following three subfields: 1) stellar evolution and rotation, 2) winds, and 3) magnetic properties. We focus on O stars.

Introduction

Massive stars are well-known to play an important role in various fields of astrophysics. Their high luminosity and effective temperature generate ionizing photons at the origin of HII regions. Their strong winds deposit huge amounts of mechanical energies in the interstellar medium, creating bubbles. Massive stars are also responsible for the chemical enrichment of galaxies since they produce most of the metals heavier than oxygen. In recent years, massive stars have also received special interest in the context of extragalactic studies as well as in cosmology. Indeed, they are identified as the progenitors of long–soft gamma–ray bursts and are often claimed to constitute the dominant population of metal free stars in the early Universe, possibly contributing to its reionization. Understanding these objects is thus important.

However, their properties and evolution are not fully constrained. Recent developments in the modelling of these object (both of their interiors and atmospheres) coupled to continuously better observational data have significantly changed our view of massive stars. The inclusion of the effects of rotation in evolutionary models (and the first observational constraints), the discovery of clumping in stellar winds and the very recent discovery of magnetic fields in massive stars are three of the main breakthroughs. In the following, we summarize some of the results related to these three fields. We restrict ourselves to O (and in Sect. 3 also Wolf-Rayet) stars, B stars being treated in other contributions of this volume.

Stellar evolution and rotation

The main improvement in stellar evolutionary models in the last decade was the inclusion of the effects of rotation. The mixing processes and transport of angular momentum and chemical elements triggered by rotation have consequences on several properties of massive stars (see Maeder & Meynet 2000 for a review). One of the main effects is the modification of surface abundances. In particular, CNO material freshly produced in the interior is brought to the surface, leading to N enrichment and C and O depletion, even when massive stars are still O stars on or close to the main sequence. Models predict a larger surface N enrichment as stars evolve off the main sequence due to mixing processes (meridional circulation, shear turbulence) caused by rotation. This prediction has been qualitatively confirmed by a number of studies (e.g. Massey et al. 05, Heap et al. 06, Mokiem et al. 06, 07). Fig. 1 illustrates this

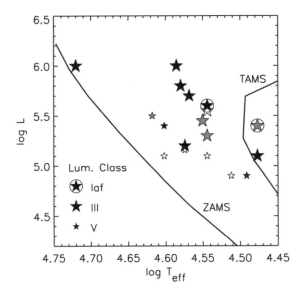

Figure 1: HR diagram of SMC stars shown by star symbols. The darker the symbol, the larger the N/C ratio. From Heap et al. 2006.

point: the larger the distance to the main sequence, the stronger the surface N/C ratio derived from tailored spectroscopic analysis. However, there are exceptions to this rule. Mokiem et al. (2006) showed that some O dwarfs close to the main sequence are He rich while some O supergiants have normal abundances. Since He is also transported to the surface by mixing processes, these observations indicate that different rotation rates lead to different degrees of surface enrichment. In practice, this means that two stars at the same position in the HR diagram can have different chemical compositions. Such a situation was nicely illustrated by Hillier et al. (2003) who studied two SMC stars and found similar T_{eff} and luminosities but different degrees of He and N enrichment. This complicates the analysis of HR diagrams: a star can be explained by various sets of evolutionary tracks with various rotational velocities intersecting at the position of the HRD (see Fig. 7 of Meynet & Maeder 2000).

Recently, Hunter et al. (2008) questioned the validity of current evolutionary models with rotation to explain the positions of two groups of SMC stars in the $\log(N/H)$–v_{sini} diagram. This question is tackled by other contributions to these proceedings (I. Brott, S. deMink, A. Maeder). We simply mention here that an alternative explanation to explain the presence of fast rotating non N–rich and slow rotating N–rich stars is binary evolution. Recently, Linder et al. (2008) conducted a spectroscopic analysis of the Plaskett's star. It is a well-known binary. Linder et al. showed that it most likely experienced mass transfer and that the primary was N-rich and slowly rotating, while the secondary was a fast rotator with initial N composition, supporting the idea that binaries can explain outliers in the $\log(N/H)$–v_{sini} diagram.

The study of surface abundances is clearly the key to test the effects of rotation in massive stars. Mokiem et al. (2006, 2007a) have studied a number of O stars in the Small and Large Magellanic Clouds. From the position of these stars in the HRD, they have derived evolutionary masses which have been compared to spectroscopic masses obtained through gravity determination with atmosphere models. These two mass estimates were generally consistent, showing that the so-called mass discrepancy problem (Herrero et al.

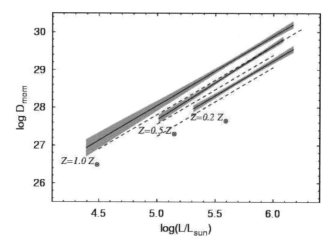

Figure 2: Modified wind momentum – luminosity relation for Galactic, LMC and SMC stars compared to theoretical predictions (dashed lines). From Mokiem et al. 2007b.

1992) was greatly reduced by the use of new evolutionary models and atmosphere models. However, a few stars still show the mass discrepancy. Mokiem et al. found a trend of a larger discrepancy for larger He content for O dwarfs. They interpreted this as a possible sign of chemically homogeneous evolution. Such an evolution occurs when fast rotation fully mixes the stellar interior, leading to very strong surface chemical enrichment. At the same time, due to the reduced opacities (the surface He content is large) the star evolves blueward in the HR diagram. Hence, using standard tracks to derive evolutionary masses leads to meaningless masses. The existence of homogeneous evolution is however difficult to prove for O stars, since homogeneous and normal evolutionary tracks are not that different near the main sequence. However, these tracks clearly differ in later evolutionary phases, normal tracks evolving redward while homogeneous tracks evolve toward the blue (see Fig. 2 of Yoon & Langer 2005). Recently, Martins et al. (2009) studied two WNh stars in the SMC and found very high surface H contents together with high temperatures typical of Wolf-Rayet stars located to the left of the main sequence. The large hydrogen mass fraction can only be accounted for by homogeneous evolution: normal tracks are H–free at the position of the two stars. This study revealed that homogeneous evolution due to fast rotation occurs in the SMC.

Winds of O stars

The winds of O stars are powered by radiative acceleration acting on thousands of metallic lines (Castor et al. 1975, hereafter CAK). One of the greatest successes of the theory is the prediction of a relation between the so-called modified wind momentum ($\dot{M} v_\infty \sqrt{(R)}$) and luminosity (MWL). This relation has been observed, confirming the validity of the radiatively driven wind theory (Puls et al. 1996, Repolust et al. 2004). Another success is the prediction of a metallicity dependence of the wind properties: line driving is mainly due to metallic lines so that lower metallicity implies lower acceleration and lower wind momentum. This prediction has been tested recently. Fig. 2 shows the results obtained by Mokiem et al. (2007b) and confirms that the MWL is shifted towards lower values for lower metallicities.

In spite of this success of the CAK theory, a new paradigm for O-star winds has emerged.

Indeed, it is now undisputed that such winds are not smooth but rather inhomogeneous, with rarefied regions separating overdensities (see the proceedings of the Potsdam workshop dedicated to wind clumping by Hamann et al. for an overview). Evidence came from different directions: either directly through observations of moving substructures in emission lines of supergiants, or indirectly from a number of indicators. Among them, one can mention the difficulty to reproduce some specific line profiles in the UV with homogeneous models (Bouret et al. 2005) or the shape of X-ray lines (Oskinova et al. 2004). Clumping is expected to happen due to the unstable nature of the acceleration mechanism, which is confirmed by hydrodynamical simulations (Runacres & Owocki 2002). In practice, clumping leads to a reduction of mass loss rates. Indeed in atmosphere models, lower global mass loss rates combined to a distribution of material in clumps and voids results in wind sensitive line profiles almost similar to the ones obtained by homogeneous models. In practice, complications arise for certain temperature and luminosity ranges because clumping also affects the ionization structure. But on average, new mass loss rates including the effects of clumping are a factor of 3 smaller than previously thought. This has important consequences on stellar evolution which for massive stars is governed by mass loss. In practice, little is known about clumping. The distribution of inhomogeneities is poorly constrained (see Puls et al. 2006 for a first attempt), the dependence (or absence of dependence) of clumping on metallicity is not known, the relation to X-ray emission (thought to be generated by shocks) remains to be established.

In addition to the emergence of clumping, massive star winds face a puzzling problem at low luminosities. Some late O dwarfs seem to have winds about 100 times weaker than predicted by the best hydrodynamical models available (Martins et al. 2004, 2005). The origin of this "weak wind" problem is not known at present. Effects of high energy radiation on the atmosphere structure or subtle driving effects in hydrodynamical models are the most attractive solutions (see Martins et al. 2005 for a discussion).

Magnetism

One of the major developments in the understanding of massive stars has been the observation of large scale magnetic fields on the surface of O stars. The young star θ^1 Ori C in the Trapezium cluster was the first O stars to reveal a magnetic field (Donati et al. 2002), soon followed by HD191612 (Donati et al. 2006), ζ Ori (Bouret et al. 2008) and HD148937 (Hubrig et al. 2008).

The origin of magnetic fields in O stars is still debated. The absence of large convective shells at the surface of the star argues against a dynamo origin, as for solar type stars. Instead, a fossil origin better explains the strengths (usually of a few hundred G to a kG) and geometry (dipole). However, ζ Ori seems to deviate from this global scenario. Its magnetic field is not a simple dipole and is quite weak (about 100 G). It is actually weaker than the field strength expected to quench MHD instabilities so that some sort of dynamo process might be at work in this star. Obviously, studies of additional stars are needed to understand the origin of magnetic fields in O stars.

The effects of magnetism on the structure, evolution and winds of massive stars have been studied theoretically. Concerning winds, the pioneering work of ud-Doula & Owocki (2002) showed that depending on the ratio of magnetic energy to wind kinetic energy (quantified by the so-called η parameter, see ud-Doula & Owocki 2002), several situations can occur. This is illustrated in Fig. 3. If the magnetic field is weak, it will be simply blown by the stellar wind which will remain basically unaffected (left panel of Fig. 3). On the contrary, if the magnetic field is strong, the outflow will be channeled along field lines, leading to wind confinement and material collision at the magnetic equator (right panel of Fig. 3). Intermediate situations occur when the magnetic and kinetic energy are similar. θ^1 Ori C is an example of strong confinement, while for ζ Ori, confinement is minimized.

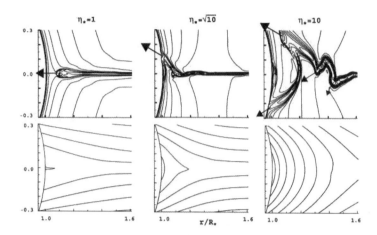

Figure 3: 2D density structure (top) and magnetic field geometry (bottom) for the case of a weak (left), medium (middle) and strong (strong) magnetic field. From ud-Doula & Owocki 2002.

Magnetism also affects stellar evolution by changing the efficiency of chemical elements and angular momentum transport (see Zahn 2009). More specifically, the presence of a magnetic field modifies the star's rotation and consequently affects the surface abundances. However, our understanding of these effects is still poor. Some simulations tend to predict larger surface chemical enrichment than in the case of normal rotating models (Maeder & Meynet 2005), while in the case of ζ Ori, the observed chemical enrichment is actually weaker than expected from rotating models *without* magnetic fields. Clearly, more studies are required to assess the interplay between winds, rotation and magnetic fields.

Conclusion

Major advances have been made in the understanding of massive stars in the last years.

- Rotation is now routinely included in evolutionary codes and its effects on stellar properties (especially surface abundances) can be tested using results of tailored spectroscopic analysis of large samples of stars. Generally, rotation explains well the observed properties. Exceptions exist for which binarity appears to be a convincing explanation.

- Magnetic fields start to be discovered among O stars. They certainly have a role to play on stellar interiors and winds. However, their effects are just beginning to be observed and quantified. More studies are needed to constrain the current theoretical predictions and to pinpoint the origin and effects of magnetic fields on massive stars.

- Winds of massive stars are radiatively driven and depend on metallicity. But winds of massive stars are also clumped. The presence of inhomogeneities translates into a reduction of mass loss rates by an average factor of 3, which affects the details of stellar evolution (governed by mass loss). Late O dwarfs have very weak winds which represent a challenge to the theory of radiatively driven winds.

Acknowledgments. F.M. thanks the organizers for the invitation and for a friendly and fruitful workshop. Thanks to an anonymous referee for helpful suggestions which improved the clarity of the text.

References

Bouret, J.-C., Donati, J.-F., & Martins, F. 2008, MNRAS, 389, 75

Bouret, J.-C., Lanz, T., & Hillier, D.J. 2005 A&A, 438, 301

Castor, J.I., Abbott, D.C., & Klein, R.I. 1975, ApJ, 195, 157

Donati, J.-F., Howarth, I.D., Bouret, J.-C., et al. 2006, MNRAS, 365L, 6

Donati, J.-F., Babel, J., Harries, T.J., et al. 2002, MNRAS, 333, 55

Herrero, A., Kudritzki, R.P., Vilchez, J.M., et al. 1992, A&A, 261, 209

Hubrig, S., Schoeler, M., Schnerr, R.S., et al. 2008, A&A, in press (astro-ph:0808.2039)

Hunter, I., Brott, I., Lennon, D.J., et al. 2008, ApJ, 676L, 29

Linder, N., Rauw, G., Martins, F., et al. 2008, A&A, 489, 713

Maeder, A., & Meynet, G. 2005, A&A, 440, 1041

Maeder, A., & Meynet, G. 2000, ARA&A, 38, 143

Martins, F, Hillier, D.J., Bouret, J.-C., et al. 2009, A&A, accepted

Martins, F, Schaerer, D., Hillier, D.J., et al. 2005, A&A, 441, 735

Martins, F, Schaerer, D., Hillier, D.J., et al. 2004, A&A, 420, 1087

Meynet, G., & Maeder, A. 2000, A&A, 361, 101

Mokiem, R., de Koter, A., Evans, C.J., et al. 2007a, A&A, 465, 1003

Mokiem, R., de Koter, A., Vink, J.S., et al. 2007b, A&A, 473, 603

Mokiem, R., de Koter, A., Evans, C.J., et al. 2006, A&A, 456, 1131

Oskinova, L.M., Feldmeier, A., & Hamann, W.-R. 2004, A&A, 422, 675

Puls, J., Markova, N., Scuderi, S., et al. 2006, A&A, 454, 625

Puls, J., Kudritzki, R.P., Herrero, A., et al. 1996, A&A, 305, 171

Repolust, T., Puls, J., & Herrero, A. 2004 A&A, 415, 349

Runacres, M.C., & Owocki, S.P. 2002, A&A, 381, 1015

ud-Doula, A., & Owocki, S.P. 2002, ApJ, 576, 413

Yoon, S.-C., & Langer, N. 2005, A&A, 443, 643

Zahn, J.-P. 2009, CoAst, 158, 27

DISCUSSION

Noels: Did you come out with an explanation for the Plaskett's star? I understood from Natacha Linder that there is a large discrepancy between the "orbital" masses and the evolutionary masses (Linder et al. 2008, A&A 489, 713).

Martins: The problem is that the stars appear underluminous for their dynamical masses. But this is the case when we take luminosities from single star evolutionary tracks. Adopting binary tracks might change the picture. Unfortunately such tracks are not available for the mass range of interest. In addition, the mass transfer history in the Plaskett star is not well constrained. Another puzzle is the He enrichment of the secondary which is not enriched in N.

Rauw: You mentioned the case of ζ Ori and you have shown the H_α line profile variability. Do you know what is the rotational period of this star?

Martins: Unpublished H_α data by O. Sthal confirm the H_α variability with a period of about 7.5 days. Besides, photospheric lines observed with NARVAL also show a very clear periodicity of $\sim$ 7.5 days.

Rauw: Stellar atmosphere models become increasingly sophisticated. They include the effects of clumping, X-rays... And these models are used to infer abundances that are compared with theory. How certain are all these results?

Martins: Mass loss rates are rather uncertain because the clumping properties are not well constrained. The current agreement is that mass loss rates have to be reduced by a factor ~ 3, but this can be a factor 10 in certain cases. As for X-rays, fuxes are provided by observations, but their implementation in atmosphere models is still crude. On the other hand, temperature can be derived with an uncertainty of ~ 1000 k, and gravities with an error of ~ 0.1 dex (for abundances, see J. Puls's answer).

Chiosi: I think that a deep revision of the scenario for forming WR stars will be needed if the rates of mass loss for O stars go down by a large factor. The only possibility I can see to accommodate lower rates and WR stars is to invoke a more efficient mass loss in the red part of the evolution than considered so far.

Comm. in Asteroseismology
Vol. 158, 2009, 38th LIAC/HELAS-ESTA/BAG, 2008
A. Noels, C. Aerts, J. Montalbán, A. Miglio and M. Briquet., eds.

Modeling the atmospheres of massive stars

J. Puls

Universitätssternwarte München, Scheinerstr. 1, D-81679 München, Germany

Abstract

In this review I summarize state-of-the-art approaches to model the atmospheres of massive stars, including their line-driven winds, and provide some examples for the potential impact of stellar pulsations on such atmospheres.

Hot star model atmospheres

Most of our knowledge about the physical parameters of hot stars[1] (e.g., effective temperatures, gravities, wind-properties, chemical composition of the outer layers) originates from *quantitative spectroscopy*, i.e., the analysis of stellar spectra by means of atmospheric models.[2] The numerical "construction" of such atmospheric models is a tremendous challenge, mostly because of the intense radiation fields of hot stars which lead to a number of effects that are not present in the atmospheres of cooler, less massive stars. In the following, I outline the basic ingredients which have to be adequately considered to allow for a reasonable description of the outer layers of massive stars.

Non-local thermodynamic equilibrium (NLTE)

In addition to the intense radiation field, massive stars have rather low densities, ρ, in their line and continuum forming regions. For objects not to close to the Eddington limit, the density in (quasi-) hydrostatic regions ("photosphere") depends almost soley on the pressure scale height, H, and the column density, m, (remember that m is roughly proportional to any reference optical depth scale, τ)

$$\rho(m) \approx \frac{1}{H}m \propto \frac{g}{T_{\text{eff}}}m \propto \frac{M_*/M_\odot}{(R_*/R_\odot)^2\, T_{\text{eff}}/T_\odot}m \qquad (1)$$

(with g the gravitational acceleration), i.e., at a given m (or τ) the density of a typical O-dwarf with 10 $M_\odot$, 10 $R_\odot$ and $T_{\text{eff}} = 30{,}000$ K is a factor of 50 lower than in the Sun. Thus collisions are less important in hot star atmospheres, at least in the upper photosphere/wind and in the UV/optical transitions.[3] In combination with the intense radiation field, this leads

---

[1] defined here to comprise OBA-stars, Luminous Blue Variables (LBVs) and Wolf-Rayet (WR) stars, and also Central Stars of Planetary Nebulae (CSPN), which have similar atmospheres as their massive O-star counterparts.

[2] For stars with $T_{\text{eff}} \gtrsim 30$ kK, photometric methods become completely unreliable discriminators of temperatures and gravites, due to the insensitivity of the Rayleigh-Jeans tail of the spectral energy distribution on temperature (e.g., Hummer et al. 1988).

[3] IR-transitions depend crucially on collisional processes, due to the lower energy separation of the involved levels.

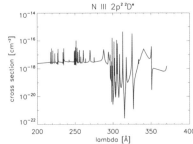
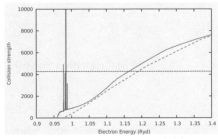

Figure 1: Examples for required atomic data. *Left:* Photo-ionization cross section of N III $2p^2$ $^2D^o$. Note the multitude of resonances (from Opacity Project, Seaton et al. 1992). *Right:* Comparison of collision strengths for the $n = 4$ to $n' = 5$ transition in hydrogen. Red (solid): Butler, in preparation; see also Przybilla & Butler (2004), a 28 state close-coupling calculation; green (dashed): Percival & Richards (1978), semi-classical approximation (note that the theory should only be accurate for $n, n' \geq 5$); blue (dotted): the van Regemorter (1962) approximation with $\bar{g} = 0.2$.

to the requirement that the occupation numbers of the atomic levels, n_i, have to be calculated from the NLTE rate equations (rather than from the Saha-Boltzmann equations),

$$n_i \sum_{j \neq i} (R_{ij} + C_{ij}) = \sum_{j \neq i} n_j (R_{ji} + C_{ji}), \qquad (2)$$

with radiative and collisional rates R_{ij} and C_{ij}, respectively, and including all bound-bound and bound-free processes. Since the radiative rates depend on the radiation field, whereas the radiation field depends on the opacities and emissivities, which themselves are functions of the occupation numbers, the rate-equations (for all levels n_i) have to be solved in parallel with the equations of radiative transfer (for all required frequencies). In order to prevent the stagnation of the ordinary Lambda-iteration, a delicate iteration scheme (*accelerated Lambda-iteration*) has to be implemented as well.

Atomic data

To calculate the radiative and collisional rates, detailed knowledge about the *individual* cross sections is required. Even though a large number of such data is available now (as calculated, e.g., within the OPAL project (Iglesias & Rogers 1996 and references therein), the *Opacity Project* (Seaton et al. 1992, for an example, see Fig. 1, left panel), and the Iron Project (Paper I: Hummer et al. 1993 until (to-date) Paper LXV: Witthoeft & Badnell 2008), most atmospheric codes do *not* contain the "best" data by default, but the individual user is responsible for compiling those data into appropriate *model atoms* (e.g., Przybilla & Butler 2001). Moreover, certain atoms/ions still lack a comprehensive description, particularly with respect to collisional data. An interesting example refers to the *hydrogen* bound-bound collision strengths, which have been (re-)considered in detail just recently (cf. Fig. 1, right panel), and significantly impact the strength of hydrogen IR-lines (Przybilla & Butler 2004). Finally, the number of levels and transitions in iron-group atoms is so large that certain simplifications need to be made in order to keep the problem treatable, mostly by packing several levels with close enough energies into one so-called *super-level* (Anderson 1985, 1989).

Line-blocking/blanketing

Just as for LTE models, also NLTE models require a careful consideration of line-blocking/ blanketing effects, due to the immense number of lines in the EUV. Various techniques are

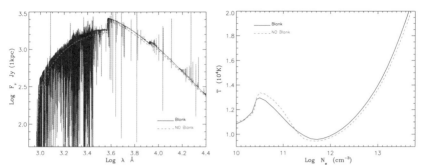

Figure 2: Effects of line blanketing (solid: blanketed model, dashed: model without blanketing) on the flux distribution (log F_ν (Jansky) vs. log λ (Å), left panel) and temperature structure ($T(10^4$ K) vs. log n_e, right panel) in the atmosphere of a late B-hypergiant. Blanketing blocks flux in the UV, redistributes it towards longer wavelengths and causes back-warming. From Puls et al. (2008).

applied to deal with the problem, using opacity distribution functions (ODFs, suitable under LTE conditions, Kurucz 1979), opacity sampling methods (Pauldrach et al. 2001), direct line-by-line calculations (involving model atoms consisting of super-levels, Hillier & Miller 1998, Hubeny 1998), and a more approximate method based on a simple statistical approach to calculate suitable means of line opacities and emissivities (Puls 2005). An example of the corresponding effects is given in Fig. 2, with the most prominent one being the reduction of the effective temperature scale for OB-stars, summarized in this volume by F. Martins. For a detailed discussion, we refer to Repolust et al. (2004).

Stellar winds

Due to their high luminosity, all massive stars display stellar winds, (mostly) driven by radiative line acceleration, since there are numerous spectral lines with high interaction probability close to the stellar flux maximum. Due to the acceleration mechanism, the mass-loss is metallicity dependent. Typical mass-loss rates range from 10^{-7} $M_\odot\mathrm{yr}^{-1}$ (with even lower values for late-O and B-dwarfs) up to 10^{-5} $M_\odot\mathrm{yr}^{-1}$ (WR-stars and LBVs), and terminal velocities scale with the photospheric escape velocity, from 200 km s^{-1} (A-supergiants) to $\gtrsim$2000 km s^{-1} (O-stars). Pioneering investigations have been performed by Lucy & Solomon (1970) and Castor, Abbott & Klein ("CAK", 1975), and important improvements with respect to a quantitative description have been provided by Friend & Abbott (1986), Pauldrach, Puls & Kudritzki (1986) and Vink, de Koter & Lamers (2000). A recent review on line-driven winds has been given by Puls, Vink & Najarro (2008; see also Kudritzki & Puls 2000).

Unified model atmospheres

With respect to the construction of atmospheric models, a consistent treatment of photosphere *and* wind is required, at least if the lines/continua are formed outside the (quasi-)hydrostatic region. Such *unified* model atmospheres were introduced by Gabler et al. (1986). Fig. 3 compares the electron-density distribution, as a function of the optical depth scale, for a hydrostatic atmosphere and unified atmospheres with a moderately dense and a thin wind. The need for using unified model atmospheres is clearly visible for the denser wind.

By means of a typical velocity law (with $\beta = 1$, see Eq. 3), one can calculate the maximum mass-loss rate for which a hydrostatic treatment is still possible, at least for UV and optical lines. ¿From the condition that unified models are required if $\tau_{\mathrm{Ross}} \geq 10^{-2}$ at the transition between the photosphere and the wind (which is roughly located at 10% of the sound-speed),

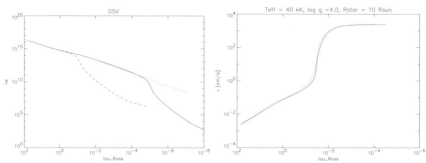

Figure 3: (Left) Electron-density as a function of the Rosseland optical depth, τ_{Ross}, for different atmospheric models of an O5-dwarf. Dotted: hydrostatic model atmosphere; solid, dashed: unified model with a thin and a moderately dense wind, respectively. In case of the denser wind, the cores of optical lines ($\tau_{Ross} \approx 10^{-1} - 10^{-2}$) are formed at significantly different densities than in the hydrostatic model, whereas the unified, thin-wind model and the hydrostatic one would lead to similar results.

Figure 4: (Right) Velocity fields in unified models of an O-star with a thin wind. Dotted: hydrodynamic solution; solid: analytical velocity law (Eq. 3) with similar terminal velocity and $\beta = 0.8$ (see text).

one finds $\dot{M} \lesssim 6 \cdot 10^{-8} M_\odot \mathrm{yr}^{-1} (R_*/10\,R_\odot)(v_\infty/1000\ \mathrm{km\,s^{-1}})$. Comparing with "observed" mass-loss rates, this limit implies that hydrostatic models are possible for the UV/optical spectroscopy of late O-dwarfs, B-stars until luminosity class II (for early sub-types) or Ib (for mid/late sub-types), and A-stars until luminosity class Ib. In any case, however, a check is required (if hydrostatic models are used), typically by comparing the observed mass-loss indicator, H_α, with the corresponding synthetic profile. If the observed core of this line is significantly shallower than the synthetic one, wind-effects *do* play a role and unified models have to be used for the analysis. Note that unified model atmospheres are computationally expensive, due to the need of accounting for the velocity-field induced Doppler shifts. Usually, this is done by solving the radiative transfer in the comoving frame (Mihalas et al. 1975, 1976). A faster method bases on the Sobolev approximation (Sobolev 1960) but is justified only for those lines which are formed predominantly in the wind[4], thus prohibiting the analysis of optical lines under typical conditions (e.g., Santolaya-Rey et al. 1996).

Unified model atmospheres may be constructed in two different ways. First, the complete (wind+photospheric) stratification is derived from a (self-) consistent approach, solving the hydrodynamic equations under the assumption of stationarity. In this case, the equation of momentum has to account for gravity and radiative acceleration (from Thomson-scattering, bound-free transitions, and line-transitions) as external forces, where the radiative acceleration follows from the NLTE-occupation numbers and the radiation field. As shown by CAK, the *line*-acceleration can be represented in the form of Thomson acceleration times a so-called force multiplier. In this description, the total line force as arising from summing up the multitude of individual contributions can be cast into one simple expression, depending on the density, velocity gradient, dilution factor, and three parameters which approximate the statistical distribution of the individual line-strengths. For a self-consistent description of the wind, these parameters have to be obtained from a multi-dimensional regression of the actual line force calculated under NLTE conditions. The force multiplier approach allows for a rather easy solution of the stationary wind dynamics, and is used within the model atmosphere code WM-basic (Pauldrach et al. 2001) to set up the density and velocity stratification throughout the complete atmosphere.

The alternative approach is based on an analytical description of the wind. In this case,

---

[4]more precisely: above the point where the velocity is larger than the thermal/turbulent speed of the considered ion.

the density and velocity field are described via

$$\rho(r) = \frac{\dot{M}}{4\pi r^2 v(r)}; \qquad v(r) = v_\infty \left(1 - \frac{R_t}{r}\right)^\beta, \tag{3}$$

where mass-loss rate $\dot{M}$, terminal velocity v_∞, and velocity field parameter β (typically in the range 0.8 - 2) are input and fit parameters. The transition radius, R_t, has to be calculated from the requirement of a smooth transition between the quasi-hydrostatic photosphere and the wind. As already mentioned, R_t is located at roughly 10% of the sound speed. The specific form of the analytical velocity law (Eq. 3) is a generalization of hydrodynamic solutions for line-driven winds from stars of different spectral types, and approximates such solutions quite well, as indicated in Fig. 4.

In this approach, the *photospheric* density stratification is calculated as for a hydrostatic atmosphere (but accounting for sphericity), from large optical depths until R_t, and can be roughly described by an exponential distribution with respect to radius in units of H. The corresponding velocity law is obtained by using the continuity equation with mass-loss rate from above and photospheric density, $\rho(r)$.

Unstable stellar winds

From early on (Lucy & Solomon 1970), there was the theoretical prediction that line-driven winds should be affected by a strong instability, inherent to the driving mechanism itself, called the line-driven or de-shadowing instability, thus rendering a stationary description at least questionable. E.g., for short-wavelength perturbations, one obtains $\delta g_{\rm rad}^{line} \propto \delta v$. First linear analyses (Owocki & Rybicki 1984, 1985) followed by a number of time-dependent hydrodynamical simulations by the groups of S. Owocki and A. Feldmeier confirmed this prediction. Recent results have been published for the 1-D case by Runacres & Owocki (2002, 2005) and for a simplified 2-D description by Dessart & Owocki (2003, 2005). The major result of these investigations - which cannot be *directly* used for model atmosphere calculations, since they are very time-consuming and based on line-statistics rather than individual NLTE-occupation numbers - can be summarized as follows: Beyond a rather stable lower wind, the outer wind ($r \gtrsim 1.3R_*$) develops extensive structure that consists of strong *reverse* shocks separating slower, dense material from high-speed rarefied regions in between. Fortunately, however, the gross quantities, such as $\dot{M}$ and v_∞, but also the density and velocity with respect to the *mass distribution*, are quite similar to the results from a stationary approach. Given the intrinsic mass-weighting of spectral formation, this suggests that at least lines with opacities proportional to the local density (e.g., UV-resonance lines) should be only weakly affected by the time-dependent structure of the wind (e.g., Puls et al. 1993).

In combination with a wealth of independent observational evidence, however, there are at least two principal features of such unstable winds which have to be transferred into the (stationary) atmospheric modeling, in order to allow for a more realistic description of occupation numbers and resulting synthetic SEDs and line profiles.

X-ray emission of hot stars.   The presence of shocks in time-dependent wind-models (with jump velocities of the order of a few hundred $\mathrm{km\,s^{-1}}$) lead to the prediction that massive stars should be X-ray emitters, and such X-ray emission has indeed been observed by EINSTEIN, ROSAT, CHANDRA, and XMM-NEWTON. For O-stars, these observations imply $L_x/L_{bol} \approx 10^{-7}$, temperatures of $T_s \approx 10^6 - 10^7$ K, and volume filling factors, f_{vol}, of a few percent. The conventional way to include this X-ray emission into unified atmospheric models is to assume a two-component wind (e.g., Hillier et al. 1993), with a small fraction (described by the volume filling factor) of shock-heated X-ray emitting gas, where the emission coefficient is described by an appropriate cooling function (depending on shock temperature and density), and a cool, X-ray absorbing wind (with density and velocity as described above). As shown by Pauldrach et al. (1994), the X-ray emission leads to significantly more flux below the He II edge (228 Å) and is an additional source of ionization (directly and via Auger-ionization) for ions of higher stages such as C IV, O IV, O V, O VI, N V, where most of these ions show strong resonance lines in the UV. Only an inclusion of X-rays can, e.g., explain the strong O VI resonance line as observed in most O-supergiants (Pauldrach et al. 1994). If included into the unified models, at least three additional parameters are required, L_x, T_s, and f_{vol}, where the last two might depend on the distance from the star.

Wind clumping.   During recent years, there have been various direct and indirect indications that hot star winds are not smooth, but clumpy, i.e., that there are small-scale density inhomogeneities which redistribute the matter into over-dense clumps and an almost void inter-clump medium. Such inhomogeneities are thought to be related to the structure formed in unstable line-driven winds (see above).

When treating wind-clumping in unified atmosphere codes, the standard assumption relates to the presence of optically *thin* clumps and a void inter-clump medium. A consistent treatment of the disturbed velocity field is still missing. The over-density (with respect to the average density) inside the clumps is described by a "clumping factor", f_{cl}. The most important consequence of such a structure is that any $\dot{M}$ derived from standard diagnostics based on processes with ρ^2-dependent opacities (such as H_α or the free-free radio excess) using *homogeneous models* needs to be scaled down by a factor of $\sqrt{f_{cl}}$.

Based on this approach, Crowther et al (2002), Hillier et al. 2003, Bouret et al. (2003, 2005) derived clumping factors of the order of 10 - 50, with clumping starting at or close to the wind base (in contradiction to theoretical expectations, but see Cantiello et al. 2009). From these values, a reduction of previous (unclumped) mass-loss rates by factors of 3 - 7 seems necessary. The *radial* stratification of the clumping factor has been studied by Puls et al. 2006, from a simultaneous modeling of H_α, IR, mm and radio observations. They found that, at least in dense winds, clumping is stronger in the lower wind than in the outer part, by factors of 4 - 6, and that unclumped mass-loss rates need to be reduced *at least* by factors 2 - 3.

Even worse, the analysis of the FUV P V-lines by Fullerton (2006) seems to imply mass-loss reductions by factors of 10 or larger, which would have an enormous impact on massive star evolution. However, as suggested by Oskinova (2007), the analysis of such optically *thick* lines might require the consideration of wind "porosity", which reduces the *effective* opacity at optically thick frequencies (Owocki et al. 2004). Consequently, the reduction of $\dot{M}$ as implied by the work from Fullerton et al. might be overestimated, and factors similar to those quoted above (around three) are more likely, particularly if considering independent arguments based on stellar evolution, such as the observed ratio of O and WR-stars (see Puls et al. 2008 and references therein).

State-of-the-art, NLTE model atmosphere codes

Table 1 enumerates and compares presently available atmospheric codes which can be used for the spectroscopic analysis of hot stars. Since the codes Detail/Surface and TLUSTY calculate occupation numbers/spectra on top of hydrostatic, plane-parallel atmospheres, they are "only" suited for the analysis of stars with negligible winds (see above). The different computation times are majorly caused by the different approaches to deal with line-blocking/blanketing. The overall agreement between the various codes (within their domain of application) is quite satisfactory, though certain discrepancies are found in specific parameter ranges, particularly regarding EUV ionizing fluxes (Puls et al. 2005, Simón-Díaz et al. 2008). Recent results of the application of these codes with respect to the determination of the stellar and wind parameters of massive stars have been summarized by Martins (2009).

Table 1: Comparison of state-of-the-art, NLTE, line-blanketed model atmosphere codes.

code	Detail/ surface[1]	TLUSTY[2]	POWR[3]	PHOENIX[4]	CMFGEN[5]	WM-basic[6]	FASTWIND[7]
geometry	plane-parallel	plane-parallel	spherical	spherical/ pl.-parallel	spherical	spherical	spherical
blanketing	LTE	yes	yes	yes	yes	yes	approx.
diagnostic range	no limitations	no limitations	no limitations	no limitations	no limitations	UV	optical/IR
major application	BA stars with negl. winds	hot stars with negl. winds	WRs	cool stars, SNe	OB(A)-stars, WRs, SNe	hot stars w. dense winds, SNe	OB-stars, early A-sgs
comments	no wind	no wind	–	no clumping no X-rays	–	no clumping	no X-rays
execution time	few minutes	hours	hours	hours	hours	1 to 2 h	few min. to 0.5 h

(1) Giddings (1981), Butler & Giddings (1985); (2) Hubeny (1998), (3) Gräfener et al. (2002), (4) Hauschildt (1992), (5) Hillier & Miller (1998), (6) Pauldrach et al. (2001), (7) Puls et al. (2005)

Massive stars, winds and pulsations

In this section, I provide some examples of the potential impact of stellar pulsations on the atmospheres of massive stars.

Stars with a luminosity to mass ratio exceeding $\sim 10^4\ L_\odot/M_\odot$ should be subject to "strange mode" oscillations (Saio, 2009). Interestingly, optical line profile variability shows the highest amplitudes just in that region of the HRD where strange mode oscillations should be present (Fullerton et al. 1996). Particularly important is the finding that the acoustic energy of such oscillations might be sufficient to initiate the mass-loss of WR-stars (Glatzel 2009), which to-date is unexplained for a large fraction of these objects.

At least B-supergiants show line profiles which require substantial extra line-broadening (in addition to rotational broadening), denoted by "macro-turbulence" and conventionally described by a supersonic, Gaussian velocity distribution in the photosphere (see Uytterhoeven et al. 2009). As shown by Aerts et al. (2009), such broadening might be *physically* explained as due to collective effects from low-amplitude, g-mode oscillations. Indeed, the spectroscopic analysis of 29 periodically variable B-supergiants[5] by Lefever et al. (2007), using unified model atmospheres as described above, revealed that most of them are located very close to the high gravity limit of the predicted pre-TAMS instability strip of high order g-modes (SPB-type, Pamyatnykh 1999) and/or within the corresponding post-TAMS instability strip of evolved stars, as predicted by Saio et al. (2006).

Non-radial pulsations (NRPs) might also be responsible for inducing *large-scale* structures in stellar winds, particularly so-called co-rotating interaction regions (Blomme 2009).

Finally, Feldmeier et al. (1997, 1998) showed that the generation of X-rays due to the self-excited line-driven instability alone results in too weak (factor of ~ 100) X-ray luminosities. To reproduce the observed values, photospheric disturbances are required to provide deep-seated seeds for clump formation, where subsequent clump-clump collisions are very effective in producing strong X-ray emission *when the photospheric excitation mechanism contains a multitude of frequencies*. This makes NRPs a prominent candidate for this process.

Acknowledgments. J.P. gratefully acknowledges a travel grant by HELAS.

---

[5] detected by Waelkens et al. (1998) from HIPPARCOS data.

References

Aerts, C., Puls, J., Godart, M., & Dupret, M.-A. 2009, CoAst, 158, 66

Anderson, L.S. 1985, ApJ, 298, 848

Anderson, L.S. 1989, ApJ, 339, 558

Blomme, R. 2009, CoAst, 158, 131

Bouret, J.C., Lanz, T., Hillier, D.J., et al. 2003, ApJ, 595, 1182

Bouret, J.C., Lanz, T., & Hillier, D.J. 2005, A&A, 438, 301

Butler, K., & Giddings, J.R. 1985, Newsl. Anal. Astron. Spectra, No. 9

Cantiello, M. 2009, CoAst, 158, 61

Castor, J.I., Abbott, D.C., & Klein, R.I. 1975, ApJ, 195, 157 (CAK)

Crowther, P.A., Hillier, D.J., Evans, C.J., et al. 2002, ApJ, 579, 774

Dessart, L., & Owocki, S.P. 2003, A&A, 406, L1

Dessart, L., & Owocki, S.P. 2005, A&A, 437, 657

Feldmeier, A., Puls, J., & Pauldrach, A.W.A. 1997, A&A, 322, 878

Feldmeier, A., Pauldrach, A.W.A., & Puls, J. 1998, ASP conf. ser., 131, 278

Friend, D.B., & Abbott, D.C. 1986, ApJ, 311, 701

Fullerton, A.W., Massa, D.L., & Prinja, R.K. 2006, ApJ, 637, 1025

Giddings, J.R. 1981, Ph.D. thesis, Univ. London

Glatzel, W. 2009, CoAst, 158, 252

Gräfener, G., Koesterke, L., & Hamann, W.-R. 2002, A&A, 387, 244

Hauschildt, P.H. 1992, J.Q.S.R.T., 47, 433

Hillier, D.J., & Miller, D.L. 1998, ApJ, 496, 407

Hillier, D.J., Kudritzki, R.-P., Pauldrach, A.W.A., et al. 1993, A&A, 276, 117

Hillier, D.J., Lanz, T., Heap, S.R., et al. 2003, ApJ, 588, 1039

Hubeny, I. 1998, ASP conf. ser., 138, 139

Hummer, D.G., Abbott, D.C., Voels, S.A., & Bohannan, B. 1988, ApJ, 328, 704

Hummer, D.G., Berrington, K.A., Eissner, W., et al. 1993, A&A, 279, 298

Iglesias, C.A., & Rogers, F.J. 1996, ApJ, 464, 943

Kudritzki, R.-P., & Puls, J. 1994, ARA&A, 38, 613

Kurucz, R.L. 1979, ApJS, 40, 1

Lefever, K., Puls, J., & Aerts, C. 2007 A&A, 463, 1093

Lucy, L.B., & Solomon, P.M. 1970, ApJ, 159, 879

Martins, F. 2009, CoAst, 158, 106

Mihalas, D., Kunasz, P., & Hummer, D.G. 1975, ApJ, 202, 465

Mihalas, D., Kunasz, P., & Hummer, D.G. 1976, ApJ, 206, 515

Owocki, S.P., & Rybicki, G.B. 1984, ApJ, 284, 337

Owocki, S.P., & Rybicki, G.B. 1985, ApJ, 299, 265

Owocki, S.P., Gayley, K.G., & Shaviv, N.J. 2004, ApJ, 616, 525

Oskinova, L.M., Hamann, W.-R., & Feldmeier, A. 2007, A&A, 476, 1331

Pamyatnykh, A.A. 1999, Acta Astronomica, 49, 119

Pauldrach, A.W.A., Puls, J., & Kudritzki, R.-P. 1986, A&A, 164, 86

Pauldrach, A.W.A., Kudritzki, R.-P., Puls, J., et al. 1994, A&A, 283, 525

Pauldrach, A.W.A., Hoffmann, T.L., & Lennon, M. 2001, A&A, 375, 161

Percival, I.C., & Richards, D. 1978, MNRAS, 183, 329

Puls, J., Owocki, S.P., & Fullerton, A.W. 1993, A&A, 279, 457

Puls, J., Urbaneja, M.A., Venero, R., et al. 2005, A&A, 435, 669

Puls, J., Markova, N., Scuderi, S., et al. 2006, A&A, 454, 625

Puls, J., Vink, J.S., & Najarro, F. 2008, AARev, in press

Przybilla, N., & Butler, K. 2001, A&A, 379, 955

Przybilla, N., & Butler, K. 2004, ApJ, 609, 1181

Repolust, T., Puls, J., & Herrero, A. 2004, A&A, 415, 349

Runacres, M.C., & Owocki, S.P. 2002, A&A, 381, 1015

Runacres, M.C., & Owocki, S.P. 2005, A&A, 429, 323

Saio, H. Kuschnig, R., Gautschy, A., et al. 2006, ApJ, 650, 1111

Saio, H. 2009, CoAst, 158, 245

Santolaya-Rey, A.E., Puls, J., & Herrero, A. 1997, A&A, 323, 488

Seaton, M.J., Zeippen, C.J., Tully, J. A., et al. 1992, Rev. Mex. AA, 23, 19

Simón-Díaz, S., & Stasińska, G. 2008, ArXiv e-prints, 805, arXiv:0805.1362

Sobolev, V.V. 1960, "Moving envelopes of stars", Cambridge: Harvard University Press

van Regemorter, H. 1962, ApJ, 136, 906

Uytterhoeven, K. 2009, CoAst, 158, 156

Vink, J.S., de Koter, A., & Lamers, H.J.G.L.M. 2001, A&A362, 295

Waelkens, C., Aerts, C., Kestens, E., et al. 1998, A&A, 330, 215

Witthoeft, M.C., & Badnell, N.R. 2008, A&A, 481, 543

DISCUSSION

Rieutord: Can the filling factor of the clumps be predicted by a stability analysis of the wind flow?

Puls: Not to my knowledge: the structure due to the line-driven instability depends (almost exclusively) on the growth-rate of the instability (from a linear analysis), and a certain damping (so-called "line-drag"), but in combination with additional effects. (i) The horizontal instability is strongly damped such that the horizontal length scale of the clumps should mostly depend on the lateral coherence length of the exciting structures in the photosphere (if there are any). (ii) The radial structure depends on when the amplitude of the non-linear flow velocity saturates (and a shock develops). Typical jump velocities are of the order of some hundred km/s. If there is no exciting mechanism, the scale of the separation of two clumps can be roughly calculated from such jump velocity and the linear growth rate. If there is a strong photospheric excitation mechanism, the corresponding frequency sets this scale (in combination with the flow speed). In conclusion, typical predicted volume filling factors are of the order of 1/10 ... 1/20, and result from the numerical models, but not directly from a stability analysis.

Comm. in Asteroseismology
Vol. 158, 2009, 38th LIAC/HELAS-ESTA/BAG, 2008
A. Noels, C. Aerts, J. Montalbán, A. Miglio and M. Briquet., eds.

Abundances of massive stars: some recent developments

T. Morel

Institut d'Astrophysique et de Géophysique, Université de Liège, Allée du 6 Août, 4000 Liège, Belgium

Abstract

The last few years, we have witnessed a large increase in the amount of abundance data for early-type stars, which have been very useful in various fields of astrophysics (e.g. mixing processes in stars, chemical evolution of galaxies). Two intriguing results emerging since the last reviews on this topic (Herrero 2003; Herrero & Lennon 2004) will be discussed: (a) nearby OB stars exhibit metal abundances generally *lower* than the solar/meteoritic estimates; (b) evolutionary models of single objects including rotation are largely unsuccessful in explaining the CNO properties of stars in the Galaxy and in the Magellanic clouds.

Introduction

This review about the chemical properties of massive stars will focus on two issues that are relevant in the context of this meeting. First, one may wonder how the chemical abundances of nearby OB stars compare to the solar values. This piece of information is, for instance, essential to model B-type pulsators properly and to draw correct inferences about their internal structure. Second, the abundances of several elements are powerful probes of mixing phenomena and, as such, can be used to improve our theoretical understanding of these processes and ultimately better model the evolution of massive stars across the HR diagram. Fascinating physical phenomena such as mass-transfer processes can affect the abundances of stars in binaries, but we refrain from discussing these systems here.

Getting the abundances

A prerequisite to obtain reliable abundances is adopting accurate atmospheric parameters (a favourable case is offered by detached eclipsing binaries; e.g. Pavlovski & Southworth 2008). The effective temperature can be derived from photometric indices or, preferably, through ionisation balance of some metals (usually Si). The surface gravity is derived by fitting the collisionally-broadened wings of the Balmer lines, while the microturbulent velocity is inferred by requiring the abundances of a given ion to be independent of the line strength. Model atmospheres (either LTE or NLTE) with adequate line-blanketing are required. Departures from LTE are significant in hot stars and a full NLTE treatment for the line formation is also needed. Plane-parallel (e.g. TLUSTY, DETAIL/SURFACE) or so-called unified codes (e.g. CMFGEN, FASTWIND) can be used depending on the strength of the stellar wind. Of course, more sophisticated analysis techniques (NLTE model atmosphere, spherical extension) are much more demanding in terms of computer resources and should be used with discernment. A hybrid approach involving hydrostatic, LTE model atmospheres coupled with an NLTE line-formation treatment is often employed for early B-type stars on the main sequence (e.g. Nieva & Przybilla 2008). At this stage, a set of model atoms as complete as possible should

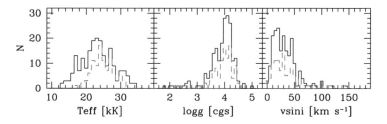

Figure 1: Distribution of the physical parameters for nearby OB stars with abundance data (either C, N, O, Mg, Al, Si, S or Fe; *solid line*) and CNO data (*dashed line*).

have also been developed. Contrary to the situation for cool stars where one can calibrate the log gf values using a solar spectrum obtained with the same instrumental set up, for hot stars one has to rely instead on theoretical calculations. The technique used for the abundance determination (classical curve-of-growth vs spectral synthesis) evidently depends on the seriousness of blending issues and hence on the $v \sin i$ of the star under study. As one might expect, most studies are strongly biased towards narrow-lined stars (see below).

Chemical composition of nearby OB stars: solar?

To estimate the baseline chemical properties of massive stars in the solar neighbourhood, abundance data from the literature have been gathered using the following criteria: (a) known, binaries are avoided; (b) supergiants are omitted to minimise evolutionary effects (they will be discussed in the following in the context of mixing); (c) only stars within $\sim$1 kpc are discussed because of the existence of a Galactic metal gradient (e.g. Daflon & Cunha 2004). Only NLTE studies based on high-resolution spectra have been selected, while carbon abundances solely based on transitions not well modelled by most line-formation codes (e.g. C II λ4267) have been excluded. Note that some bright, easily-observable stars can be represented more than once. As seen in Fig.1, most of these $\sim$150–200 stars are slowly-rotating, early B-type dwarfs/subgiants (data from Cunha & Lambert 1994, Daflon & Cunha 2004 and references therein, Gies & Lambert 1992, Gummersbach et al. 1998, Kilian 1992, 1994, Lyubimkov et al. 2005, Morel et al. 2008, Nieva & Przybilla 2008 and Thompson et al. 2008).

Before comparing the abundances of young OB stars with the solar values, two remarks are necessary. First, chemical evolution models of the Galaxy predict a small enrichment of the local ISM in metals over the past 4.6 Gyr (typically $\sim$0.1 dex; Chiappini et al. 2003) and the solar photosphere is slightly depleted in metals because of gravitational settling (typically $\sim$0.05 dex; Turcotte et al. 1998). Nearby, unevolved OB stars are hence naturally expected to appear metal rich compared to the Sun, but this should only be at moderate levels. Second, the solar abundances are indistinguishable from the mean values obtained for nearby, early G stars of the Galactic thin disk analysed in exactly the same way (Allende-Prieto 2006). The Sun therefore does not appear peculiar in terms of bulk metallicity or elemental abundances (despite hosting giant planets) and its chemical composition should be representative of the one prevailing in the local ISM at the time of its formation (see also, e.g. Gustafsson 2008).

As can be seen in Fig.2, the mean abundances of several metals (especially C, Al and Si) in nearby OB stars are found to be significantly lower than the solar values derived either using classical model atmospheres (Grevesse & Sauval 1998) or 3-D hydrodynamical simulations (Asplund et al. 2005). Although the data for OB stars are highly inhomogeneous, this result is found by most studies (if not all in case of certain elements). The results of Kilian (1992, 1994) do not appear clearly discrepant despite the fact that they are based on model

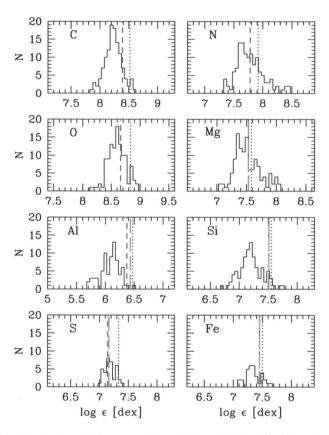

Figure 2: Distribution of the metal abundances of nearby OB stars. The logarithmic abundances are given on a scale in which log ϵ(H)=12. *Dotted line*: solar abundances of Grevesse & Sauval (1998), *short-dashed and long-dashed lines*: solar and meteoritic abundances of Asplund et al. (2005), respectively.

atmospheres that are not fully line blanketed. Low abundances are also consistently found for Mg, but the distribution depicted in Fig.2 is strongly biased by the much higher values (up to 0.5 dex above solar) found by Lyubimkov et al. (2005).[1] Nitrogen deserves special comment, as the larger spread observed is likely the consequence of deep mixing already on the main sequence (as will be discussed below). The concomitant C and O depletions are of smaller amplitude and are not expected to be readily detectable at these levels of N enrichment. Two NLTE studies are available in the literature for both S and Fe. In the former case, solar values were found (Daflon & Cunha 2004; Morel et al. 2008). For Fe, however, both subsolar (Morel et al. 2008) and solar values have been reported (Thompson et al. 2008). The fact that the mean Al, Si and Fe abundances are lower than the accurate meteoritic values supports the idea that the abundances of nearby OB stars are generally underestimated (by up to a factor 2 in the case of Si), perhaps as a result of missing physics or unaccounted systematic errors. Such a discrepancy is hard to explain on theoretical grounds (although some models have

---

[1]A number of figures showing the results of the individual studies are available from: http://www.astro.ulg.ac.be/~morel/liege_colloquium.pdf

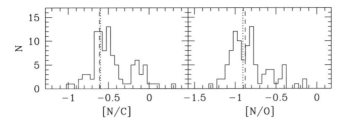

Figure 3: Distribution of the [N/C] and [N/O] ratios for nearby OB stars. The *dotted* and *short-dashed* lines indicate the solar ratios of Grevesse & Sauval (1998) and Asplund et al. (2005), respectively.

been proposed; see, e.g. Witt 2001). The neon abundance of nearby B stars is consistent with the estimate of Grevesse & Sauval (1998) based on coronal observations of the Sun and does not support the high theoretical value required to restore the past agreement between the standard solar models and the helioseismic constraints (Morel & Butler 2008, and references therein). Although the argon abundance of B stars is much higher than the solar estimate (Lanz et al. 2008), the values for these two noble gases are in excellent agreement with those inferred in the ionised gas of the Orion nebula (Esteban et al. 2004).

Deep mixing in OB stars

Mixing on the main sequence and close to it

Figure 3 shows the CNO logarithmic abundance ratios for the stars discussed above. The [N/C] ratio, which is a robust indicator of CNO-processed material dredged up to the surface, shows evidence for two subsamples of stars with either roughly solar values ([N/C]$\sim$−0.5 dex) or significantly higher ratios ([N/C]$\sim$−0.1 dex). The latter population is unexpected considering that these stars are core-hydrogen burning objects with low $v \sin i$ values (Fig.1). The [N/O] data do not clearly show evidence for two distinct populations, but rather suggest a continuum of values with a tail extending to high ratios (see also Herrero & Lennon 2004). The N-rich stars tend to be slightly more evolved than the N-normal stars ($\Delta \log g \sim$0.25 dex; Fig.4), suggesting that this nitrogen excess could arise from an evolutionary effect (however, the unphysically large gravities reaching up to $\log g$=4.6 dex found for many N-normal stars cast some doubt on this interpretation; see also Fig.6 where these stars fall well below the ZAMS). On the other hand, a higher incidence of an N excess in magnetic B stars compared to stars without a field detection is emerging (Morel et al. 2008), but more work is needed to firmly establish such a (probably statistical) link between an N enhancement and magnetic fields. Abundance data for large samples of both magnetic and non-magnetic stars are in particular needed to clearly reveal a dichotomy between the two groups. In any case, it is interesting to note in the context of models incorporating magnetic fields that some N-rich stars possess a large-scale, dipole field most likely of fossil origin (e.g. ζ Cas, β Cep).

   To examine whether this population of slowly-rotating dwarfs with an N excess is predicted by theory, we compare in Fig.5 the abundance data with the predictions of evolutionary models including rotational effects (Heger & Langer 2000). The N-rich stars have not yet evolved beyond the TAMS (Fig.4), and rotational velocities reaching $\sim$200 km s^{-1} on the ZAMS are needed to reproduce their CNO properties. This can be contrasted with their low projected, present values ($<v \sin i>\sim$30 km s^{-1}; Fig.4). Some β Cephei stars in this sample have been shown to be *intrinsically* very slowly rotating based on seismic studies (Morel et al. 2006). Assuming that the N-rich stars were rapid rotators on the ZAMS, but then dramatically spun down to the observed levels, is not supported by models (e.g. Meynet & Maeder 2003) or

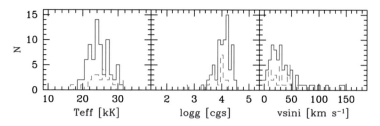

Figure 4: Distribution of the physical parameters for the N-normal ([N/C]<−0.3 dex; *solid line*) and N-rich stars ([N/C]>−0.3 dex; *dashed line*).

observations (e.g. Huang & Gies 2006) that only suggest a modest loss of angular momentum in this mass range during core-hydrogen burning. Magnetic braking is also unlikely to strongly spin down B stars with a dipole field at the few hundred Gauss level (ud-Doula et al. 2008).

Mixing in evolved objects

A broader and more complete picture of the incidence of deep mixing across the HR diagram can be gained by considering all stars irrespective of their evolutionary status or Galactocentric distance (we complement the previous data with results from Crowther et al. 2006, Kilian et al. 1994, Kilian-Montenbruck et al. 1994, Mathys et al. 2002, Przybilla et al. 2006, Schiller & Przybilla 2008, Searle et al. 2008, Smiljanic et al. 2006, Venn 1995 [with updated N abundances from Venn & Przybilla 2003], Villamariz et al. 2002, Villamariz & Herrero 2005 and Vrancken et al. 2000). The variation of the [N/C] abundance ratio across the log g-log $T_{\rm eff}$ plane clearly reveals evolutionary effects in the sense that all supergiants are N-rich by up to two orders of magnitude compared to main sequence stars of similar mass (Fig.6). The extreme enhancements observed in several cool supergiants may result from the first dredge up phase. As discussed above, two populations of normal and mildly-enriched N stars with similar masses coexist between the ZAMS and the TAMS. There is a tentative indication that the N-rich stars cluster at lower $v \sin i$ values (Fig.7; see also Fig.4), but it is unfortunately not yet possible to test rotational mixing theories thoroughly on Galactic objects as only very few CNO data for fast rotators are available. It is hoped that the data for fast-rotating stars in three Galactic open clusters obtained in the course of the VLT/FLAMES survey of massive stars will soon fill this caveat. Data for the SMC/LMC are, however, already available and have interestingly revealed two populations that cannot be accounted for by rotational mixing operating in single objects (Hunter et al. 2008; Brott, 2009): slowly-rotating dwarfs with an N excess (as in the Galaxy) and fast-rotating stars with normal nitrogen.

Summary and directions for future research

Most abundance studies are strongly biased towards early B-type stars. It would be desirable in the future to have more data for O stars. Unexpected and exciting results are also likely for late B-type stars because of the existence of diffusion effects in this $T_{\rm eff}$ range, but very few systematic studies have been undertaken up to now (e.g. Hempel & Holweger 2003).

   The metal abundances of nearby OB stars are found to be in most cases below the most recent (and likely realistic) estimates for the Sun or, more importantly, the meteoritic values. The fact that the abundances (after correction for dust depletion) of C, N, O and S in the ionised gas of the Orion nebula (Esteban et al. 2004) are in good agreement with the new solar values supports the idea that the abundances of massive stars are generally underestimated and seems to rule out scenarios such as a recent infall of metal-poor material at the Sun

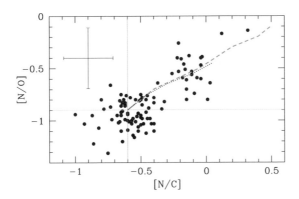

Figure 5: Comparison between the observed [N/C] and [N/O] ratios of nearby OB stars and the predictions of evolutionary models including rotation (Heger & Langer 2000). The results are shown for a 12 M$_\odot$ star (representative of this sample) and three values of the rotational velocity on the ZAMS: 99 (*solid line*), 206 (*dotted line*) and 328 km s^{-1} (*dashed line*). The loci define an age sequence with time increasing rightwards from ZAMS to TAMS. The solar ratios are indicated by horizontal and vertical solid lines.

location as an explanation for the low abundances of these objects. We should point out, however, that the values closer to solar very recently reported by Przybilla et al. (2008) for a small sample of early B-type stars suggest that improvements in the data analysis (e.g. better model atoms and/or temperature scale) could help solving this discrepancy.

The surface abundances of β Cephei stars are indistinguishable from the values found for (presumably) non-pulsating B0–B3 main sequence stars, but some objects are also N-rich and hence display the signature of deep mixing (Morel et al. 2006). Taking the results for B stars at face value would imply a higher *relative* abundance of the iron-peak elements compared to the solar mixture, and therefore that the pulsation modes are more easily excited (Montalbán et al., 2009). Until the discrepancy with the solar/meteoritic values discussed above is better understood, however, a sound assumption would be to use the solar mixture of Asplund et al. (2005) in oscillation codes. As discussed during this meeting, one of the priorities for the future is to incorporate the detailed abundances of the SMC/LMC in such theoretical codes (and not simply a scaled solar pattern) in order to account for the existence of B-type pulsators in such low-metallicity environments and to adequately model their pulsation properties. Such a work is under way (in preparation by Salmon and co-workers).

Several elements are powerful probes of mixing processes (He, B, CNO). There have been claims of an increase of the He abundance along the main sequence (Lyubimkov et al. 2004; Huang & Gies 2006), but it is still unclear whether this effect is real or is merely an artefact of the analysis (e.g. choice of microturbulence). On the other hand, boron provides precious (and complementary) information about shallow mixing close to the surface (e.g. Mendel et al. 2006). Clear evolutionary effects are observed, with all supergiants exhibiting high [N/C] ratios. Recent observations have revealed the existence of two populations challenging the very importance of rotational mixing as mechanism dredging up material from the convective core to the photosphere in massive stars: (a) slowly-rotating, N-rich dwarfs (mixing efficiency underestimated, stronger loss of angular momentum than expected and/or magnetic fields?); (b) fast-rotating stars with normal nitrogen in the LMC (binaries?). Establishing the nature of these two populations is needed for further progress. This implies that more effort should be devoted to the daunting and arduous task of deriving accurate abundances for fast rotators.

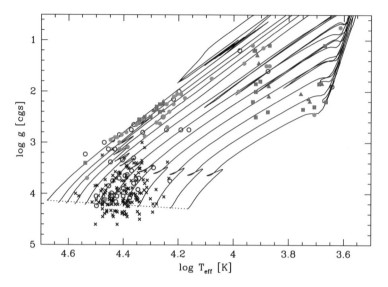

Figure 6: Variation of the [N/C] abundance ratio across the log g-log T_{eff} plane (the evolutionary tracks without rotation and for M=4.0, 5.0, 6.3, 7.9, 10.0, 12.6, 15.8, 20.0, 25.1, 31.6, 39.8 and 63.1 M$_\odot$ are taken from Claret 2004). *Crosses*: [N/C]<−0.2, *open circles*: −0.2<[N/C]<+0.2, *filled circles*: +0.2<[N/C]<+0.6, *filled squares*: +0.6<[N/C]<+1.0, *filled triangles*: [N/C]>+1.0 dex.

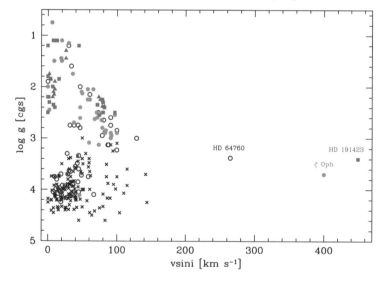

Figure 7: Variation of the [N/C] abundance ratio across the log g-$v \sin i$ plane (symbols as in Fig.6). The three fast rotators have been analysed by Searle et al. (2008), Villamariz et al. (2002) and Villamariz & Herrero (2005). Note that stars with widely different masses are shown in this figure.

References

Allende-Prieto, C. 2006, in The Metal-Rich Universe, in press (astro-ph/0612200)

Asplund, M., Grevesse, N., & Sauval, A. J. 2005, in Cosmic abundances as records of stellar evolution and nucleosynthesis, ed. T. G. Barnes III, F. N. Bash, ASP Conf. Ser., 336, 25

Brott, I., Hunter, I., de Koter, A., et al. 2009, CoAst, 158, 55

Chiappini, C., Romano, D., & Matteucci, F. 2003, MNRAS, 339, 63

Claret, A. 2004, A&A, 424, 919

Crowther, P. A., Lennon, D. J., & Walborn, N. R. 2006, A&A, 446, 279

Cunha, K., & Lambert, D. L. 1994, A&A, 426, 170

Daflon, S., & Cunha, K. 2004, ApJ, 617, 1115

Esteban, C., Peimbert, M., García-Rojas, J., et al. 2004, MNRAS, 355, 229

Gies, D. R., & Lambert, D. L. 1992, ApJ, 387, 673

Grevesse, N., & Sauval, A. J. 1998, Space Sci. Rev., 85, 161

Gummersbach, C. A., Kaufer, A., Schäfer, D. R., et al. 1998, A&A, 338, 881

Gustafsson, B. 2008, Phys. Scr., T130, 014036

Heger, A., & Langer, N. 2000, ApJ, 544, 1016

Hempel, M., & Holweger, H. 2003, A&A, 408, 1065

Herrero, A. 2003, in CNO in the Universe, ed. C. Charbonnel, et al., ASP Conf. Ser., 304, 10

Herrero, A., & Lennon, D. J. 2004, in Stellar rotation, ed. A. Maeder, P. Eenens, 209

Huang, W., & Gies, D. R. 2006, ApJ, 648, 591

Hunter, I., Brott, I., Lennon, D. J., et al. 2008, ApJ, 676, L29

Kilian, J. 1992, A&A, 262, 171

Kilian, J. 1994, A&A, 282, 867

Kilian, J., Montenbruck, O., & Nissen, P. E. 1994, A&A, 284, 437

Kilian-Montenbruck, J., Gehren, T., & Nissen, P. E. 1994, A&A, 291, 757

Lanz, T., Cunha, K., Holtzman, J., & Hubeny, I. 2008, ApJ, 678, 1342

Lyubimkov, L. S., Rostopchin, S. I., & Lambert, D. L. 2004, MNRAS, 351, 745

Lyubimkov, L. S., Rostopchin, S. I., Rachkovskaya, T. M., et al. 2005, MNRAS, 358, 193

Mathys, G., Andrievsky, S. M., Barbuy, B., et al. 2002, A&A, 387, 890

Mendel, J. T., Venn, K. A., Proffitt, C. R., et al. 2006, ApJ, 640, 1039

Meynet, G., & Maeder, A. 2003, A&A, 404, 975

Montalbán, J., Miglio, A., Morel, T. 2009, CoAst, 158, 288

Morel, T., Butler, K., Aerts, C., et al. 2006, A&A, 457, 651

Morel, T., Hubrig, S., & Briquet, M. 2008, A&A, 481, 453

Morel, T., & Butler, K. 2008, A&A, 487, 307

Nieva, M. F., & Przybilla, N. 2008, A&A, 481, 199

Pavlovski, K., & Southworth, J. 2008, MNRAS, submitted

Przybilla, N., Butler, K., & Becker, S. R. 2006, A&A, 445, 1099

Przybilla, N., Nieva, M. F., & Butler, K. 2008, ApJ, 688, L103

Schiller, F., & Przybilla, N. 2008, A&A, 479, 849

Searle, S. C., Prinja, R. K., Massa, D., et al. 2008, A&A, 481, 777

Smiljanic, R., Barbuy, B., de Medeiros, J. R., et al. 2006, A&A, 449, 655

Thompson, H. M. A., Keenan, F. P., Dufton, P. L., et al. 2008, MNRAS, 383, 729

Turcotte, S., Richer, J., Michaud, G., et al. 1998, ApJ, 504, 539

ud-Doula, A., Owocki, S. P., & Townsend, R. H. D. 2008, MNRAS, in press (arXiv:0810.4247)

Venn, K. A. 1995, ApJ, 449, 839

Venn, K. A., & Przybilla, N. 2003, in CNO in the Universe, ASP Conf. Ser., 304, 20

Villamariz, M. R., Herrero, A., Becker, S. R., et al. 2002, A&A, 388, 940

Villamariz, M. R., & Herrero, A. 2005, A&A, 442, 263

Vrancken, M., Lennon, D. J., Dufton, P. L., et al. 2000, A&A, 358, 639

Witt, A. N. 2001, Phil. Trans. R. Soc. Lond. A, 359, 1949

DISCUSSION

Grevesse: You recently redetermined the Ne abundance in some of these stars. Could you tell us why your results lead to smaller values than previously measured values?

Morel: Contrary to previous studies that were solely based on NeI lines, we used NeI and NeII lines. Forcing agreement between the mean abundances given by both ions allowed us to have a better handle on the temperature scale.

Rauw: Can you say something about binarity in your sample of B stars?

Morel: Several N-rich stars are well-known β Cephei stars and have been intensively monitored in spectroscopy both on short and long timescales without any evidence for binarity. There is no indication that they are runaways or strong X-ray sources either.

Comm. in Asteroseismology
Vol. 158, 2009, 38th LIAC/HELAS-ESTA/BAG, 2008
A. Noels, C. Aerts, J. Montalbán, A. Miglio and M. Briquet., eds.

Non-radial pulsations and large-scale structure in stellar winds

R. Blomme

Royal Observatory of Belgium
Ringlaan 3, B-1180 Brussel, Belgium

Abstract

Almost all early-type stars show Discrete Absorption Components (DACs) in their ultraviolet spectral lines. These can be attributed to Co-rotating Interaction Regions (CIRs): large-scale spiral-shaped structures that sweep through the stellar wind.

We used the Zeus hydrodynamical code to model the CIRs. In the model, the CIRs are caused by "spots" on the stellar surface. Through the radiative acceleration these spots create fast streams in the stellar wind material. Where the fast and slow streams collide, a CIR is formed. By varying the parameters of the spots, we quantitatively fit the observed DACs in HD 64760.

An important result from our work is that the spots do not rotate with the same velocity as the stellar surface. The fact that the cause of the CIRs is not fixed on the surface eliminates many potential explanations. The only remaining explanation is that the CIRs are due to the interference pattern of a number of non-radial pulsations.

Individual Objects: HD 64760, ζ Pup, ξ Per

Structure in O-type stellar winds

Many indicators exist for the presence of structure in the stellar winds of early-type stars. Here we discuss the wind structure in O and early B-type stars; the structure in Wolf-Rayet stars is discussed by Gosset et al. (2009). Two types of structure are known to coexist in the stellar wind: small-scale and large-scale. The small-scale type (clumping) has been discussed by Puls (2009); we will concentrate on the large-scale structure. The best-known indicators of this large-scale structure are the Discrete Absorption Components (DACs). These can be seen in the P Cygni profiles of some ultraviolet resonance lines. The absorption part of such a P Cygni profile covers the whole range of velocities from the base of the wind up to the terminal velocity at large distances from the star. The DACs are additional absorption features that are superposed on these P Cygni profiles. With time, the DACs move from low outflow velocities to the terminal velocity.

This behaviour is best seen on a so-called dynamic spectrum (Figure 1), which is a grey-scale plot that shows how much a single spectrum differs from the average of all spectra. (Note that other authors use the highest fluxes as a reference spectrum, rather than the average. This explains why Figure 1 shows both emission and absorption.) By stacking the spectra in time-sequence, one obtains a clear view of the outward movement of the DAC. From the DAC behaviour of a sample of OB stars, it was found that the DACs recur on timescales related to $v_{eq} \sin i$ (Henrichs et al. 1988, Prinja 1988). In addition to DACs, Figure 1 also shows rotational modulations: these are the horizontal, nearly-straight features seen in the spectrum. They cover a large range of velocities and (in this example) repeat on

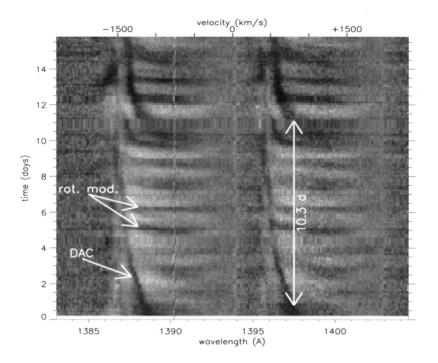

Figure 1: Grey-scale plot of the Si IV $\lambda\lambda1394,1403$ P Cygni profile of HD 64760, based on spectra extracted from the IUE archive (observers: Massa et al. 1995). At each wavelength the grey value indicates the difference between this spectrum and the average spectrum. The spectra are stacked as a function of time. DACs are clearly seen as the absorption that is slowly moving toward the terminal velocity ($v_\infty = 1500$ km s^{-1}). The recurrence time of the DACs is ~ 10.3 d. The rotational modulations move much faster than the DACs, and repeat every ~ 1.2 d.

a $\sim$ 1.2 day timescale. DACs have been detected in almost all O and early B-type stars (e.g. Kaper et al. 1996, 1997, 1999), but rotational modulations are much rarer.

The effect of structure is also seen in the Hα line, which is partly formed in the stellar wind of these early-type stars. Examples of this are given by, e.g., Morel et al. (2004).

HD 64760

Rotational Modulations

HD 64760 is a B0.5 Ib supergiant with $v_{eq} \sin i = 265$ km s^{-1} (Kaufer et al. 2006), which is quite high for a supergiant. The star is therefore probably viewed nearly equator-on, and, in what follows we will assume that it is viewed exactly equator-on as this will simplify the analysis.

Figure 1 shows the data from the IUE Mega Campaign (Massa et al. 1995) for this star. The plot is limited to the Si IV $\lambda\lambda1394,1403$ doublet. All features are seen in both components of the doublet, but we will concentrate our discussion on the (strongest) blue component to avoid contamination effects of the red component. A Fourier analysis shows that the rotational modulations are not exactly straight, but bowed. This inspired Fullerton et al. (1997) to propose a simple ad-hoc model where a spiral-shaped structure is assumed to rotate through the wind. As this spiral crosses the line of sight towards the observer, additional absorption is seen which starts at an intermediate velocity and then moves to both higher and lower velocities with time, thus creating a bow-shaped rotational modulation in the spectrum.

A more theoretical background to this ad-hoc model was provided by Cranmer & Owocki (1996) who followed up on the idea proposed by Mullan (1986) that Co-rotating Interaction Regions (CIRs) should exist in OB star winds. They used a numerical code to solve the hydrodynamical equations of the stellar wind including the line driving, which is responsible for these radiatively driven winds. They then added a bright spot on the stellar surface centred on the equator and calculated the effect of the additional line driving due to this spot. Because of the difference in radiative acceleration, the wind streaming out above the bright spot has a higher density and a lower velocity than the wind outside the spot. Due to the rotation, these streams collide and form a spiral-shaped density pattern, called a Co-rotating Interaction Region (CIR). It is important to realize that the CIR is a pattern in the wind; it is *not* the path the particles follow through the wind. Particles basically move out radially (except close to the stellar surface where some effect of the rotation is still seen, due to angular momentum conservation).

DACs

Inspired by the above success for rotational modulations, Lobel and Blomme (2008) tried to explain the DACs in a similar way. We calculated models using the Zeus hydrodynamics code, following the procedure outlined by Cranmer & Owocki (1996). A major difference in our work, however, is that we included a spot that is not fixed on the stellar surface (an idea which had already been proposed by Kaufer et al. 2006). The reason for introducing such a spot is that the recurrence timescale of the DACs is 10.3 d (see Figure 1), while the rotation period of HD 64760 is only 4.1 d. The main effect of a spot velocity different from the rotational velocity is that the CIRs spiral more for a rapidly rotating spot and less for a slower spot. An example of the resulting hydrodynamical structure is shown in Figure 2.

For a given hydrodynamical model, we calculate the P Cygni profiles using the 3D radiative transfer code Wind3D developed by A. Lobel (see Lobel & Blomme 2008). In this way, we can create artificial grey-scale plots and compare their DACs to the observed ones. By varying the parameters of the 2 spots we put in the model (brightness enhancement and opening angle of the spot), we obtained a very good fit to the shape and width of the DACs observed in the HD 64760 Si IV resonance line (Lobel & Blomme 2008).

As has already been pointed out by Cranmer & Owocki (1996), the velocity plateaus formed due to the CIRs are more important than the density enhancements in determining the effect on the P Cygni profiles. This can easily be seen from an extreme simplification of the radiative transfer, where we just consider the Sobolev approximation to the optical depth (τ) and limit ourselves to the radial direction only. We then have:

$$\tau \propto \frac{\rho}{dv/dr},\tag{1}$$

from which it follows that both density (ρ) and velocity gradient (dv/dr) can provide additional absorption. By plotting these variables for our set of models, we found that in most cases it is the nearly-zero velocity gradient that is responsible for the DACs.

The density contrast we find for the best-fit CIRs is quite small (20 to 30 %) and the total mass loss is increased by less than 1 % compared to a smooth wind. These CIR structures are therefore quite delicate, and it is surprising that they are not destroyed by the instability mechanism that is responsible for the clumping in the wind. Owocki (1998) has made some

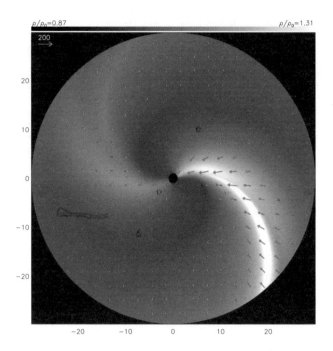

Figure 2: Model CIRs in the stellar wind of HD 64760. The grey-scale indicates the density contrast (with respect to a smooth wind) and the velocity vectors the deviation from the smooth wind velocity. The spots causing the CIRs have a brightness enhancement of 20 % and 8 % and an opening angle of 20° and 30°, respectively. Both spots rotate at 1/5 of the equatorial velocity. This model gives the best fit to the observed HD 64760 DACs seen in the Si IV line.

calculations combining CIRs and the instability mechanism. He found that, depending on the strength of the instability mechanism, the CIRs can be completely destroyed. However, observations tell us that DACs are seen in nearly all O and early B-type stars (Prinja & Howarth 1986). In principle, we can therefore use the presence of DACs to constrain the amount of clumping that can be present in a stellar wind.

Non-Radial Pulsations (NRPs)

The fact that the CIRs rotate more slowly than the stellar surface eliminates many potential explanations. A slower rotation timescale would be possible in a differentially rotating star, if the spots were located at higher latitude. However, in that case, the nearly-radial outflow of the wind would prevent the CIRs from crossing in front of the stellar disc, and would therefore not explain the DACs. The only remaining explanation is that the CIRs are due to NRPs (or a beat pattern of NRPs). From the spectra of photospheric lines, Kaufer et al. (2006) found NRPs in HD 64760 with three different periods (P_1=4.810 h, P_2=4.672 h, P_3=4.967 h). They also found variability in the Hα profile (a line which is formed partly in the stellar wind), but with a 2.4-day period. They tried to attribute this period to a beat period between P_1

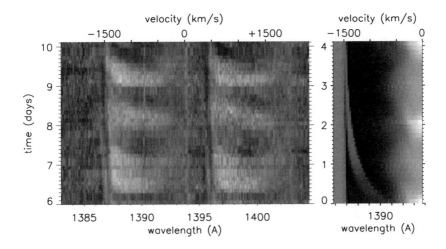

Figure 3: Attempted hydrodynamical spot model (*right*) to explain the observed rotational modulations (*left*). It is clear that the curvature of the model (toward the terminal velocity) is much too high to provide a good explanation for the observation. None of the models we tried gave a satisfactory fit to the data.

and P_2, but found that this beat period is 6.8 d rather than 2.4 d.

However, if we look at the power spectrum they published (their Figure 5), we see that the NRP periods are not that well-determined. A small shift of the P_1 and P_2 periods (by only 0.0235 hrs, which is well within the uncertainty) would result in a beat pattern of 10.3 d, which is exactly the recurrence timescale of the DACs. It would be very interesting to see if the NRP periods of HD 64760 could be determined to such a precision that the beat period can indeed be shown to correspond with the DAC recurrence time.

Rotational Modulations revisited

The claim that rotational modulations are well explained by CIRs is based on a kinematical model (Owocki et al. 1995), i.e. a model that uses a pre-specified density wave superposed on a smooth wind. When we try to make a hydrodynamical spot model, however, we never achieve a rotational modulation that is sufficiently flat to compare well with the observations (see Figure 3).

The curvature of the rotational modulation is mainly determined by the ratio v_{eq}/v_∞. Introducing a spot with a velocity lower than v_{eq} might suggest itself as a possible solution, but in that case the timescale needs to be stretched (because the spot takes a longer time to cross the line of sight as it moves more slowly). This stretching basically compensates for the effect of the slower spot velocity and we again end up with a curvature incompatible with the observed one. We thus have considerable problems in finding a *hydrodynamical* model based on surface spots that can explain the rotational modulations.

Other stars

We also started modelling the O7.5III(n)((f)) star ξ Per, which has an earlier spectral classification than HD 64760. For this star it is not clear if the spots responsible for the DACs are fixed on the stellar surface or not. The rotational period is less well-determined (there is some discussion in the literature about the radius of this star) and it is possible that the DAC recurrence time scale of 2.1 or 4.2 day is compatible with the (uncertain) rotation period of 3.1 day. An interesting feature of ξ Per is that it sometimes shows two DACs at the same time that (apparently) cross one another. We have succeeded in finding theoretical models that also show such crossing DACs. It should be stressed that the crossing of the DACs is only a line-of-sight effect; the corresponding CIRs do not physically cross one another.

Another well-known star is ζ Pup (spectral type O4I(n)f), which has both DACs and rotational modulations. Compared to HD 64760, the ζ Pup rotational modulations are more slanted, straight lines (Howarth et al. 1995). The complete absence of curvature will present a further challenge to hydrodynamical modelling on top of those already presented by HD 64760.

Conclusions

Different types of structure are present in a stellar wind. CIRs are one form of large-scale structure, which can very well explain the DACs, but not the rotational modulations (contrary to what had been previously thought). Explaining the rotational modulations therefore remains a challenge for the hydrodynamical modelling. Specifically for HD 64760, the CIRs are related to a beat pattern of NRPs. We caution, however, that it is not certain if we can extrapolate this conclusion to other stars, such as ξ Per, where a magnetic field might explain the observations as well. The additional mass loss rate required for the CIRs is very small. It is surprising therefore that these structures are not destroyed by clumping. This fact can, in principle, be used to constrain the amount of clumping in the stellar winds of early-type stars.

References

Cranmer, S.R., & Owocki, S.P. 1996, ApJ, 462, 469

Fullerton, A.W., Massa, D.L., Prinja, R.K., et al. 1997, A&A, 327, 699

Gosset, E., & Rauw, G. 2009, CoAst, 158, 138

Henrichs, H.F., Kaper, L., & Zwarthoed, G.A.A. 1988, in A Decade of UV Astronomy with the IUE Satellite, ESA SP-281, Vol. 2, 145

Howarth, I.D., Prinja, & R.K., Massa, D. 1995, ApJ, 452, L65

Kaper, L., Henrichs, H.F., Nichols, J.S., et al. 1996, A&AS, 116, 257

Kaper, L., Henrichs, H.F., Fullerton, A.W., et al. 1997, A&A, 327, 281

Kaper, L., Henrichs, H.F., Nichols, J.S., & Telting, J.H. 1999, A&A, 344, 231

Kaufer, A., Stahl, O., Prinja, R.K., & Witherick, D. 2006, A&A, 447, 325

Lobel, A., & Blomme, R. 2008, ApJ, 678, 408

Massa, D., Fullerton, A. W., Nichols, J. S., et al. 1995, ApJ, 452, L53

Morel, T., Marchenko, S.V., Pati, A.K., et al. 2004, MNRAS, 351, 552

Mullan, D.J. 1986 A&A, 165, 157

Owocki, S.P. 1998, in Cyclical Variability in Stellar Winds, Eds. L. Kaper & A.W. Fullerton, ESO Astrophysics Symposia, Springer-Verlag, 325

Owocki, S.P., Cranmer, S.R., & Fullerton, A.W. 1995, ApJ, 453, L37

Prinja, R.K. 1988, MNRAS, 231, P21

Prinja, R.K., & Howarth, I.D. 1986, ApJS, 61, 357

Puls, J. 2009, CoAst, 158, 113

DISCUSSION

Rauw: We heard about the origin of structure in the winds. Beating of NRPs, magnetic fields, small-scale structure all seem to produce observational signatures. What I found puzzling is the small density contrast between the Corotating Interaction Regions and the rest of the wind. This really brings up the question whether these large scale features can survive in the presence of clumps with a larger density contrast. What I would like to ask you is whether you can imagine having all these 3 features at the same time in a given star ?

Blomme: I would invert the question. We know from observations that all these types of structure co-exist. This therefore puts substantial constraints on the modeling of these structures. Note that the Owocki's (1997) calculations do not say that clumping and CIRs are incompatible. You can have the radiative instability and CIRs at the same time, but the radiative instability should not be too large (otherwise the CIRs get destroyed).

Meynet: What are the physical characteristics of the spot giving birth to the CIR region ? What is, for instance, its luminosity with respect to the rest of star ? Its dimension ?

Blomme: In the best-fitting model, we have 2 spots, one with 20 % more flux than the stellar surface, the other with 8 % more flux. These spots are on opposite sides, centered on the equator. They are circularly symmetric and have angular diameters of $20°$ and $30°$ respectively. The only effect they have is the way they change (locally) the radiative acceleration in the stellar wind.

Comm. in Asteroseismology
Vol. 158, 2009, 38th LIAC/HELAS-ESTA/BAG, 2008
A. Noels, C. Aerts, J. Montalbán, A. Miglio and M. Briquet., eds.

Spectroscopic and photometric variability of O and Wolf-Rayet stars

E. Gosset[1,2], G. Rauw[1,2]

[1] Institut d'Astrophysique et de Géophysique, Université de Liège, Allée du 6 Août, Bât B5c,
4000 Liège, Belgium
[2] Research Associate FRS/FNRS (Belgium)

Abstract

Low-level line profile variability in the optical domain is a ubiquitous feature of O, Of and Wolf-Rayet type stars. This variability can arise from pulsations, from magnetic fields or wind structures (both small and large scales). For main-sequence O-type stars, the spectra display many absorption lines that provide stringent diagnostics of photospheric features. However, as the stars evolve off the main-sequence towards the Of and Wolf-Rayet stage, the stellar wind densities (and hence the wind optical depths) increase dramatically. The wind eventually dominates the formation of the entire spectrum rendering the investigation of photospheric structures more ambiguous. We discuss the observational analyses of the spectroscopic and photometric variability of massive stars of spectral type O, Of and WR. In particular, we highlight the search for a connection between the photospheric and the wind variability.

Optical spectra of O, Of and Wolf-Rayet stars

Main sequence O-type stars have optical spectra that are dominated by rotationally broadened photospheric absorption lines (mainly H I, He I and He II as well as some weaker metallic lines). The stellar winds of O V stars are usually not sufficiently dense to produce strong and broad emission lines in the optical domain. On the other hand, Of-type objects are defined as O-stars showing some selective lines in emission (N III $\lambda\lambda$ 4634-41 and He II λ 4686). These Of stars are believed to be somewhat more evolved supergiants with denser stellar winds. As the stars evolve, they become Wolf-Rayet stars of the WN sequence with spectra dominated by broad helium and nitrogen emission lines formed in a very dense expanding wind. Later on, these stars evolve towards the WC stage where the optical spectra are dominated by helium, carbon and oxygen lines. The O $\rightarrow$ Of $\rightarrow$ WN $\rightarrow$ WC sequence not only reflects a progression in the evolutionary stage of the star, but also in its wind parameters. Indeed, typical values of the stellar and wind parameters and wind densities are given in Table 1.

An immediate consequence of the progression in stellar wind density concerns the observability of the stellar surface. In fact, in the case of O V stars, the stellar photosphere can be observed directly in the optical domain and deformations of this photosphere, e.g. due to pulsations, can be Doppler mapped through the investigation of the variability of rotationally broadened photospheric absorptions. For Wolf-Rayet stars, on the contrary, the winds are optically thick so that there are no purely intrinsic photospheric absorptions left in their spectra. As a result, any information from the stellar core is 'filtered' by the wind before it reaches the observer. Actually, the situation is even more complex since these stellar winds are highly clumped media (see below). An intermediate situation is observed in Of stars where the photosphere still produces an observable signature in the spectrum whilst the denser regions of the stellar wind produce optical emission lines such as Hα and He II λ 4686.

Table 1: Typical wind parameters of various categories of early-type stars. In this table, v_∞ is the wind's terminal velocity, R_* the stellar radius, β the exponent of the wind's velocity law and $\dot{M}$ the mass loss rate of the star. $\rho(2\,R_*)$ is the wind density at $2\,R_*$. The parameters listed here refer to clumped model atmosphere fits to the spectra of HD 96715 (O4 V), HD 190429A (O4 If$^+$) (both from Bouret et al. 2005), WR 78 (WN7, Hamann et al. 2006) and Br 43 (WC4, Hamann & Koesterke 1998).

Spectral type	v_∞ ($\mathrm{km\,s^{-1}}$)	R_* ($R_\odot$)	β	$\dot{M}$ ($M_\odot\,\mathrm{yr^{-1}}$)	$\rho(2\,R_*)$ ($\mathrm{g\,cm^{-3}}$)
O4 V	3000	12.0	1.0	0.25×10^{-6}	3×10^{-15}
O4 If$^+$	2300	19.6	0.8	1.80×10^{-6}	9×10^{-15}
WN7	1385	16.7	1.0	79.0×10^{-6}	1×10^{-12}
WC4	2800	1.06	1.0	95.0×10^{-6}	1.5×10^{-10}

The spectroscopic variability of single O-type stars

Spectroscopic variability is a widespread phenomenon among presumably single early-type stars. On the one hand, long time series of UV spectra of OB and Wolf-Rayet (WR) stars obtained with the *IUE* satellite revealed a strong variability of those lines that are preferentially formed in the wind (see e.g. Prinja et al. 1998). This variability usually takes the form of irregularly recurring discrete absorption components (DACs) or periodic global modulations of the intensity of the absorption components. Both kinds of features are interpreted as being due to large-scale structures in the stellar wind. The most intensively monitored O-stars display probably cyclic variability on time scales of a few days, i.e. exceeding the typical flow times of the winds (see Blomme 2009). On the other hand, the monitoring of a sample of 30 O-type stars with spectral types between O4 and O9.7 and luminosity classes V to I in optical spectroscopy revealed that 77% of the targets display significant line profile variability on time scales ranging from several hours to about one week (Fullerton et al. 1996). In their campaign, Fullerton et al. (1996) found all supergiants to display line profile variability (lpv), whilst non-variable stars were preferentially found among the main-sequence objects. Although a large fraction of the variability must arise in the wind, Fullerton et al. found a good agreement between the location of the stars showing lpv and the theoretically expected pulsational instability strip. This finding suggests the existence of a link between the variability at the level of the photosphere and in the wind.

As long as the *IUE* observatory was available, a powerful approach was to organize coordinated multi-site observations that combined UV resonance lines (e.g. Si IV, C IV, N V) and H I, He I, He II optical lines. These campaigns revealed that in many cases, the cyclical wind variability can be traced down from the outer wind regions (where the P Cygni troughs of the UV resonance lines are formed) to the inner wind region where Hα is formed (see e.g. the cases of ξ Per, O7.5 III(n)((f)) and 68 Cyg, O7.5 III:n((f)) discussed by Kaper et al. 1997 and de Jong et al. 2001). Since such campaigns are no longer possible, the current focus is mainly on the optical domain. For O-type stars, the optical spectra include photospheric absorption lines as well as some emission lines (Hα, He II λ 4686,...) that provide an important diagnostics of the inner regions of stellar winds. In fact, these emission lines are formed through radiative recombination which has a ρ^2 dependence. These lines are thus very sensitive to the high density, strongly accelerating part of the wind close to the stellar surface.

Whilst the DAC features discussed in the previous paragraphs are believed to be due to large-scale wind structures, additional emission line profile variations can also occur as a result of stochastic fluctuations of the number of small-scale turbulent clumps propagating outwards with the wind. Evidence for the existence of such a process in O-stars was found in the emission lines of ζ Pup (O4 Ief) by Eversberg et al. (1998) and HD 93129A (O3 If*) by

Lépine & Moffat (2008) in the form of low-level (a few percent) emission features that move away from the line core to the blue or red wing of the line.

The search for a photospheric connection?

As we have seen above, there is overwhelming evidence that the large-scale structures in the winds of early-type stars arise from, or extend into, the deepest layers of these winds. Also, from an observational point of view, it is found that a detectable photospheric variability is a sufficient condition for the presence of detectable wind variability, whilst the reverse is not true. These considerations then lead to the question whether there exists a connection between the wind features and the photospheric variability. The search for this *photospheric connection* thus became a kind of quest for the holy grail in early-type stars' astrophysics.

Such a photospheric connection could result

- either from magnetic fields rooted in the photosphere that would force part of the wind into corotation with the star (Babel & Montmerle 1997, ud-Doula & Owocki 2002),

- or from nonradial pulsations (NRPs) that perturb the stellar wind and combine with the instability of the latter to produce structures (Owocki & Cranmer 2002).

On theoretical grounds, it is expected that a perturbation of the photospheric conditions leads to the formation of structures in the wind. Indeed, in the commonly accepted radiation pressure wind driving paradigm for O-star winds, the mass loss per unit surface area is directly proportional to the surface radiation flux F. Hence, in a pulsating star with spatial or temporal variations in F, modulations of the overlying stellar wind are expected. The base variations induce a wind structure with fast rarefactions that ram into slower flows inducing a shock that compresses material into a dense shell (Owocki & Cranmer 2002).

Searching for such a connection is not an easy task. As we have seen above, the optical domain (offering simultaneous access to photospheric absorption lines and emissions or P Cygni features from the deeper wind layers) is ideally suited for such a research. However, the variability of O-type stars can be quite complex, involving many different time scales. This brings up one of the major problems of the NRP scenario to explain the wind variability. Indeed, whilst several cases exist where DACs and NRPs have been observed (e.g. λ Cep, O6 Ief and ξ Per, de Jong et al. 1999, 2001), DACs or global modulations of UV absorption troughs have recurrence time scales of several days ($\sim$ 2 d for λ Cep, 2.087 d for ξ Per) whilst NRPs occur with much shorter periods of several hours ($\sim$ 6.6 and 12.3 h for λ Cep and 3.45 h for ξ Per).

Probably the most secure examples of NRPs in O-type stars are provided by the two rapidly rotating O9.5 V stars ζ Oph (Walker et al. 2005) and HD 93521 (Rauw et al. 2008). For HD 93521, two NRP periods (1.75 and 2.89 h) are clearly detected in the He I features, including the He I λ5876 line. However, the latter line as well as Hα[1] are dominated by variations at much longer time scales (of the order of one day, see Fig. 1). It must be stressed that *IUE* observations did not reveal any variability of the wind lines on the pulsational time scales (Howarth & Reid 1993). A similar conclusion applies to ζ Oph (O9.5 V) where Howarth et al. (1993) found no evidence for a direct role of the well-established NRPs in the triggering of the DAC events. The DACs have a recurrence time of about 20 h, much longer than the periods of the NRPs which are 3.34 h and several other hourly periods (Kambe et al. 1997, Walker et al. 2005).

Another issue with the NRP model is that fast rotators have shorter DAC recurrence times and it has been suggested that the recurrence time scales with the rotational period of the star. Such a scaling is not expected if the DACs are triggered by single NRP modes

---

[1] The Hα line displays emission wings that are likely formed in an equatorial wind.

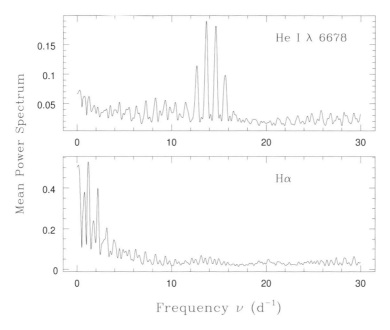

Figure 1: Comparison between the power spectrum of the variations of the He I λ 6678 and Hα lines of HD 93521. Whilst the former is dominated by two pulsational frequencies at 13.7 and 8.3 d^{-1} along with their neighbouring one-day aliases, the latter is dominated by longer time scale variations (Rauw et al. 2008).

or by beating of several pulsation modes. This argument would make magnetic fields more serious candidates since an oblique magnetic rotator model could force part of the wind into corotation with the star.

However, the magnetic field scenario also faces a number of problems. One of the most embarrassing issues is the rarity of direct detections of strong magnetic fields. Indeed, to date, only two O-stars (θ^1 Ori C, O7 V, and HD 191612, O6.5-8 f?p; Donati et al. 2002, 2006) are known to display a strong magnetic field of the order of 1 kG. In both cases, the measured magnetic field strengths are sufficient to significantly affect the dynamics of the stellar wind. Both stars display strong periodic line profile variability, but on very different time scales: 15 d for θ^1 Ori C and 528 d for HD 191612 (see Nazé et al. 2008 for a full comparison between the properties of these two stars). According to the oblique magnetic rotator model, these periods would have to be interpreted as the rotational periods of the stars, but this result would then imply that, at least in the case of HD 191612, the width of the absorption lines must be unrelated to the actual rotational velocity. In the majority of the cases, attempts to directly detect magnetic fields responsible for the large-scale wind structures failed (e.g. de Jong et al. 2001, Schnerr et al. 2008). As a conclusion, strong ($\geq$ 500 G) fields are certainly not widespread among normal O stars although weaker (more difficult to detect) fields might actually still be sufficient to modulate the stellar winds.

Another issue comes from multi-epoch observations of Oef stars (Rauw et al. 2003, De Becker & Rauw 2004). These are rapidly rotating Of stars with double-peaked He II λ 4686 emission lines (the most prominent examples of this sparse category are ζ Pup and λ Cep). In the case of BD+60° 2522 (O6.5 ef), variations were detected on time scales of 2 to 3 d.

However, both the time scales and the variability pattern were found to be epoch dependent. Moreover, in some cases, the pattern of variability (restricted to a limited range in velocity) is not consistent with a corotating structure.

A possible breakthrough in the search for a photospheric connection was obtained by Kaufer et al. (2006). These authors investigated the line profile variability of the photospheric absorption lines of the B0.5 I supergiant HD 64760. The photospheric absorptions of Si III $\lambda 4553$ and He I $\lambda 6678$ display travelling pseudo absorptions and emission features with a peak-to-peak amplitude of 1%. The Fourier analysis revealed three significant periods at 4.810, 4.672 and 4.967 h likely associated with p mode nonradial pulsations. The observed periods are closely spaced in time in the observer's frame of reference. Hα is dominated by longer time scale variations of the order of days (including the 2.4 d period seen in the UV data of this star, see Blomme (2009) and references therein). The beating of the two strongest of the closely spaced NRP pulsation modes leads to a retrograde beat pattern with two regions of constructive interference diametrically opposite on the stellar surface with a beat period of 6.8 d ($\nu_{beat} = \nu_2 - \nu_1$). While it must be stressed that the beat period of 6.8 d does not match the 2.4 d periods (it is actually closer to the DAC recurrence time scale), it nevertheless provides a promising avenue to search for a link between NRPs and large-scale wind structures.

The spectroscopic variability of single WR stars

In the case of WR stars, the entire spectrum is generally formed in the wind and any possible star-surface signal is necessarily filtered by the part of the wind below the zone of formation. The situation is rather complex because the winds of these stars present a marked structure: low ionization lines being statistically broader and formed in outer layers. The emission lines themselves are formed over a volume corresponding to a large range of depths in the wind (or distance from the hydrostatic core). According to the standard model of Hillier (1988, 1989), the formation region, in a WC5 star, of the He II $\lambda 4542$ line ranges from 3 to 8 stellar radii whereas for He I $\lambda 10830$, it corresponds to 10 to 100 stellar radii. In a WN5 star, the N IV $\lambda 4058$ line is formed between 1.8 and 5 stellar radii. Each wavelength in the profile of a typical emission line corresponds to one projected line of sight velocity and thus to layers of different depths in the wind. So, in addition to the filtering phenomenon, some blurring of the signal is also to be expected.

From the observational point of view, variability of WR emission lines is quite widespread, particularly among WC stars. It mainly consists in the presence of numerous, relatively narrow emission subpeaks on top of the main emission profiles; these subpeaks tend to move away from the center of the line. They are interpreted as originating in small-scale wind structures of higher density propagating radially outwards. These features, sometimes called blobs, are stochastic in time and are often associated with the small-scale clumping of the wind. They have been noticed in WR stars well before the two examples identified among O stars, most probably due to a better marked contrast. In the majority of the stars, no determinism has been detected in this phenomenon. It is clear that this additional behaviour is a source of stochastic variability that will further contribute to hide any possible deterministic signal coming from the stellar surface (Robert 1994; Lépine & Moffat 1999 and references therein).

In the UV domain, the absorption components of P Cygni profiles of most of the lines are saturated and no flux is emitted in the absorption trough which is at zero residual intensity. Therefore, the equivalents of O-star DACs are much more difficult to detect. Very often, the variability is only visible at or near the terminal velocity. Actually, the only report of DACs in WR stars is for WR 24 (WN6ha, Prinja & Smith 1992) where recurrent (periodic?) wide features are seen moving outwards in the unsaturated He II $\lambda 1640$ line. The similarity with the case of O stars is only partial and another interpretation could apply. It is interesting to note that we have been able to outline the existence of inhomogeneities in the wind of

WN8 stars on the basis of particularly narrow transient dips superimposed on the absorption component of the He I P Cygni profiles (see Fig. 2 of Gosset et al. 2005).

Although line-profile variations are usually considered as stochastic (in single WR stars), a couple of WN stars present a different behaviour. Two of the best studied WN stars seem to have a semi-deterministic behaviour. Actually, the line profile exhibits a pattern of subpeaks that look randomly distributed and evolving in a stochastic way. However, after a typical time-scale, the very same profile is displayed again by the line. It recurs in a periodic way a few times but after several cycles, the structure slowly evolves. The coherency is only lasting typically 10 cycles or so. The phenomenon is explained by large-scale structures distributed over the envelope of the star and corotating with the star. These structures are slightly evolving with time on a longer time-scale than the rotation period. There are clues indicating that they are deeply rooted close to the stellar surface. Such corotating features are claimed to be present in WR 6 (WN4, Morel et al. 1998) and in WR 134 (WN6, Morel et al. 1999). Some authors also proposed WR 1 (WN4), but the discussion is not closed (Flores et al. 2007; Chené et al. 2008). The latter authors proposed to use this property to investigate the rotational velocity of WR stars. In any case, these slowly moving features constitute a further disturbing factor in the analysis of the spectroscopic variability of WR stars. One still cannot reject the idea that the large-scale structure is related to pulsations (through beating) on the surface.

The photometric variability of single WR stars

For a long time, the WR stars have been known to be photometrically variable and this phenomenon was studied with the hope to unveil the exact nature of these stars. Various limited surveys concluded that the WR stars are presenting a global photometric dispersion of the order of $\sigma = 0.003$–0.030 mag (intrinsic to the star, see Moffat & Shara 1986, Lamontagne & Moffat 1987, Gosset et al. 1994, Marchenko et al. 1998). Very few WRs are not to be listed as variable but, from the ground, differential photometry has difficulties to be more precise than $\sigma \sim 0.003$ mag. The WR variability presents a distribution of the power on the various time scales that is reminiscent of white (or slightly red) noise, i.e. a predominantly stochastic behaviour. The presence of some determinism (periods?) has been claimed in a few cases but full agreement has never been reached on a single WR star. Later spectral types are more variable than early-type ones. Stars from the WC sequence are slightly less variable than those of the WN sequence. In addition, the latter includes the WN8 subclass which contains by far the most variable objects (0.1 mag peak to peak). It is interesting to notice that WR 135 (WC5), one of the less photometrically variable ones, is highly variable from the spectroscopic point of view (blobs).

The flux in a visible broadband filter is essentially determined by the photons from the continuum (formed in deep wind layers but usually not down to the surface; Hamann et al. 1995). It could thus be sensitive to the signal coming from the surface of the star. For this reason, astronomers scrutinized the WR lightcurves to discover short time variations. Actually the fundamental mode of pulsation is expected to have a period of 10-60 m whereas oscillation modes with periods in the range 0.2-1.0 d could also be present (Maeder 1986, Noels & Scuflaire 1986, Scuflaire & Noels 1986, Cox & Cahn 1988, Glatzel et al. 1993). Table 2 is a critical compendium of the WR stars having been convincingly claimed to exhibit short-term variations. Discarding the possible binary systems, there are very few good examples. Most of the time, the periodic behaviours were ephemeral. The variations around 10 m reported for WR 40 (WN8h) have never been confirmed but were refuted on various occasions (Bratschi & Blecha 1996, Martinez et al. 1994, Marchenko et al. 1994). The short periodicity attributed to WR 86 could actually be due to a pair of periods and originate in a previously unresolved (0.''2) B-type visual companion belonging to the β Cephei class (Paardekooper et al. 2002).

Table 2: A compendium of WR stars claimed to have exhibited a short-period variation.

Object	Period	References	Remarks
Cyg X-3	4.8h	–	WR+compact comp.
WR46 (WN3p)	7h	Veen et al. 2002	Probable binary
WR40 (WN8h)	627s	Blecha et al. 1992	Never confirmed
WR6 (WN4)	20-30m	Bratschi & Blecha 1996	During 1 night
WR78 (WN7h)	25m	Bratschi & Blecha 1996	During 1 night
WR111 (WC5)	20m	Bratschi & Blecha 1996	During 2 nights
WR86 (WC7)	3.5h	van Genderen et al. 1990	With a B companion
		Paardekooper et al. 2002	B = β Cephei type ?
WR66 (WN8(h))	3.5-4h	Antokhin et al. 1995	Due to a companion ?
		Rauw et al. 1996	
WR 123 (WN8)	9.8h	Lefèvre et al. 2005	MOST era

Probably the best case in Table 2 is WR 66 (WN8(h)) which has been detected as peri-odically variable by two independent teams. However, in the case of WR 66 also, there is a companion at 0."4 that could be responsible for the periodicity. However, this is not the most likely hypothesis. The final conclusion from all these surveys was that ground-based variability studies alone present difficulties to get rid of the atmospheric variations at the necessary low level (0.003 mag) and at the time scales of tens of minutes. Therefore, space-borne measurements were claimed to be much more promising. Indeed, very recently, the MOST satellite discovered the first indubitable case of the presence of a short period in a WR star (the case of WR 123, WN8; Lefèvre et al. 2005 and Lefèvre 2009). The same satellite observed a very stable WR star....WR 111 (WC5; Moffat et al. 2008). The recent discovery (to be confirmed) of the presence of the same periodicity in the P Cygni line profile variations of WR 123 (see Lefèvre 2009) provides the very first plausible proof that the signal at the surface could be propagated outwards by the wind without being completely annihilated.

Conclusion

Spectroscopic and photometric variabilities of O and WR stars are widespread. At least part of the variations are due to the winds that present small- and large-scale structures. The connection between pulsations or magnetic fields and the wind structures are likely but not yet firmly established. With present space-borne facilities, we are certainly on the edge to outline the possible missing link between putative pulsations and observed variations.

References

Antokhin, I., Bertrand, J.F., & Lamontagne, R., et al. 1995, AJ, 109, 817

Babel, J., & Montmerle, T. 1997, ApJ, 485, L29

Blecha, A., Schaller, G., & Maeder, A. 1992, Nature, 360, 320

Blomme, R. 2009, CoAst, 158, 131

Bouret, J.-C., Lanz, T., & Hillier, D.J. 2005, A&A, 438, 301

Bratschi, P., & Blecha, A. 1996, A&A, 313, 537

Chené, A.N., St-Louis, N., & Moffat, A.F.J. 2008, ASP Conf. Ser., 388, 157

Cox, A.N., & Cahn, J.H. 1988, ApJ, 326, 804

De Becker, M., & Rauw, G. 2004, A&A, 427, 995

de Jong, J.A., Henrichs, H.F., Schrijvers, C., et al. 1999, A&A, 345, 172

de Jong, J.A., Henrichs, H.F., Kaper, L., et al. 2001, A&A, 368, 601

Donati, J.-F., Babel, J., Harries, T.J., et al. 2002, MNRAS, 333, 55

Donati, J.-F., Howarth, I.D., Bouret, J.-C., et al. 2006, MNRAS, 365, L6

Eversberg, T., Lépine, S., & Moffat, A.F.J. 1998, ApJ, 494, 799

Flores, A., Koenigsberger, G., Cardona, O., et al. 2007, AJ, 133, 2859

Fullerton, A.W., Gies, D.R., & Bolton, C.T. 1996, ApJS, 103, 475

Glatzel, W., Kiriakidis, M., & Fricke, K.J. 1993, MNRAS, 262, L7

Gosset, E., Rauw, G., Manfroid, J., et al. 1994, NATO ASI Series C, 436, 101

Gosset, E., Nazé, Y., Claeskens, J.F., et al. 2005, A&A, 429, 685

Hamann, W.R., & Koesterke, L. 1998, A&A, 335, 1003

Hamann, W.R., Koesterke, L., & Wessolowski, U. 1995, A&A, 299, 151

Hamann, W.R., Gräfener, G., & Liermann, A. 2006, A&A, 457, 1015

Hillier, D.J. 1988, ApJ, 327, 822

Hillier, D.J. 1989, ApJ, 347, 392

Howarth, I.D., & Reid, A.H.N. 1993, A&A, 279, 148

Howarth, I.D., Bolton, C.T., Crowe, R.A., et al. 1993, ApJ, 417, 338

Kambe, E., Hirata, R., Ando, H., et al. 1997, ApJ, 481, 406

Kaper, L., Henrichs, H.F., Fullerton, A.W., et al. 1997, A&A, 327, 281

Kaufer, A., Stahl, O., Prinja, R.K., & Whiterick, D. 2006, A&A, 447, 325

Lamontagne, R., & Moffat, A.F.J. 1987, AJ, 94, 1008

Lefèvre, L., Marchenko, S.V., Moffat, A.F.J., et al. 2005, ApJ, 634, L109

Lefèvre, L. 2009, CoAst, 158, 297

Lépine, S., & Moffat, A.F.J. 1999, ApJ, 514, 909

Lépine, S., & Moffat, A.F.J. 2008, AJ, 136, 548

Maeder, A. 1986, A&A, 147, 300

Marchenko, S.V., Antokhin, I., Bertrand, J.F., et al. 1994, AJ, 108, 678

Marchenko, S.V., Moffat, A.F.J., Eversberg, T., et al. 1998, MNRAS, 294, 642

Martinez, P., Kurtz, D., Ashley, R., & Tripe, P. 1994, Nature, 367, 601

Moffat, A.F.J., & Shara, M.M. 1986, AJ, 92, 952

Moffat, A.F.J., Marchenko, S.V., Zhilyaev, B.E., et al. 2008, ApJ, 679, L45

Morel, T., St-Louis, N., Moffat, A.F.J., et al. 1998, ApJ, 498, 413

Morel, T., Marchenko, S.V., Eenens, P.R.J., et al. 1999, ApJ, 518, 428

Nazé, Y., Walborn, N.R., & Martins, F. 2008, RevMexAA, 44, 331

Noels, A., & Scuflaire, R. 1986, A&A, 161, 125

Owocki, S.P., & Cranmer, S.R. 2002, ASP Conf. Ser., 259, 512

Paardekooper, S.J., Veen, P.M., van Genderen, A.M., & van der Hucht, K.A. 2002, A&A, 384, 1012

Prinja, R.K., & Smith, L.J. 1992, A&A, 266, 377

Prinja, R.K., Massa, D., Howarth, I.D., & Fullerton, A.W. 1998, MNRAS, 301, 926

Rauw, G., Gosset, E., Manfroid, J., et al. 1996, A&A, 306, 783

Rauw, G., De Becker, M., & Vreux, J.-M. 2003, A&A, 399, 287

Rauw, G., De Becker, M., van Winckel, H., et al. 2008, A&A, 487, 659

Robert, C. 1994, Ap.&Sp. Sci, 221, 137

Schnerr, R.S., Henrichs, H.F., Neiner, C., et al. 2008, A&A, 483, 857

Scuflaire, R., & Noels, A. 1986, A&A, 169, 185

ud-Doula, A., & Owocki, S.P. 2002, ApJ, 576, 413

van Genderen, A.M., Larsen, I., & van der Hucht, K.A. 1990, A&A, 229, 123

Veen, P.M., van Genderen, A.M., & van der Hucht, K.A. 2002, A&A, 385, 619

Walker, G.A.H., Kuschnig, R., Matthews, J.M., et al. 2005, ApJ, 623, L145

Comm. in Asteroseismology
Vol. 158, 2009, 38th LIAC/HELAS-ESTA/BAG, 2008
A. Noels, C. Aerts, J. Montalbán, A. Miglio and M. Briquet., eds.

Atmospheric parameters and chemical composition of β Cephei stars in the open cluster NGC 3293

E. Niemczura[1], Rodler, F.[2,3], Müller, A.[3,4]

[1] Instytut Astronomiczny, Uniwersytet Wrocławski, Kopernika 11, 51-622 Wrocław, Poland; [2] Instituto de Astrofísica de Canarias, C/Via Láctea s/n, 38205 La Laguna, Spain; [3] Max Planck Institut für Astronomie, Königstuhl 17, D-69117 Heidelberg, Germany; [4] European Southern Observatory, Karl-Schwarzschild-Strasse 2, D-85748, Garching bei München, Germany

Abstract

We introduce a new project concerning the spectroscopic analysis of β Cephei stars in open clusters. The high-resolution, high signal-to-noise observations were collected for β Cephei stars and comparison non-pulsating stars in NGC 3293, NGC 4755, NGC 6231, NGC 6910 and NGC 884. These data allow us to obtain atmospheric parameters (effective temperature, surface gravity, microturbulence), chemical abundances and rotational velocities of all analysed objects. Here we present the preliminary results for six pulsating stars in NGC 3293.

Individual Objects: NGC 3293, V 380 Car, V 400 Car, V 401 Car, V 403 Car, V 404 Car, V 406 Car

Introduction

Asteroseismology enables studying interiors of pulsating stars by means of matching observed and modelled frequency spectra. So far, seismic modelling has been successfully applied to four field β Cephei stars with rich pulsation spectra containing some identified modes: HD 129929 (Aerts et al. 2003), ν Eridani (see Jerzykiewicz et al. 2005 and references therein), θ Ophiuchi (Handler et al. 2005, Briquet et al. 2005) and 12 Lacertae (Handler et al. 2006). For HD 129929 and ν Eri the presence of differential rotation in the interior and the small convective overshooting parameter have been determined. Additionally, for ν Eri the mean metallicity in the interior has been estimated.

All these stars have similar masses. The general applicability of seismic results needs to be checked by examining stars with masses in a wider range. The obvious step after the successful seismic modelling of field β Cephei stars is to perform asteroseismology on such objects in the open cluster. Studying β Cephei variables in the same environment has some advantages. Seismic modelling requires the knowledge of stellar parameters. For cluster stars, we can assume some of them, because age, reddening, and chemical composition are nearly the same for each star. Moreover, the spread of stellar distances is small in comparison with the average distance of the cluster. The position of stars in the colour-magnitude diagram can be used to gain information on their evolutionary status. However, the reddening can be different for different stars in very young clusters, because of the remnants of clouds of which the cluster was formed from.

The variability content of the cluster is defined by its metallicity, initial mass function, but first of all by its age. This is the consequence of stellar evolution and the fact that pulsating stars occupy well-defined instability strips in the H-R diagram. The β Cephei variables are not expected to be found in clusters older that 30 Myr. Clusters in the range of ages between

5 – 25 Myr are best suited for studying such variables. There are many open clusters with at least one β Cephei variable. Among them, NGC 3293 is one of the richer in pulsating stars of this type.

The open cluster NGC 3293 ($\alpha = 10^h 35^m 48.8^s$, $\delta = -58°13'00''$, J2000) is a young object located in the Carina region, north-west of the open clusters Trumpler 14/16. It is a compact and well-populated object. The cluster distance, 2750 ± 250 pc, and age, 8 ± 1 Myr, were determined by Baume et al. (2003) from photometric data. NGC 3293 contains eleven known β Cephei variables, discovered by the analysis of photometric data (see Balona et al. 1997 and references therein). Many β Cephei stars in NGC 3293 are among the most massive members of the group. The multisite extensive CCD photometric campaign on NGC 3293 was performed in 2006. Preliminary results indicate that none of β Cephei stars in this cluster is singly periodic and several objects have at least a few modes excited (Handler et al. 2008).

Observations and methods of analysis

The study is based on observations carried out with the MPG/ESO 2.2-m telescope at La Silla Observatory in Chile. High-resolution (R = 48000), high signal-to-noise (S/N $\sim$ 100) spectra were obtained in April 2004, May 2007 and May 2008 with the FEROS spectrograph, as a part of the 073.D-0291(A), 079.A-9008(A) and 081.A-9006(A) programs, respectively. The exposure times ranged from 900 to 1800 seconds. The spectral range extended from 3600 to 9200 Å. The pipeline reduction was performed using the MIDAS echelle package. For each analysed star at least four spectra were collected. All the spectra for individual objects were normalised, cross-correlated and co-added in order to eliminate the effect of pulsations and to increase the quality of the spectrum. The normalisation was performed using the IRAF package.

The synthetic spectra are computed with a hybrid non-LTE approach. First, hydrostatic, plane-parallel, line-blanketed LTE models of the atmospheres were calculated with the AT-LAS 9 code (Kurucz 1996). Then, NLTE population numbers and synthetic spectra were derived using the updated version of DETAIL (Giddings 1981, Butler 1984) and SURFACE codes (Butler and Giddings 1985). The DETAIL code solves radiative transfer and statistical equilibrium equations, while the SURFACE code calculates the spectrum. The analysis is based on a grid of synthetic spectra obtained by Thierry Morel (for the detailed description see Morel et al. 2006).

The adopted spectrum synthesis method consists of comparison of high-resolution spectrum (f_{obs}) with theoretical spectra (f_{theo}). The shape of the spectrum depends on many parameters, like effective temperature, T_{eff}, surface gravity, $\log g$, chemical abundances of elements, $\log \epsilon(X)$, radial velocity, V_{rad}, rotational velocity, $V \sin i$, and microturbulence, ξ. The value of effective temperature was determined from the analysis of Si lines in ionisation stages III and IV. There are at least seven Si III features and only one Si IV (4116 Å) line available for every star. The average surface gravity was calculated from hydrogen $H\beta$ and $H\gamma$ lines for each object. Microturbulence was derived from O II lines of different strength. From 13 to 27 unblended oxygen lines were used, depending on the stellar rotation. Microturbulence velocities obtained in this way are consistent with values determined from the Si III lines. In order to optimise the other parameters, we used classical least-squares fitting method (LS-method), i.e. we minimised the differences between observed and theoretical spectra, $\text{RMS} = \left[\sum (f_{obs}^2 - f_{theo}^2)\right]^{1/2}$. In the LS-method, the corrections to analysed parameters were determined in each iteration step. The process was repeated until these corrections were close to zero.

The total errors of all abundances include line-to-line scatter (σ_i) and variations of abundance arising from errors of effective temperature ($\sigma_{T_{eff}}$), surface gravity ($\sigma_{\log g}$) and microturbulence (σ_ξ). The total errors were calculated as follows: $\sigma_{Tot} = (\sigma_{T_{eff}}^2 + \sigma_{\log g}^2 + \sigma_\xi^2 + \sigma_i^2)^{1/2}$.

Table 1: Atmospheric parameters and rotation velocities of the analysed β Cephei pulsators in NGC 3293.

Star	T_{eff} [K]	log g [dex]	ξ [km s^{-1}]	V sin i [km s^{-1}]
V 406 Car	24800	3.75	6	26
V 403 Car	24300	3.65	9	42
V 404 Car	25800	3.60	8	55
V 400 Car	25000	3.80	5	60
V 380 Car	22600	3.60	9	82
V 401 Car	22500	3.70	8	110

Results and discussion

We investigated six β Cephei stars in the open cluster NGC 3293. All considered objects: V 380 Car, V 400 Car, V 401 Car, V 403 Car, V 404 Car and V 406 Car are stars with moderate rotational velocities, allowing for detailed abundance analysis. We excluded five other β Cephei pulsators, also members of this cluster. V 405 Car is a double line spectroscopic binary. V 381 Car is an eclipsing system with a β Cephei component, and was analysed in detail by Freyhammer et al. (2005). Three other variables, V 412 Car, V 378 Car and V 440 Car, are fast rotators with V sin i higher than 200 km s^{-1}.

The obtained effective temperatures, surface gravities, microturbulences and rotational velocities of analysed β Cephei variables are presented in Table 1. The errors of these parameters were not determined here. In order to estimate the uncertainties of derived abundances we adopted the errors of $T_{eff} = 1000$ K and log $g = 0.1$ dex, resulting from steps in our grid of theoretical fluxes. The error of microturbulence comes from the comparison of ξ values determined from O II and Si III lines and is equal to 2 km s^{-1}. For all stars, the standard deviation of V sin i derived from all analysed lines never exceeded 5 km s^{-1}.

We were able to determine the abundances of C, N, O, Mg, Al, Si, S, Fe and He. The total errors, σ_{Tot}, are smaller than 0.25 dex. For all but one element (magnesium) more than one spectral line was available for each object. For all stars with rotation velocity higher than 50 km s^{-1} Mg II line is blended by Al III feature. In such cases, the abundance of Al was determined from other lines before the analysis of Mg and the additional sub-grids were calculated for given atmospheric parameters and Al abundances. These grids were used for the determination of abundance of Mg. The remaining considered features were not blended by lines of different element. The derived CNO abundances are in agreement with the results obtained by Mathys et al. (2002) for a few stars from NGC 3293 cluster. In order to obtain the average carbon abundance we used all available lines but 4267.0/2, 6578.0 and 6582.9 Å. The abundances derived from these features were significantly different in comparison with the others (for more information see Nieva and Przybilla 2008).

In Fig. 1, the average NLTE abundances, log ϵ(X), (by convention, log ϵ(H) = 12) of C, N, O, Mg, Al, Si, S and Fe are compared with the average values derived for B-type stars by Przybilla et al. (2008) and with the average abundances of field β Cephei stars by Morel et al. (2006). Przybilla et al. (2008) analysed chemical composition of unevolved early-B stars in the solar neighbourhood and found homogeneous abundances, consistent with the results for the interstellar medium. Their values are also close to the solar abundances found by Asplund et al. (2005). Almost all the abundances obtained for pulsating β Cephei stars in NGC 3293 are considerably lower than values derived by Przybilla et al. (2008). It can be a real effect or the result of differences in atmospheric and atomic models. In this paper we applied grid of fluxes calculated and described by Morel et al. (2006) in the paper concerning the determination of chemical abundances of field β Cephei variables. It is clear from Fig. 1, that our results are more consistent with the average values obtained by Morel et al. (2006).

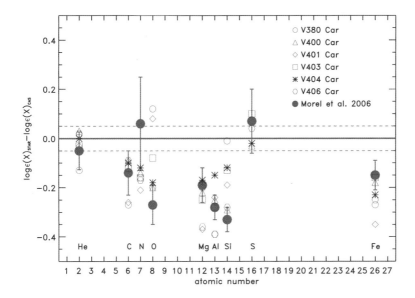

Figure 1: The obtained chemical abundances of analysed β Cephei variables (log $\epsilon(X)_{STAR}$) in comparison with values for B-type stars from Przybilla et al. (2008) (log $\epsilon(X)_{CAS}$, where CAS – cosmic abundance standards). For two elements, S and Al, there is no results in the paper of Przybilla et al. (2008). In these cases, we used as standards solar values from Asplund et al. (2005). The dashed lines are plotted for ± 0.05 dex, which is the highest error of abundances determined by Przybilla et al. (2008). The average values derived by Morel et al. (2006) for field β Cephei stars are also included.

However, there are some differences. First of all, there is no nitrogen enhancement found by Morel et al. (2006) for field β Cephei stars. Moreover, the obtained oxygen and silicon abundances are higher for the stars from NGC 3293. The largest effect is for V 380 Car and V 401 Car, two stars with the highest rotation velocity. On the other hand, the iron abundances for cluster variables are lower than for field β Cephei pulsators.

The results presented here are preliminary. In the next step of our analysis the obtained atmospheric parameters and abundances of elements will be fine-tuned. Especially, the effect of different atmospheric and atomic models on the determined abundances will be checked. Moreover, the macroturbulence related with pulsations will be accounted for in the line fitting. We will also compare the results derived for β Cephei variables and for non-pulsating stars from NGC 3293.

Acknowledgments. I am very grateful to Thierry Morel for making the grid of theoretical NLTE fluxes accessible for me, and for all useful comments. I acknowledge the financial support of the HELAS Consortium (No. 026138). This paper has been partially supported by the Wrocław University grant 2945/W/IA/08.

References

Aerts, C., Thoul, A., Daszyńska, J., et al. 2003, Science, 300, 1926

Asplund, M., Grevesse, N., & Sauval, A.J. 2005, 2005, ASP, 336, 25

Balona, L.A., Dziembowski, W., & Pamyatnykh, A. 1997, MNRAS, 289, 25

Baume, G., Vázquez, R.A., Carraro, G., & Feinsein, A. 2003, A&A, 402, 549

Briquet, M., Lefever, K., Uytterhoeven, K., & Aerts, C. 2005, MNRAS, 362, 619

Butler, K. 1984, Ph.D. Thesis, University of London, UK

Butler, K., & Giddings, J.R. 1985, Newsletter of Analysis of Astron. Spectra, No. 9, Univ. of London, UK

Freyhammer, L.M., Hensberge, H., & Sterken, C., et al. 2005, A&A, 429, 631

Giddings, J.R. 1981, Ph.D. Thesis, University of London, UK

Handler, G., Shobbrook, R.R., Mokgwetsi, T., et al. 2005, MNRAS, 362, 612

Handler, G., Jerzykiewicz, M., Rodríguez, E., et al. 2006, MNRAS, 365, 327

Handler, G., Tuvikene, T., Lorenz, D., et al. 2008, CoAst, 150, 193

Jerzykiewicz, M., Handler, G., Shobbrook, R.R., et al. 2005, MNRAS, 360, 619

Kurucz, R. 1996, CD-ROM No. 13

Mathys, G., Andrievsky, S.M., Barbuy, B., et al. 2002, A&A, 387, 890

Morel, T., Butler, K., Aerts, C, et al. 2006, A&A, 457, 651

Nieva, M.F., & Przybilla, N. 2008, A&A, 481, 199

Przybilla, N., Nieva, M.F., & Butler, K. 2008, ApJ, 688, L103

Comm. in Asteroseismology
Vol. 158, 2009, 38th LIAC/HELAS-ESTA/BAG, 2008
A. Noels, C. Aerts, J. Montalbán, A. Miglio and M. Briquet., eds.

The solar abundance of Oxygen

N. Grevesse[1,2]

[1] Centre Spatial de Liège, Université de Liège,
Avenue Pré Aily, B-4031 Angleur-Liège, Belgium
[2] Institut d'Astrophysique et de Géophysique, Université de Liège, Allée du 6 Août, 17, B-4000 Liège,
Belgium

Abstract

With Martin Asplund (Max Planck Institute of Astrophysics, Garching) and Jacques Sauval (Observatoire Royal de Belgique, Brussels) I recently published detailed reviews on the solar chemical composition (Asplund et al. 2005, Grevesse et al. 2007). A new one, with Pat Scott (Stockholm University) as additional co-author, will appear in Annual Review of Astronomy and Astrophysics (Asplund et al. 2009).

Here we briefly analyze recent works on the solar abundance of Oxygen and recommend a value of 8.70 in the usual astronomical scale.

Introduction

The new abundances mentioned in the abstract result essentially from the use of a 3D hydrodynamic model of the solar atmosphere instead of the classical 1D hydrostatic models, accounting for departures from LTE when possible, and the use of improved atomic and molecular data for many different indicators of the abundances including atomic as well as molecular lines. These new abundances are significantly smaller than previously recommended values at least for the most abundant metals, resulting in a decrease of the solar metallicity. While resolving a number of longstanding problems, the new solar abundances pose a very serious problem with the Standard Solar Model. Models computed with the new abundances disagree with the very precise measurements of the sound speed profile, the abundance of He in the convection zone and the depth of this convection zone as derived from helioseismology. The abundance of O does play a key role as a major contributor to the opacity in the layers just below the convection zone.

The solar abundance of oxygen

New works on the solar abundance of Oxygen have recently been published; we discuss them briefly hereafter.

Asplund et al. (2004) and Scott et al. (2006) have recently redetermined the solar abundance of oxygen from all the indicators available, permitted and forbidden atomic lines as well as the CO vibration-rotation lines and the OH vibration-rotation and pure rotation lines in the infrared solar spectrum. The new solar abundance of O is A(O)=8.66, in the usual astronomical scale. This value is much lower, by about 0.2 dex, than previously recommended values.

Ayres et al. (2006) also analyzed the CO lines with quite a different result. Their analysis used various empirical 1D models and they found a high value of the abundance of O. This is quite normal because the CO lines are extremely sensitive to temperature. The 3D effects are very large on such lines as shown by Scott et al. (2006), decreasing the abundance found when using 1D models, by rather large amounts.

Socas-Navarro & Norton (2007) confirmed the low abundance of O. They constructed an empirical 3D model from spatially resolved spectropolarimetric observations of spectral lines and applied it to the analysis of the infrared triplet of O I at 777nm.

However Centeno & Socas-Navarro (2008) recently analyzed the forbidden O I line at 630nm in sunspot spectra, a line that is blended with a Ni I line. They derived a much larger abundance of O, A(O)=8.86. If this result is updated for a better transition probability of the [O I] line (-0.06 dex), for a better solar abundance of Ni (-0.06 dex) and for a more precise effect of the formation of CO in sunspots(-0.08 dex), as shown by Scott et al. (2009), the resulting abundance of O decreases to A(O)=8.66, in agreement with our low O value.

Very recently, Ayres (2008) reanalyzed the same forbidden O I line, using profile fits with a 3D model atmosphere. He got a high abundance of O, A(O)=8.81. But the best fits of the profiles are obtained if the Ni I blend is treated as a free parameter. This results in too small a contribution of Ni to the blend. If this contribution is increased to its normal value, obtained with the new abundance of Ni we have derived from a new analysis (to be published soon), the abundance of O can decrease to A(O)=8.74 i.e. very near to our value (see here after).

The most important new analysis of the solar abundance of O has been recently done by Caffau et al. (2008). These authors analyze very carefully the permitted and forbidden lines of O I on various solar atlases observed at the center of the solar disc and on the whole disc. They use different 1D models as well as their own 3D original model (CO5BOLD). They recommend a rather high abundance of O, A(O)=8.79. One of the main differences between this work and previous works is to be found in the equivalent widths of the O I lines. The values from Caffau et al. are always larger than any other measurement by other authors. We remeasured very carefully these equivalent widths and always found values lower than Caffau et al.. We therefore normalized the Caffau et al. (2008) results as well as our own results (Asplund et al. (2004)) to the new equivalent widths and found about the same abundances of O, A(O)= 8.71 versus A(O)=8.70.

It is also this low value which is found by Meléndez & Asplund (2008) from the analysis of the third forbidden line of O I at 557.7nm. This line is however heavily perturbed by two lines of C_2.

Very recently, Asplund et al. (2009) have revisited the solar O abundance with all the indicators, the 3 blended forbidden lines of O I and the highly excited O I lines, as well as the molecular lines of OH and CO in the infrared, taking NLTE effects into account for the permitted O I lines and with a new 3D model. All the indicators now converge to a value A(O)= 8.70±0.05.

Conclusions

This solar abundance of O, A(O)=8.70, combined with the new abundances of Asplund et al. (2009) for the other elements, for example A(C)=8.42 and A(Ne)=7.94, leads to a present day solar metallicity, Z=0.0135. These solar abundances, about 10% larger than the values of Asplund et al. (2005) and Grevesse et al. (2007), are still much too low to solve the problem of the disagreement between the Standard Solar Models and helioseismology.

References

Asplund, M., Grevesse, N., & Sauval, A. J. 2005, in Cosmic Abundances as Records of Stellar Evolution and Nucleosynthesis, ed. T.G. BarnesIII, F.N. Bash, ASP Conf. Ser. 336, 25

Asplund, M., Grevesse, N., Sauval, A. J., et al. 2004, A&A, 417, 751

Asplund, M., Grevesse, N., Sauval, A. J., & Scott, P. 2009, ARA&A, in press

Ayres, T. R. 2008, ApJ, 686, 731

Ayres, T. R., Plymate, C., & Keller, C. U. 2006, ApJS, 165, 618

Caffau, E., Ludwig, H.-G., Steffen, M., et al. 2008, A&A, 488, 1031

Centeno, R., & Socas-Navarro, H. 2008, ApJ, 682, 61

Grevesse, N., Asplund, M., & Sauval, A. J. 2007, Space Sci. Rev., 130, 105

Meléndez, J., & Asplund, M. 2008, A&A, 490, 817

Scott, P., Asplund, M., Grevesse, N., & Sauval, A. J. 2006, A&A, 456, 675

Scott, P., Asplund, M., Grevesse, N., & Sauval, A. J. 2009, ApJ, 691, L119

Socas-Navarro, H. & Norton, A. A. 2007, ApJ, 660, L153

Session 4
Observed frequencies in pulsating massive stars

Comm. in Asteroseismology
Vol. 158, 2009, 38th LIAC/HELAS-ESTA/BAG, 2008
A. Noels, C. Aerts, J. Montalbán, A. Miglio and M. Briquet., eds.

Ground-based observations of O and B stars

K. Uytterhoeven[1,2]

[1] Instituto de Astrofísica de Canarias, Calle Via Láctea s/n, E-38205 La Laguna (TF), Spain
[2] INAF-Osservatorio Astronomico di Brera, Via E. Bianchi 46, I-23807 Merate, Italy

Abstract

Ground-based observations are a strong tool for asteroseismic studies and even in the era of asteroseismic space missions they continue to play an important role. I will report on the activities of the CoRoT/SWG Ground-Based Observations Working Group, discuss the observational efforts of the Open Cluster campaigns and the search for the origin of extra line-broadening in massive OB stars.

Individual Objects: HD 180642, HD 50209, HD 181231, HD 49330, NGC 3293, NGC 6910, NGC 884, NGC 1893, NGC 869

The role of ground-based observations in asteroseismic studies

Over the last decades large observational efforts have been undertaken to carefully monitor the pulsational variability of massive B stars, resulting in a breakthrough in seismic modelling. Also, several observational initiatives are taken to open up new horizons, such as the study of B-type pulsators in clusters and pulsational studies of more evolved massive OB stars. The important key ingredients for an asteroseismic study are precise pulsational frequencies, accurate identified pulsation modes, and strong constraints on physical parameters. To obtain these, a large amount of observing time is required as continuous time-series are needed to unravel beat-frequencies. The best we can do from the ground are multi-site campaigns. Data from space have the additional advantage of a good time sampling and phase coverage, as well as a higher precision. However, the ground-based observations continue to play an important role as simultaneous ground-based data are complementary to the 'white light' space data. Multi-colour photometry provides information on amplitude ratios and phase shifts, which are used to identify the degree l, while high-resolution spectroscopy allows the detection of high-degree modes and the identification of both l and m values.

The ground-based CoRoT support observations

The CoRoT/SWG Ground-based Observations Working Group has played an important role in preparing the CoRoT satellite mission. Several mid- and high-resolution spectrographs, multi-colour photometers, and a spectropolarimeter at different observatories[1] were involved (see

---

[1]CAHA: Calar Alto Astronomical Observatory (E); KO: Konkoly Observatory (HU); ESO: European Southern Observatory, La Silla (CL); MJUO: Mount John University Observatory (NZ); OHP: Observatoire de Haute Provence (F); OPM: Observatoire du Pic du Midi (F); ORM: Observatorio Roque de los Muchachos (E); OT: Observatorio del Teide (E); SAAO: South African Astronomical Observatory (ZA); SLN: Osservatorio Serra La Nave (I); SNO: Sierra Nevada Observatory (E); SPMO: San Pedro Mártir Observatory (MX)

Table 1: Observatories, instruments and telescopes involved in the CoRoT preparatory and simultaneous ground-based observations.

preparatory		simultaneous	
Obs.	Instrument(s)	Obs.	Instrument
ESO	FEROS@1.52m	CAHA	FOCES@2.2-m
KO	V@0.5m	ESO	FEROS@2.2m
OHP	Elodie@1.93m		HARPS@3.6m
	Aurelie@1.52m	MJUO	HERCULES@McLellen
OPM	NARVAL@TBL	OHP	Sophie@1.93m
ORM	P7@Mercator	OPM	NARVAL@TBL
	SARG@TNG	SNO	$uvby$@0.9m
OT	CCD@STARE	SPMO	$uvby$@1.5m
SAAO	UBV@(0.5m,0.75m)		
SLN	FRESCO@0.91m		
SNO	$uvby$@0.9m		
SPMO	$uvby$@1.5m		

left panel of Table 1) to characterise and select suitable CoRoT targets (Poretti et al. 2003, 2005).

Now the CoRoT satellite is in successful operation, huge efforts are being made to guarantee the simultaneous monitoring of a handful of selected β Cep, δ Sct, γ Dor and Be CoRoT targets from the ground. Large Programme and normal observing proposals have been applied for, and have been approved, with high-resolution spectrographs at ESO, OHP, CAHA, and MJUO. Strömgren multi-colour information is provided for by telescopes at SPMO and SNO, and spectropolarimetry is done with NARVAL@TBL at OPM (right panel Table 1). Table 2 gives an overview of the targets and the amount of spectra that have been collected in the framework of the simultaneous ground-based observations campaign during CoRoT's first one-and-a-half years of operation[2].

First ground-based results of B-type CoRoT targets

The β Cep star HD180642 (B1.5III, M $10M_\odot$) was one of the asteroseismic targets in CoRoT's LRc1. Analysis of a dataset consisting of 507 high-quality multi-colour photometric data, obtained with the Mercator telescope and with photometers at KO, SPMO, and SNO, and 280 high-resolution high S/N spectra (FEROS, SOPHIE, Aurelie) confirmed the presence of a dominant radial mode, and revealed evidence for at least two additional non-radial pulsations modes, including a possible high-order g-mode (Uytterhoeven et al. 2008a; Briquet et al. 2009).

No frequencies were found in the 68 FEROS spectra of the late-B type Be star HD 50209 (B8IV, vsini= 209km s^{-1}), observed with CoRoT in LRa1 (Gutiérrez-Soto & Neiner, private comm.). On the other hand, photometric data, consisting of HIPPARCOS satellite data, ASAS3 data and Strömgren $uvby$ data obtained at SNO, reveals one frequency 1.47(2) $\pm$ 1d^{-1} (Gutiérrez-Soto et al. 2007).

No variable signals were detected in the HIPPARCOS data and SNO Strömgren $uvby$ magnitudes of HD 81231 (B5IV) (Gutiérrez-Soto et al. 2007). The Be star was observed with CoRoT in LRc1. A line-profile analysis, based on 72 FEROS spectra, leads to the detection and identification of one $l = 3$ mode (Gutiérrez-Soto & Neiner, private comm.).

Gutiérrez-Soto et al. (2007) report the detection of frequencies in the 2-3d^{-1} domain in HIPPARCOS and SNO $uvby$ data for the early-B type Be star HD 49330. From a total

[2]IR: Initial Run (Jan.- Feb. 2007); LRc1,LRc2: first (May-Oct. 2007) and second (Apr.- Sep. 2008) Long Run in the center direction; LRa1: first Long Run in the anti-center direction (Oct. 2007 - Mar. 2008)

Table 2: Overview of the amount of spectra obtained from December 2006 until August 2008 in the framework of the simultaneous ground-based observational campaign of CoRoT targets with the FEROS, SOPHIE, FOCES and HERCULES spectrographs . The last column indicates the CoRoT run.

type	target	FEROS	SOPHIE	FOCES	HERCULES	
δ Sct	HD 50844	216				IR
Be	HD 50209	68				IR
δ Sct	HD 181555	343	66	285		LRc1
β Cep	HD 180642	213	35			LRc1
Be	HD 181231	72				LRc1
δ Sct	HD 174966			134		LRc1
γ Dor	HD 49434	409	711	75	194	LRa1
Be	HD 49330	127				LRa1
δ Sct	HD 172189	176				LRc2
γ Dor	HD 171834	193	447	401	55	LRc2
Ap	HD 171586	12				LRc2

of 127 FEROS and 41 NARVAL spectra it seems that this CoRoT target, observed in LRa1, pulsates in several short-period variations (frequencies $> 11d^{-1}$), associated to high-degree modes ($5 < \ell < 7$), as well (Floquet, Gutiérrez-Soto & Neiner, private comm.). Longer-term variations of the order of 50 days are also detected and can be associated with the envelope of the Be system (Floquet, private comm.).

Future

Analyses of the ground-based time-series in combination with the CoRoT data are in progress, and the preliminary results look promising. To continue the huge observational ground-based effort, there is a need for (time on) high-resolution spectrographs. Even with the current Large Programmes in progress, we need to keep convincing time allocation committees of the need for strings of consecutive nights and a huge amount of observing time, which is not always straightforward given the competition and the small amount of available high-resolution spectrographs. Moreover, two of the instruments intensively used for the ground-based support observations so far, FOCES (R=40000) and FEROS (R=48000), will not be available anymore to the community soon. This currently leaves us with the high-resolution spectrographs HARPS (R=80000), SOPHIE (R=70000), HERCULES (R=35000) and FIES (R= 46000 or 67000). Despite the two new spectrographs that will be operational in the near future, HERMES (R=40000–90000) and SONG (R=100000), it would be wise to look for alternatives to secure continuous ground-based high-resolution spectroscopic time-series in the future.

The open cluster campaigns

After successful multi-site campaigns on isolated β Cep stars, a new challenge was found in performing asteroseismic studies on β Cep stars in open clusters. A multi-site campaign using CCD photometry was set up by Andrzej Pigulski et al. The advantages of such a study are clear: the cluster members are supposed to have the same age and metallicity, which puts serious constraints on the asteroseismic models. Moreover, as the CCD field contains several cluster members, several pulsators can be studied simultaneously. The Open Cluster campaigns are being carried out in two phases: observations in 2005–2007 were dedicated to NGC 3293, NGC 6910 and NGC 884 (χ Per), while NGC 1893 and NGC 869 (h Per) are being observed in 2007-2009.

Preliminary results

The Southern cluster NGC 3293 has been observed in January-May 2006 from seven sites. When combining all the data, a detection level of 0.5mmag is expected. Handler et al. (2007) reported on the first results obtained from the SAAO data only: the cluster contains at least ten multi-periodic β Cep stars and a few faint variable B stars with periods between 8-12 hours. These variables are very interesting from a theoretical point of view, as their periods are shorter than typical Slowly Pulsating B(SPB)-type pulsations and longer than the typical periods of β Cep stars.

For the multi-site campaign on NGC 6910 and χ Per, twelve sites in ten countries on three continents were involved. The main telescopes were the 0.6m telescope at Białkow (PL), the Mercator telescope at ORM (E) and the 0.8m telescope at the Observatory of Vienna University (A). We refer to Saesen et al. (2008; 2009) for a detailed description of all the telescopes involved, as well as for the preliminary results on the cluster χ Per. Analysis of the Białkow data of NGC 6910 resulted in the confirmation of four known β Cep stars and the detection of five new β Cep variables (Pigulski et al. 2007; Pigulski 2008). For some of them (e.g. NGC 6910 18 and NGC 6910 16) up to ten or more modes are observed (see Fig. 1). Interestingly, both p- and g-modes seem to be present in NGC 6910 38 (Pigulski 2008).

Sites involved in the multi-site campaigns of the open clusters NGC 1893 and h Per, executed in 2007-2009, are Xinglong (CN), Białkow (PL), Baja (HU) and ORM (E, Mercator telescope) observatories. Preliminary results based on Białkow data, enlarged with older KO data, of h Per include the detection of several variable stars, including three known and six new β Cep stars, some of them seemingly showing g-modes, and eight Be/SPBe stars (Majewska-Świerzbinowicz et al. 2008).

Discussion

The open cluster campaign is one of the largest asteroseismic observational ground-based efforts currently executed, with many sites and small telescopes involved. Observations, reduction and analysis are in progress. So far, the campaign has been very successful with several β Cep, SPB and Be stars already discovered from isolated datasets. The combined multi-site, multi-colour time-series, free from aliasing effects, promise the detection and identification of several pulsation modes. Moreover, the discovered pulsators are promising targets for future seismic modelling, given the constraints on age and chemical composition dictated by the clustership member.

It seems to be not uncommon to observe high-order g-modes in β Cep stars. This follows from recent results on the cluster data, e.g. NGC 6910 38 (Pigulski 2008) and some β Cep variables in h Per (Majewska-Świerzbinowicz et al. 2008). Also other examples are known (e.g. the CoRoT target HD 180642, see above). Recently, Miglio et al. (2007) theoretically predicted high-order modes to be unstable in β Cep stars, which leaves room for nice prospects in asteroseismic studies, as the simultaneous detection of g- and p-modes enables the mapping of stellar interiors from the core to the surface.

Time-scales of extra line-broadening in massive OB stars

Spectroscopic studies show that a non-negligible extra line-broadening, often called macro-turbulence, plays an important role in the line broadening of OB giants and supergiants. The origin of the broadening, however, is still unknown (e.g. Simón-Díaz & Herrero 2006). One of the possible explanations is in terms of stellar oscillations (e.g. Aerts et al. 2009). Currently a new observational project is set up by Simón-Díaz et al. to investigate this explanation, to explore other origins, to study whether the macro-turbulence in massive OB stars show variability, and if so, to investigate on what time-scales it does. Five nights have been awarded in November 2008 with FIES@Nordic Optical Telescope in low-resolution mode (R=25000). Eight OB targets have been selected and will be observed in different time-intervals, ranging from a few minutes to 3 days. This project is a good step in the direction of exploring possibilities of pulsational and asteroseismic studies of more evolved OB stars.

Acknowledgments. The CoRoT ground-based support observations benefit greatly from time on ESO Telescopes at the La Silla Observatory under the ESO Large Programmes LP178.D-0361 and LP182.D-0356 (PI: E. Poretti). KU acknowledges Andrzej Pigulski, Sophie Saesen, Juan Gutiérrez-Soto, Coralie Neiner, Michele Floquet, Conny Aerts, and Artemio Herrero for their input on the latest results on the projects described in this paper. She also thanks Ennio Poretti for careful reading of the manuscript. Part of this work has been carried out at INAF-OA Brera (Merate, Italy) in the framework of a Marie Curie Intra-European Fellowship, contract number MEIF-CT-2006-024476, and of the Italian ESS project ASI/INAF I/015/07/0, WP 03170.

References

Aerts, C., Puls, J., Godart, M., & Dupret, M.-A. 2009, CoAst, 158, 66

Briquet, M., Uytterhoeven, K., Aerts, C., et al. 2009, CoAst, 158, 292

Gutiérrez-Soto, J., Fabregat, J., Suso, J., et al. 2007, A&A, 476, 927

Handler, G., Tuvikene, T., Lorenz, D., et al. 2007, CoAst, 150, 193

Majewska-Świerzbinowicz, A., Pigulski, A., Szaboó, R., & Csubry, Z. 2008, J. Phys.: Conf. Ser., 118, 012068

Miglio, A., Montalbán, J., & Dupret, M.-A. 2007, MNRAS, 375, 21

Pigulski, A., Handler, G., Michalska, G., et al. 2007, CoAst, 150, 191

Pigulski, A. 2008, J. Phys.: Conf. Ser., 118, 012011

Poretti, E., Garrido, R., Amado, P.J., et al. 2003, A&A, 406, 203

Poretti, E., Alonso, R., Amado, P.J., et al. 2005, AJ, 129, 2461

Saesen, S., Pigulski, A., Carrier, F., et al. 2008, J. Phys.: Conf. Ser., 118, 012071

Saesen, S., Carrier, F., Pigulski, A., et al. 2009, CoAst, 158, 179

Simón-Díaz, S., & Herrero, A., 2006, A&A 468, 1063

Uytterhoeven, K., Poretti, E., Rainer, M., et al. 2008a, J. Phys.: Conf. Ser., 118, 012077

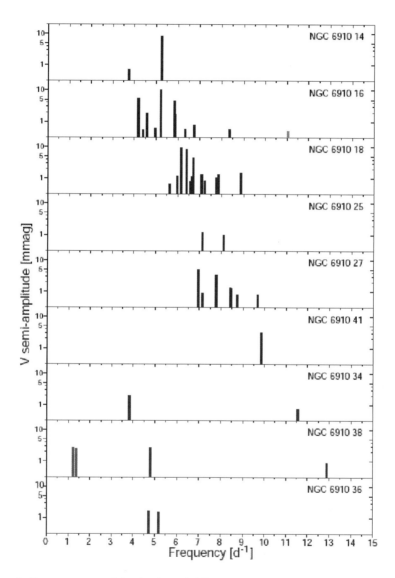

Figure 1: Frequency spectra of the nine detected β Cep variables in the open cluster NGC 6910 (taken from Pigulski 2008).

Comm. in Asteroseismology
Vol. 158, 2009, 38th LIAC/HELAS-ESTA/BAG, 2008
A. Noels, C. Aerts, J. Montalbán, A. Miglio and M. Briquet., eds.

Space observations of O and B stars with MOST

R. Kuschnig

Institut für Astronomie, Türkenschanzstrasse 17, A-1180 Vienna, Austria

Abstract

The MOST (Microvariability and Oscillation of STars) satellite mission has observed more than 1100 stars during its first 5 years of operation. While the primary application was to probe bright (0-6 mag) solar type stars and red giants for p mode oscillations, other main targets were roAp stars, planet hosting stars including known transiting systems, pulsating Pre-Main-Sequence (PMS) stars and also Wolf-Rayet objects. In addition, photometric data from about 120 O and B stars have also been obtained. The majority of those as secondary targets (stars near primary targets), in the so-called 'Guide Star Photometry' mode. This led to the discovery of numerous SPB, pulsating Be (SPBe), pulsating B super giants (SPBsg), spotted stars with rotational light modulation and eclipsing binaries. More recently O and B stars in young clusters have been observed with MOST.

Individual Objects: zeta Oph, beta CMi, Spica, delta Ceti, Procyon, NGC2244, HD163830, WR123, WR103, WR111, WR110, WR124

MOST instrument and photometry

The MOST satellite, launched on June 30th 2003 in a 820km sun-synchronous low earth orbit, hosts a 15cm aperture Rumak-Maksutov telescope feeding a frame transfer CCD camera. Details about the instrument are found most comprehensively in Walker & Matthews et al. (2003). MOST photometry data are obtained in three different modes:

Fabry Imaging was designed to be the primary method, in which the entrance pupil of the telescope is imaged onto the CCD via micro-lens optics. This ring shaped image comprises about 1500 pixels which is optimally exposed by stars ranging 0 to 4th mag with integration times $\leq$ 30 sec. Fabry images rendered within a 60x60 pixel segment of the CCD are transmitted with a low binning factor (2x2) to ground. About 1/3 of the MOST science detector is covered by the micro-lens array.

Direct Imaging provides a PSF of about 2 pixels FWHM (plate scale 3 arcsec/pixel) in the area not covered by the micro-lens array. In general images (20x20 pixels segments with the PSF in the center) from 5 stars in any given field are obtained in this mode. While Direct Imaging was originally meant to be conducted for a handful of stars near a bright primary target (which was placed on a micro-lens), it is by now the most frequently used mode for observation. It provides high precision photometry for fainter stars 6-13 mag.

Guide Star Photometry is used for the vast majority of objects. It is based on a simple onboard data reduction and delivers a single signal value from a set of exposures. Only signal values from pixels exceeding a threshold value, which is determined by an instantaneously measured background value plus a fixed signal offset, are recorded and summed up across the PSF. Due to data transfer volume constraints photometry from no more than 5 stars in any given field can be obtained in Direct Imaging mode. For the remaining objects the data

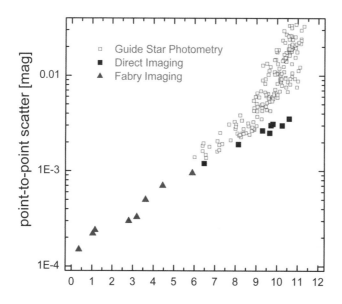

Figure 1: MOST point-to-point scatter

reduction is conducted onboard as descibed above. MOST can obtain Guide Star Photometry for up to 60 stars simultaneously.

In all modes a total of more than 1100 stars have been observed over the past 5 years of operations.

MOST was designed to perform ultra high precision photometry for very bright stars, collected for up to 60 days with high duty cycle (90 %), to achieve a sensitivity of 1-2 ppm noise amplitude for periods of 30 minutes and shorter. For the best cases like Procyon, 0.4 mag(V), the point-to-point scatter was as low as 120 ppm with a 15 sec sampling rate and about 300ppm for a 3.0 mag star. Direct Imaging can deliver 1-2 mmag scatter for objects with 6-8 mag(V) and still 0.03 mag for stars as faint as 12th magnitude. The rms noise from Guide Star Photometry is higher compared to Direct Imaging photometry. The difference is greater for fainter objects. However even Guide Star Photometry provides respectable performance values. Figure 1 shows typical scatter values for the MOST photometry modes.

Primary O and B stars

This is a brief description of the results obtained from the hot primary target stars observed with MOST up to June 2008.

ζ Oph, this β Cephei variable was the first early type target star that MOST observed (June 2004). The collected data including ground based simultaneous spectroscopy unambiguously proved that multimode nonradial pulsations are excited in this star (Walker et al. 2005a).

β CMi, a cool (B8Ve) Be type star was observed for 41 days starting Dec 2005 till the end of January 2006. The search for variability was successful with the discovery of low amplitude pulsations (Saio et al. 2007).

Spica, is a β Cephei primary element of an ellipsoidal binary system. MOST obtained unprecedented high precision photometry for that object during 30 days in March and April 2007. While the analysis is still in progress, preliminary results are presented in this volume (Desmet et al.).

δ Ceti was the first star which MOST observed for a prolonged period of time, during the commissioning phase of the mission in Sept 2003. Past measurements of this β Cephei star revealed one frequency hence it was considered a monoperiodic member of this class of variable stars. Even MOST photometry in first order appeared to confirm this simple nature. However, when comparing residual variations, after subtracting the main mode, with previously obtained high precision RV observations more pulsation modes were identified (Aerts et al. 2006a).

O stars in clusters

Initially O stars were not specifically selected as targets and even serendipitous observations of only a few objects in the fields around other stars did not reveal any variability. In January 2008 one of two main objectives of observing the young open cluster NGC2244 was to probe O stars members for variability. However, the six O type stars we monitored did not show any intrinsic brightness changes at levels from 200ppm and above.

Wolf-Rayet targets

Up to now 5 Wolf-Rayet stars have been observed with MOST, all of them as primary targets in the Direct Imaging mode since their brightness ranges only from 7.7-11.4 mag(V). Those were, in chronological order: WR123 (WN8), WR103 (WC9d), WR111 (WC5), WR110 (WN5) and most recently WR124 (WN8h). For WR123, belonging to the Nitrogen-rich class, a pulsation period with 9.8h was found (Lefevre et al., 2005). In contrast WR111, an object ascribed to the Carbon-rich stars did not reveal any coherent variations above a level of 50ppm for periods ¡2.4 hr (Moffat et al., 2008) and only a very modest 1/f noise spectrum below frequencies 10c/d.

Serendipitous discoveries of variable B stars

The majority of newly discovered variable B stars came from the measurements conducted from objects in close proximity to the primary targets. In particular fields around Wolf-Rayet stars and young clusters have showed a high abundance of B stars naturally. A significant amount revealed variability caused by pulsation, spot modulation and eclipsing binarity. In total about 60 percent of the B stars observed by MOST appeared variable. Among the pulsating stars we have had 'classical' SPB stars, B emission (SPBe), SPB super-giants (SPBsg) and β Cephei objects. In fact the SPBe and SPBsg subgroups have been introduced based on MOST photometry and the subsequent analysis (Walker et al. 2005, Saio et al. 2006). In particular the amplitude spectra of a sample of Be stars in combination with model calculations, see Saio et al. in this volume, seem to provide direct information on the rotation rates of those stars. The largest group are the newly discovered 'classical' SPB stars. About 30 have been detected by MOST. The first one HD163830, a B5 II/III, published in Aerts et al. 2006b, revealed more than 20 pulsation modes. Unfortunately for most of the SPB candidates in our archive, spectroscopic information is scarce or missing hence a lack of knowledge about the fundamental parameters is hindering the analysis. The aim is to gather that information via ground-based follow-up observations at least for the stars with the richest amplitude spectra.

As an example MOST photometry from a newly discovered SPB candidate star is presented in Figure 2, showing multimode variation with high signal-to-noise. However no fundamental

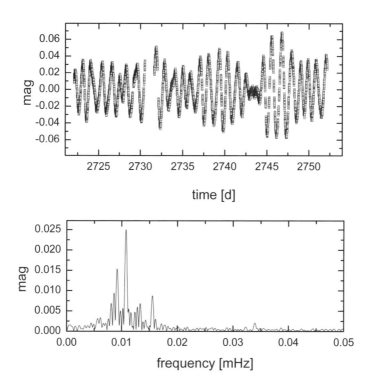

Figure 2: One of many SPB candidate stars observed by MOST. Light curve (top) and amplitude spectrum (bottom).

parameters are known for that object.

Acknowledgments.   This project is funded by the Canadian Space Agency (CSA) and the Natural Sciences and Engineering Research Council (NSERC) of Canada. Support is also provided by the University of Vienna and the Austrian Science Promotion Agency (FFG-ARL).

References

Aerts, C., Marchenko, S.V, Matthews, J.M. et al. 2006a, ApJ, 642, 470

Aerts, C., De Cat, P., Kuschnig, R. et al. 2006b, ApJ, 642, L165

Lefevre, L., Marchenko, S.V, Moffat, A.F.J. et al. 2005, ApJ, 634, L109

Moffat, A.F.J., Marchenko, S.V, Zhilyaev, B.E. et al. 2008, ApJ, 679, L45

Saio, H., Kuschnig, R., Gautschy, A. et al. 2006, ApJ, 650, 1111

Saio, H., Cameron, C., Kuschnig, R. et al. 2007, ApJ, 654, 544

Walker, G.A.H., Matthews, J.M., Kuschnig, R. et al. 2003, PASP, 115, 1023

Walker, G.A.H., Kuschnig, R., Matthews, J.M. et al 2005, ApJ, 623, L145

DISCUSSION

De Ridder: Did MOST observe hybrid B-type pulsators, i.e. B-type pulsators that show both p-modes and g-modes ?

Kuschnig: Not that I am aware of. But I will check the data we have more carefully for those.

Rauw: Concerning the O stars where you have not detected any photometric variability, do you know whether these objects show spectroscopic variability ?

Kuschnig: No, we did not check that in advance.

Comm. in Asteroseismology
Vol. 158, 2009, 38th LIAC/HELAS-ESTA/BAG, 2008
A. Noels, C. Aerts, J. Montalbán, A. Miglio and M. Briquet., eds.

Space observations of B stars with CoRoT

P. Degroote[1], A. Miglio[2], J. Debosscher[1], J. Montalbán[2], J. Cuypers[3], M. Briquet[1],P. De Cat[3], A. Thoul[2], T. Morel[2], E. Niemczura[4], L. Balaguer-Núñez[5], C. Maceroni[6], I. Ribas[7], A. Noels[2], C. Aerts[1,8], M. Auvergne[9], A. Baglin[9], C. Catala[9], M. Deleuil[10], E. Michel[9], M. Ollivier[11], L. Jorda[10], R. Samadi[9]

[1] Institute of Astronomy, K.U.Leuven, Celestijnenlaan 200D, 3001 Leuven, Belgium
[2] Institut d'Astrophysique et de Géophysique, Université de Liège, Allée du 6 Août 17, 4000 Liège, Belgium
[3] Koninklijke Sterrenwacht van België, Ringlaan 3, 1180 Brussels, Belgium
[4] Astronomical Institute of the Wroclaw University, ul. Kopernika 11, 51-622 Wroclaw, Poland
[5] Departament d'Astronomia i Meteorologia, Universitat de Barcelona, Av. Diagonal, 647, 08028 Barcelona, Spain
[6] INAF-Osservatorio di Roma, via Frascati-33, Monteporzio Catone (RM), Italy
[7] Institut de Ciències de l'Espai (CSIC-IEEC), Campus UAB, Facultat de Ciències, Torre C5 parell, 2a pl, 08193 Bellaterra, Spain
[8] Department of Astrophysics, IMAPP, Radboud University Nijmegen, PO Box 9010, 6500 GL Nijmegen, The Netherlands
[9] Observatoire de Paris, LESIA, CNRS (UMR8109), UPMC, Université Paris Diderot, 5 Place Jules Janssen, 92190 Meudon, France
[10] Laboratoire d'Astrophysique de Marseille, CNRS UMR 6110, Traverse du Siphon, 13376 Marseille, France
[11] Institut d'Astrophysique Spatiale, UMR8617, Université Paris XI, Bâtiment 121, 91405 Orsay Cedex, France

Abstract

We present the preliminary results of the exploration of pulsating B stars observed with the CoRoT[1] space mission. The previously known group of Slowly Pulsating B stars gains a substantial amount of new candidates, offering the opportunity to test stellar models beyond individual cases. Besides these well-defined stars, the analysis of other B star candidate pulsators hints towards the presence of different variability behaviour, co-existing in the same space in terms of the timescale of the variations and location in the (T_{eff},$\log g$) diagram.

Introduction

The CoRoT space mission (see "The CoRoT Book", Fridlund et al. 2006) is a great opportunity for main sequence B-star asteroseismology. Two different kinds of datasets are available, each measured with an unprecedented duty cycle and precision. One type of dataset, the EXO field, contains tens of thousands of stars, and will continue to grow in the following months and years. Its main goal is to detect planetary transits, but this data also results in the discovery of new pulsators, where the B-star candidates can be selected from. The main advantage of this dataset is the wealth of stars, which will be used to constrain statistical properties of the target classes as a whole. The other dataset from CoRoT (SISMO field)

---

[1] The CoRoT space mission, launched on 27 December 2006, has been developed and is operated by CNES, with the contribution of Austria, Belgium, Brazil , ESA, Germany and Spain.

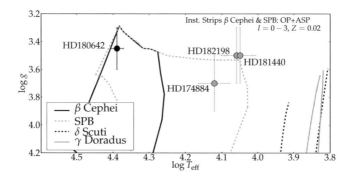

Figure 1: (T_{eff},log g) diagram, showing the position of the B stars in the initial run's SISMO channel. HD180642 is a known, bright β Cephei star. The CoRoT photometry clearly showed the binary nature of the former SPB candidate HD174884. HD181440 and HD182198 are two cool SPB star candidates. The instability strips are taken from Miglio et al. (2007).

is specifically aimed at asteroseismology, and focuses on few stars, but with an even higher precision and better sampling of 32 sec. The goal of this dataset is to do in-depth case studies of individual stars.

Non-emission main sequence B-star asteroseismology is restricted to two classes of stars so far: the Slowly Pulsating B stars (SPB stars hereafter; Waelkens 1991) are stars with masses between 3 and 8 $M_\odot$, showing high-order g modes with periods roughly between 1 and 3 days. The second class is that of the hotter β Cephei stars, with a mass range between 7 and 20 $M_\odot$. These stars usually have no surface convection zone, or at most a thin convective layer near the surface. The mechanism that drives these pulsations is the κ mechanism (e.g. Dziembowski & Pamyatnykh 1993).

SISMO field

In the CoRoT field dedicated to asteroseismology of the initial, first short and long runs, four non-emission B stars have been targeted (Fig. 1). The optical spectra obtained for these stars (Solano et al. 2005) have been used to derive the abundances and fundamental parameters (in preparation by Niemczura and collaborators). The main B target HD180642 is a known β Cephei star target, its dominant mode was already identified as radial by Aerts (2000). Because of the huge amplitude of this mode, non-linear effects have to be taken into account to model this star (Briquet et al., 2009). The three other secondary targets are candidate SPB stars. The achieved precision of the light curves showed that one of these candidates is actually a binary, and will be subject to a separate analysis. The two other targets are cool SPB candidates or cool B stars, which could hopefully help to determine the red boundaries of the SPB instability strip.

EXO field

The CoRoT satellite continues to transmit light curves of thousands of stars, of which most have never been observed outside the CoRoT program. Because so little is known about these stars, there is no other choice than to rely on automated classification methods based on CoRoT's white light photometry to unravel their nature (Debosscher et al. 2007, Sarro et al. 2008). Our main interest is to extract the B-type pulsators from the sample (β Cephei and SPB), while the cooler δ Scuti stars are in our case only used as a reference to limit the red edge of the SPB instability strip. We used the CoRoT Variability Classifier (in preparation by Debosscher and collaborators), which uses 2 independent supervised classification methods. In essence, the code compares the observed light curves from CoRoT with other targets that are independently observed and whose pulsational properties are determined with high reliability. Next, the code labels the CoRoT light curves with its 'best match'.

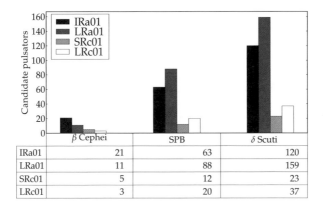

	β Cephei	SPB	δ Scuti
IRa01	21	63	120
LRa01	11	88	159
SRc01	5	12	23
LRc01	3	20	37

Figure 2: Results from the CoRoT Variability Classifier on all available runs, showing only the most probable candidates. There are significantly more β Cephei, SPB and δ Scuti star candidates detected in the fields directed at the galactic anticentre (IRa01,LRa01) than the fields towards the galactic centre (SRc01,LRc01).

Initially, we restricted the selection to those stars assigned to the same class with both methods. The results from this extraction can be seen in Fig.2. For all these stars, an auto-mated complete frequency analysis was performed on which we will report in a forthcoming paper (in preparation by Degroote and collaborators), after some basic piecewise detrending to correct for the largest (instrumental) jumps still occurring in the light curves. A model of the form $F(t) = C + \sum_{i=1}^{n_f} A_i \sin[2\pi(f_i t + \phi_i)]$ with A_i and ϕ_i free parameters for each $i = 1, \ldots, n_f$, is fitted to each light curve, with frequency values f_i determined by the highest peak in the Scargle periodogram (Scargle 1982) of each prewhitening stage.

Some post-processing has been done to build a reliable frequency list: identification of candidate harmonics and combination frequencies and rare window frequencies. Frequency regions with known instrumental effects are also avoided.

In a next step, Strömgren photometry was obtained with INT-WFC for the stars in the initial run and used to determine the effective temperature T_{eff} and gravity $\log g$ of each star, in order to be able to roughly compare the stars with the predicted instability strips. This was done following the method of Balona (1994) and allowed us to place the new suspected B-type pulsators in a $(T_{\mathrm{eff}}, \log g)$ diagram, and compare their position with theoretical predictions (Fig. 3).

It is immediately clear that almost all SPBs are situated well within the predicted area, as well as most of the δ Scuti stars. However, the picture is entirely different for the β Cephei candidates. Although these stars are supposed to be the hottest in the sample, several appear to exist with much lower effective temperature than the SPBs, in some cases even comparable to the T_{eff} typical of δ Scuti stars. Because of this discrepancy, the classification of the subsample was redone using a simple, but also more flexible cluster algorithm. Instead of comparing each light curve with light curves from already known class members, we compare

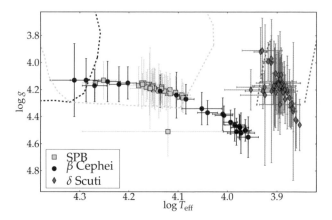

Figure 3: Position of the most probable β Cephei, δ Scuti and SPB stars in the (T_{eff},log g) diagram. The SPB and δ Scuti candidates are located well within their instability strips (dashed lines), while the β Cephei stars are spread out over the diagram.

the light curves to each other, taking also the temperature and gravity information into account in the process. The most important choice we have to make here is the number of groups we are willing to divide the subsample into. This arbitrariness is resolved by the observation that the clusters are stable with respect to these changes; introducing more groups tends to divide existing groups instead of mixing them. Raising the number of groups therefore means resolving more differences between observations. The result is a confirmation of the tight clustering of SPB and δ Scuti stars, and the breaking up of the β Cephei candidates in a hot and cool part.

To obtain a global picture of every group of stars in the (T_{eff},log g) diagram, we graphically depict the distribution of the highest-amplitude frequency and average frequency, and the main amplitude of pulsation modes of the stars in the four groups we found (Fig. 4). In terms of frequencies, both the δ Scuti candidates and SPB candidates confirm their expected position in the (T_{eff},log g) diagram. The stars on the hot side of the SPB candidates, (formerly β Cephei candidates and indicated as group U1 in Fig. 4), appear to have frequencies between these two classes. For this subgroup, two possible explanations come to mind when investigating the light curves and their frequencies (some examples can be found in Fig. 5): either they are candidate Be stars, which would explain some of the more erratic behaviour in the light curves (e.g. Neiner et al 2005), or they are in fact true candidate SPB-stars, but are rapid rotators, shifting their frequencies to higher values. The second unidentified group of stars (indicated as U2) lies in between the δ Scuti and SPB group, both in temperature and frequency value. They are difficult to separate from the β Cephei stars without any additional colour information. This group's main amplitudes are low, thus more difficult to detect from ground-based observations than the main amplitudes of the other three groups. This collection of U2 stars seems rather heterogeneous in nature.

An example of one of these stars, is CoRoT 102771057 (Fig. 5). We see from the fit and the power spectrum that this star pulsates with many frequencies. A few frequencies clearly stand out, and are spread out over a wide range: the first frequency is in the β Cephei p-mode range; $f_1 = 6.0872 \pm 0.0002\,\mathrm{d}^{-1}$, the second frequency equals $f_2 = 3.2834 \pm 0.0002\,\mathrm{d}^{-1}$, in between the p- and g mode range, while the third frequency, $f_3 = 0.5141 \pm 0.0003\,\mathrm{d}^{-1}$, enters the g mode range. On top of that, some power excess is detected around $\approx 7.9\,\mathrm{d}^{-1}$ and

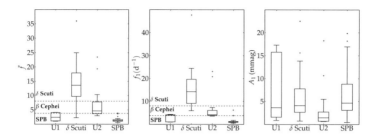

Figure 4: Properties of reclustered groups (box and whiskers denote the 50% and 75% interval, the horizontal line is the median value). Reclustering divided the β Cephei candidates in two groups of undetermined type (U1 and U2). Dotted lines show the frequency regions where SPB, β Cephei and δ Scuti pulsations are to be expected. The distribution of average (*left*) and first (*middle*) frequencies are shown. The amplitudes of U2 stars are clearly smaller than the amplitudes of the other groups. (*right*).

$\approx 9.3\,\mathrm{d}^{-1}$, which could very well be multiplets with $\Delta f \approx 0.16\,\mathrm{d}^{-1}$. This wealth and broad range of frequencies for U2 stars is striking, because, at present, no star has been predicted to pulsate in this region of the HR diagram. Theory, however, does predict the existence of some type of B star pulsations in the hot part of this region (Townsend 2005, Savonije 2005).

If we focus on the unidentified group of stars on the hot side of the SPB group (U1), we see that, in contrast to the previous group, these stars form a fairly homogeneous group. Their dominant mode amplitude is on the low side of the β Cephei range, while their frequencies are on the high side of the SPB frequency range. They show small groups of peaks, around $2\,\mathrm{d}^{-1}$, $4\,\mathrm{d}^{-1}$, $6\,\mathrm{d}^{-1}$ and even $8\,\mathrm{d}^{-1}$. The collection of peaks is a signature of slightly variable frequencies or amplitudes, making them good pulsating Be star candidates.

One of the stars that is behaving as a classical SPB star, is shown in Fig. 6. Around 60 frequencies are identified between $0\,\mathrm{d}^{-1}$ and $3\,\mathrm{d}^{-1}$. There are a few peaks above $3\,\mathrm{d}^{-1}$, but they can all be reasonably well explained by combination frequencies. A candidate period spacing of $\Delta P \approx 0.08\mathrm{d}$ is also detected.

Finally, although no typical β Cephei stars are confidently identified in the Initial Run's subsample of stars with available ground-based photometry, there are certainly good candidates available. An example is shown in Fig. 6: a star with frequencies between 3 and $10\,\mathrm{d}^{-1}$.

Conclusion

From the exploration of pulsating B stars in the SISMO field, we can conclude that they are compatible with current theoretical models. Rotational velocities and different evolutionary stages will be further exploited in upcoming runs.

In the initial run's EXO field, we have encountered a large number of new pulsating B stars. For the SPB candidates, the T_{eff} is as expected, but for the candidate β Cephei stars, there are two groups, one of which the T_{eff} is not compatible with theory. There are several likely scenarios to explain the behaviour of these two groups; they could be binary pulsators, fast rotating SPBs or δ Scuti with retrograde modes. For the hottest stars (U1), it is also possible that they are pulsating Be-stars. A thorough frequency analysis and theoretical interpretation of the results is ongoing and will be presented in a forthcoming paper. To further discriminate between the different types, spectroscopic observations from the VLT will be used in the future.

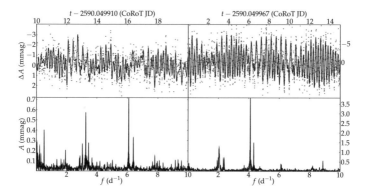

Figure 5: Example light curves and frequency spectra of a star of unknown type U1 having a rich spectrum similar to the one of a β Cephei star but a lower temperature (*left*) and U2 showing both frequencies too high for an SPB star and too low to be a β Cephei or δ Scuti star (*right*).

Acknowledgments. The Belgian authors are supported by the Belgian Federal Space Policy Office (PRODEX). PD, JD, and CA also acknowledge support from the Research Council of Leuven university (grant GOA/2008/04). MB is postdoctoral researcher of the Fund for Scientific Research of Flanders (FWO) and AT Chercheur Qualifié du Fonds de la Recherche (FRS).

References

Aerts, C. 2000, A&A, 361, 245

Balona, L. A. 1994, MNRAS, 268, 119

Briquet, M., Uytterhoeven, K., Aerts, C., et al. 2009, CoAst, 158, 292

Debosscher, J., Sarro, L. M., Aerts, C., Cuypers, J., et al. 2007, A&A, 475, 1159

Dziembowski, W. A. & Pamiatnykh, A. A. 1993, MNRAS, 262, 204

Fridlund, M., Baglin, A., Lochard, J., & Conroy, L. (Eds), 2006, The CoRoT Mission, ESA Publications Division, ESA Spec. Publ. 1306

Miglio, A., Montalbán, J. & Dupret, M.-A. 2007, MNRAS, 375, L21

Neiner, C., Floquet, M., Hubert, A. M., Frémat, Y., et al. 2005, A&A, 437, 257

Savonije, G. J. 2005, A&A, 443, 557

Sarro, L. M., Debosscher, J., Lopez, M., Aerts, C. 2009, A&A, 494, 739

Scargle, J. D. 1982, ApJ, 263, 835

Solano, E., Catala, C., Garrido, R., Poretti, E. et al. 2005, AJ, 129, 547

Townsend, R. H. D. 2005, MNRAS, 364, 573

Waelkens, C. 1991, A&A, 246, 453

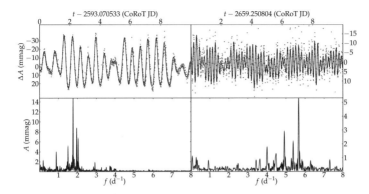

Figure 6: An SPB candidate (*left*) and a β Cephei candidate (*right*). Plotted are a zoom on the light curve (*top*) and part of the frequency spectrum (*bottom*). The SPB candidate shows strong pulsations around $f \approx 2\mathrm{d}^{-1}$, while the β Cephei candidate shows clear variations with frequencies above $f > 4\mathrm{d}^{-1}$

DISCUSSION

Chiosi: Studying the evolution of Pop III low mass stars, Marigo et al. (2001) suggested that old stars of about 0.9 M$_\odot$ with age of about 12 Gyr could intersect the instability strip in the region of δ Scuti while burning hydrogen in the core (turn-off). Therefore if Pop III stars of low mass could form, we expect a new class of δ Scuti-like pulsators with roughly the same colors and luminosity but longer periods (P $\sim$ M$^{-0.5}$) because of the lower masses involved. In there any way to check whether the new class of pulsators you have presented might correspond to this prediction? It would be nice because it would prove the existence of Pop III.

Comm. in Asteroseismology
Vol. 158, 2009, 38th LIAC/HELAS-ESTA/BAG, 2008
A. Noels, C. Aerts, J. Montalbán, A. Miglio and M. Briquet., eds.

Massive B-type pulsators in low-metallicity environments

C. Karoff[1,2], T. Arentoft[2], L. Glowienka[2], C. Coutures[3], T. B. Nielsen[2], G. Dogan[2],
F. Grundahl[2], and H. Kjeldsen[2]

[1] School of Physics and Astronomy, University of Birmingham, Edgbaston, Birmingham B15 2TT, UK
[2] Danish AsteroSeismology Centre (DASC), Department of Physics and Astronomy, University of Aarhus,
DK-8000 Aarhus C, Denmark
[3] Institut d'Astrophysique de Paris, CNRS, Universite Pierre et Marie Curie UMR7095, 98bis Boulevard
Arago, 75014 Paris, France

Abstract

Massive B-type pulsators such as β Cep and slowly pulsating B (SPB) stars pulsate due to layers of increased opacity caused by partial ionization. The increased opacity blocks the energy flux to the surface of the stars which causes the layers to rise and the opacity to drop. This cyclical behavior makes the star act as a heat engine and the star will thus pulsate. For β Cep and SPB stars the increased opacity is believed to be caused by partial ionization of iron and these stars should therefore contain non-insignificant quantities of the metal. A good test of this theory is to search for β Cep and SPB stars in low-metallicity environments. If no stars are found the theory is supported, but, on the other hand, if a substantial number of β Cep and SPB stars are found in these environments then the theory is not supported and a solution is needed. With a growing number of identified β Cep and SPB stars in the low-metallicity Magellanic Clouds we seem to be left with the second case. We will in this context discuss recent findings of β Cep and SPB stars in the Magellanic Clouds and some possible solutions to the discrepancy between these observations and the theory. We also describe an ambitious project that we have initiated on the Small Magellanic Cloud open cluster NGC 371 which will help to evaluate these solutions.

Individual Objects: NGC 371

The Problem

Though the number of β Cep and SPB stars is predicted to be very limited in the Magellanic Clouds due to the reduced metallicity in these environments a growing number of studies are identifying β Cep and SPB in the Magellanic Clouds. Pigulski & Kołaczkowski (2002) used OGLE II and MACHO data to identify 3 β Cep in the Large Magellanic Cloud (LMC). This study was later updated by Kołaczkowski et al. (2006) who identified 92 β Cep and 6 SPB stars in the LMC and 59 β Cep and 11 SPB stars in the Small Magellanic Cloud (SMC). Diago et al. (2008) have reanalyzed MACHO data of 186 absorption-line B stars and identified 1 β Cep and 8 SPB stars in the SMC. Recently, Karoff et al. (2008) have identified 29 candidate SPB stars in the young open SMC cluster NGC 371. This result is particularly interesting because (if confirmed) it indicates that the population fraction of SPB stars in this cluster is probably larger than or equal to the population fraction of SBP stars in the Galaxy.

The metallicity of Magellanic Clouds is found to be around $Z = 0.007$ and $Z = 0.002$ for the LMC and SMC, respectively (Maeder et al., 1999) . Though no instability domains have been published for such low metallicities there are indications that pulsation can only be driven down to 0.01 in β Cep stars and down to 0.005 in SPB stars using standard physics (solar abundances from Grevesse & Noels 1993) which means that the increasing number of observed β Cep and SPB stars in the Magellanic Clouds cannot be explained by standard stellar models. Pamyatnykh (1999) has calculated instability domains down to $Z = 0.01$ and Miglio et al. (2007a,b) have calculated instability domains down to $Z = 0.005$ where the SPB domain seems to have vanished, but as no domains were calculated for lower metallicities it cannot be concluded that the SPB domain disappears for lower metallicities. There is clearly an urgent need for calculations of instability domains of SPB stars at the metallicity of the SMC ($Z = 0.002$), as it has not been proven that pulsation in SPB stars cannot be driven at this metallicity using standard stellar models, though it seems doubtful.

The Solutions

Though the discrepancy between the lowest metallicity at which standard stellar models predict pulsation in β Cep and SPB stars and the metallicity of the β Cep and SPB stars in especially the SMC is quite large we can identify three possible extensions to the standard models which can make the low-metallicity β Cep and SPB stars pulsate:

- The stars that pulsate have higher metallicity than the average metallicity of the Magellanic Clouds.

- The new (solar) abundances might explain why pulsation can be driven at lower metallicity.

- Local iron enhancement might drive pulsation at lower metallicity in these stars.

Metallicity

Of course not all the stars in the Magellanic Clouds have the same metallicity and as indicated by the wide range of published estimates of the metallicity ($Z = 0.004$ to 0.01 for LMC and 0.001 to 0.003 for SMC (Maeder et al., 1999) there are differences in the values of the metallicities within the two clouds. This probably reflects the fact that the Magellanic Clouds are still subject to reasonable amounts of star formation, which is also seen in the large correlation between age and metallicity for the two clouds (see e.g. Pagel & Tautvaišienė, 1999). Another important issue to remember when using the total metallicity Z (defined as the mass fraction of heavy elements to hydrogen) to evaluate a star's ability to drive pulsation is that Z is not a direct measurement of the iron content in the star. In fact the most important elements for calculating Z are O, C and Ne and then Fe. This means that a low Z does not necessarily reflect a low iron content. It could also reflect e.g. a low oxygen content. The problem is of course that we only have detailed abundance estimates for a limited number of stars in the Magellanic Clouds and that standard stellar models are calculated using the total metallicity Z, and then assuming solar abundances instead of using the individual abundances as e.g. spectroscopic metallicity [Fe/H].

The New Solar Abundances

In this way the abundance of the Sun becomes important for the ability of β Cep and SPB stars in the Magellanic Clouds to drive pulsation. Miglio et al. (2007a,b) have shown that the instability strip in the Hertzsprung-Russell diagram is increased for both β Cep and SPB stars, by using the new abundances of Asplund et al. (2005). Especially it is shown by Miglio et al. (2007a,b) that the new abundances and the new opacities can drive radial oscillations in stars with a metallicity as low as $Z = 0.01$, which cannot be done with the old abundances and opacities. It is also seen that the new abundances have the effect of extending the excited frequencies towards higher overtones.

Though there are contradictory interpretations of the effect of the new abundances on the excitations of β Cep and SPB stars (see for example Pamyatnykh & Ziomek, 2007) it might be possible to test the new abundances and opacities on oscillations in low-metallicity β Cep and SPB stars.

Local Iron Enhancement

The last solution to the problem with the low-metallicity massive B-type pulsators is the same solution that was found to solve the problem of pulsating subdwarf B (sdB) stars. To begin with, the oscillations in these stars were believed to be driven by the He_{II} – He_{III} convection zone; however, it was soon realized that the driving was negligible in the He_{II} – He_{III} convection zone as this region contains only very little mass, and therefore it carries practically no inertia to drive the pulsations. Instead it was shown by Charpinet et al. (1996, 1997) that pulsation in sdB stars could be driven by an opacity bump due to a local iron enhancement in the envelope of these stars. The enhancement was shown to be caused by gravitational settling and radiative levitation of heavy elements. A similar mechanism for β Cep and SPB stars was suggested by Pamyatnykh et al. (2004) and has been investigated for β Cep and SPB stars in low-metallicity environments by Bourge et al., (2006, 2007), Bourge & Alecian (2006) and Miglio et al. (2007c). Though it is still not clear if this is the solution to the possible low-metallicity β Cep and SPB stars, one way to test this would be to look for chemical peculiarities at the surfaces of these stars, especially Si enrichment, which could reflect local enhancement of different elements (i.e. iron) by diffusion processes if the stars are not fast rotators.

Case Study – NGC 371

In order to test the prediction that β Cep and SPB stars are rare in the Magellanic Clouds we have initiated an ambitious project which includes: 1) a survey for candidate β Cep and SPB stars in the open SMC cluster NGC 371; 2) spectroscopic determination of important physical parameters; 3) determination of precise eigenmode frequencies based on a multisite campaign and; 4) detailed modeling of the stars. The result of the first part of this project was the discovery of 29 candidate SPB stars in the cluster (amplitude spectra for the 29 candidates are shown in Fig. 1; see Karoff et al. 2008 for details), a discovery that clearly contradicts the theoretical predictions. The next step is therefore to obtain high-resolution spectra for these stars in order 1) to discriminate between binaries and bona fide SPB stars; 2) to determine cluster membership and; 3) to obtain important physical parameters of the candidates. This step in now underway as stellar spectra are now being obtained from the Gemini Multi-Object Spectrographs ($R = 4\,400$) as part of variable star one-shot project (Dall et al. 2007) We have also applied for VLT time with the Fibre Large Array Multi Element Spectrograph ($R = 18\,470$) for this cluster.

Acknowledgments.   CK acknowledges financial support from the Danish Natural Science Research Council and CK, FG, GD and TA also acknowledge support from the Danish AsteroSeismology Centre.

References

Asplund, M., Grevesse, N., Sauval, A. J., et al. 2005, A&A 431, 693

Charpinet, S., Fontaine, G., Brassard, P., et al. 1996, ApJ, 471, L103

Charpinet, S., Fontaine, G., Brassard, P., et al. 1997, ApJ, 483, L123

Cox, A. N., Morgan, S. M., Rogers, F. J., & Iglesias, C. A. 1992, ApJ, 393, 272

Basu, S., & Antia, H. M. 2008, Phys. Rep., 457, 217

Bourge, P.-O., Théado, S., & Thoul, A. 2007, CoAst, 150, 203

Bourge, P.-O., Alecian, G., Thoul, A., et al. 2006, CoAst, 147, 105

Bourge, P.-O., & Alecian, G. 2006, ASP, 249, 201

Dall, T. H., Foellmi, C., Pritchard, J., et al. 2007, A&A, 470, 1201

Diago, P. D., Gutirrez-Soto, J., Fabregat, J., et al. 2008, A&A, 480, 179

Grevesse, N., & Noels, A. 1993, Origin and Evolution of the Elements, 15

Karoff, C., Arentoft, T., & Glowienka, L. 2008, MNRAS, 386, 1085

Kołaczkowski, Z., Pigulski, A., Soszyński, I., et al. 2006, Mem. Soc. Astron. Ital., 77, 336

Maeder, A., Grebel, E. K., & Mermilliod, J.-C. 1999, A&A, 346, 45

Miglio, A., Montalbán, J., & Dupret, M.-A. 2007a, MNRAS, 375, L21

Miglio, A., Montalbán, J., & Dupret, M.-A. 2007b, CoAst, 151, 48

Miglio, A., Bourge, P.-O., Montalbán, J., & Dupret, M.-A. 2007c, CoAst., 150, 209

Pamyatnykh, A. A. 1999, MNRAS, 49, 119

Pamyatnykh, A. A., Handler, G., & Dziembowski, W. A. 2004, MNRAS, 350, 1022

Pamyatnykh, A. A., & Ziomek, W. 2007, CoAst, 150, 207

Pigulski A., & Kołaczkowski, Z. 2002, A&A, 388, 88

Pagel, B. E. J., & Tautvaišienė, G. 1999, Ap&SS, 265, 461

DISCUSSION

Dziembowski: Did you, in your SPB sample, see objects showing pairs of modes with near (but not exactly !) 2:1 period ratios? Such ratios have been found by Kolaczkowski (Ph.D. thesis) in a number of SPB stars in the SMC.

Karoff: I am sorry, but I do not think that the quality of these single-site data is good enough to see such an effect. In order to answer your question we would need a multisite campaign.

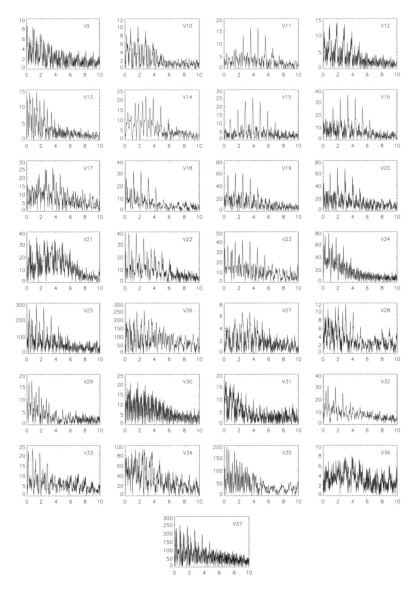

Figure 1: Amplitude spectra of the 27 stars pulsating in the upper part of the main sequence of NGC 371. The y-axes are in mmag (in B) and the x-axes are in c/d. Note the different scaling of the y-axes.

Comm. in Asteroseismology
Vol. 158, 2009, 38th LIAC/HELAS-ESTA/BAG, 2008
A. Noels, C. Aerts, J. Montalbán, A. Miglio and M. Briquet., eds.

Asteroseismology of massive stars in the young open cluster NGC 884: a status report

S. Saesen[1], F. Carrier[1], A. Pigulski[2], C. Aerts[1], G. Handler[3], A. Narwid[2], J. N. Fu[4], C. Zhang[4],
X. J. Jiang[5], G. Kopacki[2], J. Vanautgaerden[1], M. Steślicki[2], B. Acke[1], E. Poretti[6],
K. Uytterhoeven[6], W. De Meester[1], M. D. Reed[7], Z. Kołaczkowski[2], G. Michalska[2],
E. Schmidt[3], R. Østensen[1], C. Gielen[1], K. Yakut[8,9], A. Leitner[3], B. Kalomeni[10], S. Prins[1],
V. Van Helshoecht[1], W. Zima[1], R. Huygen[1], B. Vandenbussche[1], P. Lenz[3], D. Ladjal[1],
E. Puga Antolín[1], T. Verhoelst[1], J. De Ridder[1], P. Niarchos[11], A. Liakos[11], D. Lorenz[3],
S. Dehaes[1], M. Reyniers[1], G. Davignon[1], S.-L. Kim[12], D. H. Kim[12], Y.-J. Lee[12], C.-U. Lee[12],
J.-H. Kwon[12], E. Broeders[1], H. Van Winckel[1], E. Vanhollebeke[1], G. Raskin[1], Y. Blom[1],
J. R. Eggen[7], P. Beck[3], J. Puschnig[3], L. Schmitzberger[3], G. A. Gelven[7], B. Steininger[3], and
R. Drummond[1]

[1] Instituut voor Sterrenkunde, Katholieke Universiteit Leuven, Belgium
[2] Instytut Astronomiczny, Uniwersytet Wrocławski, Poland
[3] Institut für Astronomie, Universität Wien, Austria
[4] Beijing Normal University, China
[5] National Astronomical Observatories, Chinese Academy of Sciences,Beijing, China
[6] INAF-Osservatorio Astronomico di Brera, Merate, Italy
[7] Department of Physics Astronomy and Material Science,
Missouri State University, USA
[8] Institute of Astronomy, University of Cambridge, UK
[9] Department of Astronomy & Space Sciences, University of Ege, Izmir, Turkey
[10] Izmir Institute of Technology, Department of Physics, Izmir, Turkey
[11] Department of Astrophysics, Astronomy and Mechanics,
University of Athens, Greece
[12] Korea Astronomy and Space Science Institute, Daejeon, South Korea

Abstract

To improve our comprehension of the β Cephei stars, we set up a photometric multi-site campaign on the open cluster NGC 884 (χ Persei). Thirteen telescopes joined the 2005-2007 campaign which resulted in almost 78 000 CCD frames. We present an up-to-date status of the analysis of these data, in which several interesting oscillating stars are pointed out. We end with the future prospects.

Individual Objects: NGC 884

Introduction

Recent progress in the seismic interpretation of selected β Cephei stars was remarkable in the sense that standard stellar structure models are unable to explain the oscillation data for the best-studied stars: HD 129929 (Aerts et al. 2003), ν Eridani (Pamyatnykh et al. 2004, Ausseloos et al. 2004, Dziembowski & Pamyatnykh 2008) and 12 Lacertae (Ausseloos 2005, Handler et al. 2006, Dziembowski & Pamyatnykh 2008). Non-rigid internal rotation and core convective overshoot are needed to fit the measured oscillation frequencies and the standard models have now been upgraded to include these effects, albeit in a crude parametrised way. Pamyatnykh et al. (2004) have suggested to include in future models radiative diffusion processes as well, in an attempt to resolve the yet unsolved excitation problem encountered for some of the modes detected in ν Eridani.

A next step in asteroseismology of β Cephei stars was undertaken recently, with the study of these stars in clusters. Indeed, with the current CCD cameras we are able to obtain simultaneous measurements of thousands of stars. Another big advantage is the cluster membership of the stars: this gives us much tighter constraints when modelling their observed and identified oscillation modes.

Krzesiński & Pigulski (1997, 2000) discovered one candidate and two bona fida β Cephei stars in NGC 884. The variability study on this cluster conducted by Waelkens et al. (1990) showed that at least half of the brighter stars are variable, while most of them seem to be Be stars.

Observations

To perform ensemble-asteroseismolgy of NGC 884, we needed time-resolved multi-colour differential photometry of a selected field of this cluster, for which we organised a large-scale multi-site campaign. An international team monitored the cluster with 13 telescopes in the northern hemisphere in the filters U, B, V, I. The data were taken between 2005 and 2007, spread over three observational seasons, spanning in total 800 days. 77 900 CCD images and 92 hours of photo-electric measurements were collected, representing 1290 hours of data taken by about 60 observers. A world map indicating all participating sites and their characteristics (telescope diameter and filters used) and the distribution of the observations in time can be found in Saesen et al. (2008).

The effect of the multi-site character of the campaign is best seen when comparing the spectral windows of the data (see Fig. 1). The spectral window of the Polish data, the site that has the largest contribution to the whole data set, has a high one-day-alias: its amplitude is 90% of the main frequency peak. If we add one site, China, situated at a different latitude, the one-day-alias falls down to 70% of the central peak. Taking all sites into account makes the alias drop to only 55%, which makes it much easier to identify the correct frequency peak.

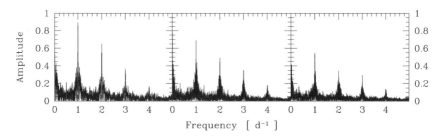

Figure 1: Spectral windows of the Polish data (left panel), the combined Polish and Chinese data (middle panel) and all data (right panel).

To transform the CCD frames into interpretable light curves, we extracted the fluxes of the stars with Daophot (Stetson 1987), in which we combined PSF and aperture photometry. We performed differential photometry in which we correct for atmospheric extinction by taking several reference stars distributed over the CCD frame into account. Currently we are detrending the data with SysRem (Tamuz et al. 2005) to remove the linear systematic effects which are present in a lot of stars. For sites with many data points we obtain an overall

V accuracy of 3-5 mmag over the entire campaign, depending on the telescope. The error on the frequency is smaller than $0.000\,14\,\mathrm{d}^{-1}$ and the detection threshold at high frequencies for Polish data is about 0.3 mmag. Frequencies in the β Cephei range can be accepted if their amplitudes are larger than 1 mmag, but this limit will go down once all data are put together.

Detected variable stars

A preliminary analysis on single-site data led to the confirmation of the two known β Cephei stars, Oo 2246 and Oo 2299, and to the discovery of numerous new pulsators of this type, among which several are multi-periodic (Pigulski et al. 2007, Saesen et al. 2008). However, spectroscopy has shed additional light on two of these newly classified β Cephei stars, Oo 2085 and Oo 2566. They show variations on both short ($\sim$hours) and long ($\sim$days) time scales and turn out to have Hα emission. As a consequence, Oo 2085 and Oo 2566 are also categorised as Be stars. Oo 2444, Oo 2488 and Oo 2572 remain accepted as β Cephei stars.

The β Cephei stars show at least a double-mode behaviour, except Oo 2299, which seems to show a single, predominant mode. Fig. 2 shows the periodograms in subsequent stages of prewhitening based on dual-site data of two of these oscillators. We clearly detect two frequencies for each of them (first two panels) and after subtracting these variations from the data, there is still power present in the residuals (lower panels). We expect that additional frequencies will peak above the frequency acceptance level after detrending and merging the data of all sites. The case of Oo 2572 also points out that multi-site observations are important to pinpoint the correct frequency peak. In Saesen et al. (2008), which is only based on single-site data, an alias frequency was mistaken for the correct peak.

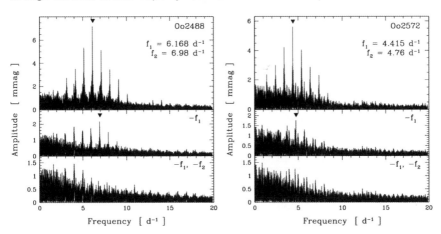

Figure 2: Periodograms of subsequent stages of prewhitening based on Polish and Chinese data of the β Cephei stars Oo 2488 and Oo 2572.

Besides these five established β Cephei stars, we have five more candidates. Three of them show evidence for low frequencies and might turn out to be hybrid oscillators. We have detected seven SPB candidates, amongst which is Oo 2253: three significant frequencies are extracted and shown in the subsequent periodograms of Fig. 3. In addition, the observed field of NGC 884 contains several eclipsing binaries (see Fig. 4 for two newly discovered cases) and other variable stars.

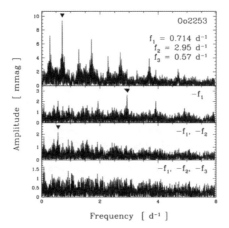

Figure 3: Periodograms of subsequent stages of prewhitening based on Polish and Chinese data of the SPB star Oo 2253.

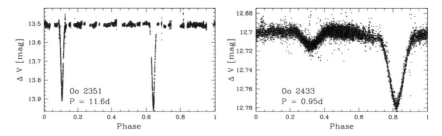

Figure 4: Phase plots of two newly discovered eclipsing binaries Oo 2351 and Oo 2433.

Future prospects

First of all, we will conduct a detailed search for variable stars in the cluster NGC 884. For these variable stars we will do a frequency analysis. We will especially look for B-type pulsators, for which we will perform a mode identification to determine the degree ℓ. The well-known photometric method, which compares the theoretical amplitude ratios with the observed ones at different wavelengths, will be used (Dupret et al. 2003). For this purpose we have observations in different bands at our disposal. The photometric amplitude ratios are going to be combined with radial-velocity amplitudes deduced from simultaneous NOT spectra for the three brightest β Cephei stars. This will make the mode identification more conclusive than only the amplitude ratios (Daszyńska-Daszkiewicz et al. 2005).

In the end we will fit theoretical frequencies to the observed ones and their mode identification, simultaneously for all pulsating cluster members. Indeed, as the stars in a cluster were born out of the same cloud, we can assume that they have the same age and had the same chemical composition at birth. The Liège stellar evolution code CLÉS (Scuflaire et al. 2008) and the non-adiabatic oscillation code MAD (Dupret et al. 2002) will be applied in this process. Only models that fulfill additional criteria, such as the position in the HR diagram derived from photometry, and the abundances of the stars obtained by NOT spectroscopy,

will be retained.
In any case, the first results are very promising for our future analysis of all campaign data. In-depth evaluation of stellar evolution models seems therefore within reach, now that the technique of asteroseismology has been extensively tested on single field stars.

Acknowledgments. S. Saesen is an Aspirant Fellow and F. Carrier is a Postdoctoral Fellow of the Fund for Scientific Research, Flanders (FWO). K. Uytterhoeven acknowledges financial support from a *European Community Marie Curie Intra-European Fellowship*, contract number MEIF-CT-2006-024476.

References

Aerts, C., Thoul, A., Daszyńska, J., et al. 2003, Science, 300, 1926

Ausseloos, M., Scuflaire, R., Thoul, A., & Aerts, C. 2004, MNRAS, 355, 352

Ausseloos, M. 2005, dissertation, K.U.Leuven, Belgium

Daszyńska-Daszkiewicz, J., Dziembowski, W.A., & Pamyatnykh, A.A. 2005, A&A, 441, 641

Dupret, M.-A., De Ridder, J., De Cat, P., et al. 2003, A&A, 398, 677

Dupret, M.-A., De Ridder, J., Neuforge, C., et al. 2002, A&A, 385, 563

Dziembowski, W.A., & Pamyatnykh, A.A. 2008, MNRAS, 385, 2061

Handler, G., Jerzykiewicz, M., Rodríguez, E., et al. 2006, MNRAS, 365, 327

Krzesiński, J., & Pigulski, A. 1997, A&A, 325, 987

Krzesiński, J., & Pigulski, A. 2000, ASPC, 203, 496

Pamyatnykh, A.A., Handler, G., & Dziembowski, W.A. 2004, MNRAS, 350, 1022

Pigulski, A., Handler, G., Michalska, G., et al. 2007, CoAst, 150, 191

Saesen, S., Pigulski, A., Carrier, F., et al. 2008, JPhCS, in press

Scuflaire, R., Théado, S., Montalbán, J., et al. 2008, Ap&SS, 316, 83

Stetson, P.B. 1987, PASP, 99, 191

Tamuz, O., Mazeh, T., & Zucker, S. 2005, MNRAS, 356, 1466

Waelkens, C., Lampens, P., Heynderickx, D., et al. 1990, A&AS, 83, 11

Comm. in Asteroseismology
Vol. 158, 2009, 38th LIAC/HELAS-ESTA/BAG, 2008
A. Noels, C. Aerts, J. Montalbán, A. Miglio and M. Briquet., eds.

More on pulsating B-type stars in the Magellanic Clouds

P.D. Diago[1], J. Gutiérrez-Soto[1,2,3], J. Fabregat[1,2], C. Martayan[2,4]

[1] Observatori Astronòmic de la Universitat de València, Ed. Instituts d'Investigació,
Polígon La Coma, 46980 Paterna, València, Spain
[2] GEPI, Observatoire de Paris, CNRS, Université Paris Diderot,
Place Jules Janssen 92195 Meudon Cedex, France
[3] LESIA, Observatoire de Paris, CNRS Université Paris Diderot,
Place Jules Janssen 92195 Meudon Cedex, France
[4] Royal Observatory of Belgium, 3 Avenue Circulaire, B-1180 Brussels, Belgium

Abstract

We present here the results of our research for B-type pulsators in low metallicity environments, searching for short-term periodic variability in a large sample of B and Be stars in the Magellanic Clouds (MC), for which the fundamental astrophysical parameters were accurately determined. A significant number of β Cephei and SPB-like pulsators at low-metallicity have been detected, conflicting with the current theoretical models of pulsation. In addition, we have placed these pulsating stars in the HR diagram mapping the observational instability regions for the MC metallicities. The large sample of B and Be stars analysed allows us to perform a reliable statistical analysis on the pulsating B-type stars in the MC. Finally, we have made a comparison between pulsational theory and observations in low metallicity environments.

Introduction

A significant fraction of main-sequence B-type stars are variable. The whole main-sequence in the B spectral domain is populated by two classes of pulsators: the β Cephei stars and the Slowly Pulsating Stars (SPB). Pulsations in β Cephei and SPB stars are due to the κ-mechanism activated by the metal opacity bump . β Cephei stars do pulsate in low-order p- and g-modes with periods similar to the fundamental radial mode. SPB stars are high-radial order g-mode pulsators with periods longer than the fundamental radial one. Pamyatnykh (1999) showed that the β Cephei and SPB instability strips practically vanish at $Z < 0.01$ and $Z < 0.006$, respectively. The metallicity of the Magellanic Clouds (MC) has been measured to be around $Z = 0.002$ for the Small Magellanic Cloud (SMC) and $Z = 0.007$ for the Large Magellanic Cloud (LMC) (see Maeder et al. 1999 and references therein). Therefore, it is expected to find a very low occurrence of β Cephei and SPB pulsators in the LMC and no pulsating B-type stars in the SMC. Recently, new B-type pulsators have been found in low-metallicity environments (e.g. Karoff et al. 2008 or Diago et al. 2008), suggesting that pulsations are still driven by the κ-mechanism even in low metallicity environments.

The new models provided by Miglio et al. (2007a,b) based on OPAL and updated OP opacities have shown that the blue border of the SPB instability strip is displaced at higher effective temperatures at solar metallicity, and that a SPB instability strip exists at metallicities as low as $Z = 0.005$. Their calculations however, do not predict β Cephei pulsations at $Z = 0.005$.

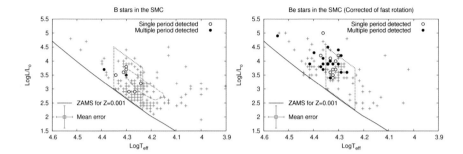

Figure 1: Location of the B (left) and Be (right) samples of the SMC in the the HR diagram: single crosses represent stars in our sample, the empty circles represent single period detection and the filled ones multiple period detection. The dashed line delimits the suggested SPB instability strip for the SMC. In the left panel we have depicted with dash-point-dashed line the SPB instability strip computed by Miglio et al. (2007b) at $Z = 0.005$ with OP opacities.

Another large class of stars populating the B-type main-sequence are the Be stars. They are defined as non-supergiant B stars whose spectrum has displayed at least once emission lines mainly in the Balmer series. Emission lines come from a circumstellar disk created by episodic matter ejections from the central star. In the Milky Way (MW), they show short-term variations like β Cephei or SPB stars. Be stars are also known to be fast rotating stars.

Here we present the analysis of a sample of 128 B and Be stars from the LMC and 313 B and Be stars from SMC, for which Martayan et al. (2006, 2007) provided accurate fundamental astrophysical parameters.

Results

The search for periodic variability has been done by analysing the photometric time series provided by the MACHO project. The frequency analysis was performed with the self-developed code pasper. The criterion used to determine whether the frequencies are statistically significant is the *signal-to-noise amplitude ratio requirement* described in Breger et al. (1993). Moreover, once we had the significant frequencies, we performed a visual inspection of the phase diagrams for each star folded with the detected frequencies. All the details concerning data and methods used can be found in Diago et al. (2008).

Many of the short-period variables have been found multiperiodic and some of them show a beating phenomenon due to the presence of close frequencies. This effect is a signature of nonradial pulsations.

Small Magellanic Cloud

In the left panel of Fig. 1 we show the position in the HR diagram for the 9 short-period variable B stars found. All pulsating B stars are restricted to a narrow range of temperatures. Moreover, all stars but one have periods longer than 0.5 days, characteristic of SPB stars. Thus, we suggest an observational SPB instability strip for the SMC metallicity, that it is shifted towards higher temperatures than for the MW. We propose the hottest pulsating star in our sample to be a β Cephei variable, because it has two close periods in the range of p-mode Galactic pulsators. If it is indeed a β Cephei star, this would be an unexpected result, as the current stellar models do not predict p-mode pulsations at the SMC metallicities (see Miglio et al. 2007a,b).

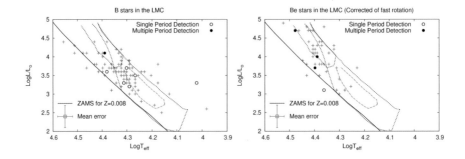

Figure 2: Location of the B (left) and Be (right) samples of the LMC in the the HR diagram: single crosses represent stars in our sample, the empty circles represent single period detection and the filled ones multiple period detection. The dashed line delimits the β Cephei and SPB instability strips at $Z = 0.01$ and the dash-point-dashed line the SPB instability strip at $Z = 0.005$ computed by Miglio et al. (2007b) with OP opacities.

A puzzling circumstance regarding our sample of SPB stars is that only one has been detected as multiperiodic. So, the variability in some of these stars is perhaps not caused by pulsations, but by other phenomena like eclipsing binaries or ellipsoidal binarity. However, none of these stars have been detected as binary in the spectroscopic survey of Martayan et al. (2007). Thus, we consider our figure of eight stars as an upper limit to the number of bona-fide SPB stars in our sample.

We have represented in the right panel of Fig. 1 the 32 pulsating Be stars in the HR diagram (using $\Omega/\Omega_c = 95\%$). In the case of Be stars it is more difficult to distinguish the pulsational type by using the detected frequencies, because the fast rotation effects affect the observed periods. We have included the suggested SPB region as described above in the figure and since most of the Be stars are located inside or very close to this region, this suggests that they are g-mode SPB-like pulsators. Three stars are significantly outside the strip towards higher temperatures, all of them multi-periodic, with periods lower than 0.3 days. Therefore, we propose that these stars are probably β Cephei-like pulsators.

The detected frequencies with their amplitudes and phases, the phase diagrams, beating of close frequencies and the detailed discussion for the SMC results are published in Diago et al. (2008).

Large Magellanic Cloud

In the LMC we have found 7 short-period variables among the B star sample (one multi-periodic). Their positions are displayed in the HR diagram in the left panel of Fig. 2. The periods obtained for these stars are typically of Galactic β Cephei stars except one. Concerning Be stars, we found 4 short-period variables (3 multiperiodic), depicted in the right panel of Fig. 2 (using $\Omega/\Omega_c = 85\%$). The periods obtained for these pulsating Be stars are compatible with Galactic β Cephei, but as in the SMC, the high rotational rates prevent us from distinguishing the pulsational type using the observed periods. It is remarkable that, as in the SMC, the hottest stars are multiperiodic. Our work is ongoing and it will be published by Diago et al., in preparation.

Table 1: Percentages of short-period variables in the MC and in the MW compared with their rotational velocity rates and the metallicity of the stellar environment.

	MW	LMC	SMC
Metallicity	0.020	0.007	0.002
Ω/Ω_c (B stars)	40%	37%	58%
Pulsating B stars	16%	6.9%	4.9%
Ω/Ω_c (Be stars)	88%	85%	95%
Pulsating Be stars	74%	15%	25%

Discussion

As mentioned in the introduction, one expects a lower occurrence of B-type pulsators in the low metallicity environments of the MC. On the other hand, according to Maeder & Meynet (2001), high rotational velocities favour the metal enrichment of the surface of fast rotating stars due to the rotational mixing. In this way, this metal enrichment could feed the pulsational mechanism on the fast rotating stars. Moreover, Zorec et al. (2005) and Frémat et al. (2005) have deduced from observed Be stars that these fast rotation effects appear in stars with a ratio of the angular rotational velocity with respect to the critical angular rotational velocity (Ω/Ω_c) of much larger than 60%.

Concerning our sample of B stars in the MC, the ratios of angular velocities are measured to be about $\Omega/\Omega_c = 37\%$ for the LMC and $\Omega/\Omega_c = 58\%$ for the SMC (see Martayan et al. 2007). These values are very close to the one for the Galactic B-type stars, which is about $\Omega/\Omega_c \sim 40\%$. In both cases, the ratios are lower than 60%. Therefore, we conclude that the B star samples of the LMC and SMC are not affected by fast rotation effects which can excite the pulsational mechanism. The B star samples are, consequently, only affected by the decreasing trend in metallicity that makes a lower fraction of observed B-type pulsators as we can see in our results (3rd row of Table 1).

In the case of Be stars, the decreasing trend in metallicity is also present (see last row of Table 1). However, Be stars are fast rotators and in addition they can rotate faster in lower metallicity environments, since the radiative winds are less efficient. The values of rotational velocity rates are about $\Omega/\Omega_c = 85\%$ for the LMC and $\Omega/\Omega_c = 95\%$ for the SMC (see Martayan et al. 2007). They are very close to the one for the Galactic Be stars, which ranges from 83% to 88%, depending on the authors. Consequently, the percentages of pulsating Be stars in our sample suggests that the fast rotation enhances the nonradial pulsations or amplifies the existing modes in the pulsating Be stars between MW/LMC and SMC. A similar result was obtained by Gutiérrez-Soto et al. (2007) for Be stars in the MW.

Conclusions

The most important result in our investigation is the detection of β Cephei and SPB-type pulsators in low metallicity environments, in contrast with the predictions of the current theoretical models. Pulsations in B stars seem to be damped by the decreasing trend of metallicity, so the lower the metallicity the lower the pulsations observed in these stars. For Be stars, the rapid rotation seems to enhance the presence of the nonradial pulsations or to amplify the existing modes, making them easier to be detected. As an alternative explanation, the prevalence of nonradial pulsations could be related to the yet unknown nature of the Be phenomenon, being in fact Z-enriched stars due to rotational mixing.

References

Breger, M., Stich, J., Garrido, R., et al. 1993, A&A, 271, 482

Diago, P. D., Gutiérrez-Soto, J., Fabregat, J., & Martayan, C. 2008, A&A, 480, 179

Frémat, Y., Zorec, J., Hubert, A.-M., & Floquet, M. 2005, A&A, 440, 305

Gutiérrez-Soto, J. , Fabregat, J., Suso, J., et al. 2007, A&A, 476, 927

Karoff, C., Arentoft, T., Glowienka, L., et al. 2008, MNRAS, 386, 1085

Maeder, A., Grebel, E. K., & Mermilliod, J.-C. 1999, A&A, 346, 459

Maeder, A., & Meynet, G. 2001, A&A, 373, 555

Martayan, C., Floquet, M., Hubert, A.-M., et al. 2006, A&A, 452, 273

Martayan, C., Floquet, M., Hubert, A.-M., et al. 2007, A&A, 462, 683

Miglio, A., Montalbán, J., & Dupret, M.-A. 2007, MNRAS, 375, L21

Miglio, A., Montalbán, J., & Dupret, M.-A. 2007, CoAst, 151, 48

Pamyatnykh, A. A. 1999, Acta Astron., 49, 119

Zorec, J., Frémat, Y., & Cidale, L. 2005, A&A, 441, 235

DISCUSSION

Karoff: I am relieved to see that you have made the same mistake as I did. The papers by Miglio et al. 2007a,b do not conclude that pulsation is not driven in SPB stars with $Z<0.005$. The papers only show that the instability strip for the SPB stars is really small at $Z=0.005$, but no models have been computed for $Z<0.005$. A. Miglio agreed on this comment.

Aerts: It is clear that we need spectroscopic data to better understand the pulsating B stars in the SMC and the new class that seems to emerge from the CoRot exoplanet data. But be aware that those stars are faint and that we have to take into account error bars for vsini and for the abundances.

Alecian: Atomic diffusion can enhance iron at the right depth in β Cep stars to explain observed pulsations. This enhancement will occur even if the initial metallicity is weak. The process of enhancement will take more time but atomic diffusion will try to reach the same Fe/H. Of course, atomic diffusion will compete with other processes and could hardly work for high rotational velocities (due to mixing).

Noels: I understand that there is an urgent call for detailed abundance determinations, especially the relative importance of Fe at low Z. But the problem could come from an underestimated opacity in the iron bump region. We need an increase of about 10 % only. When switching from Los Alamos to OPAL opacities, the increase was a factor 2 or 3 in the iron bump region.

Cantiello: It is true. Los Alamos opacities did lack the correct inclusion of spin-orbit coupling in the atomic calculations. This effect is important for high atomic number, hence resulted in an increase of the iron bump in the OPAL opacities (Iglesias, Rogers & Wilson 1992). The next order approximation, from an atomic physics point of view, would be the inclusion of spin-spin interaction. It would be interesting to ask experts in atomic physics if the inclusion of this effect could be important in this respect.

Baglin: Can you give statistical significances of the percentages you present? Are they 10% or 50%?

Diago: The samples analysed consist of 313 stars in the SMC and 128 stars in the LMC, so the size of both samples is large enough to consider the differences in Table 1 to be significant, although we have not performed formal calculations of the statistical significance.

Comm. in Asteroseismology
Vol. 158, 2009, 38th LIAC/HELAS-ESTA/BAG, 2008
A. Noels, C. Aerts, J. Montalbán, A. Miglio and M. Briquet., eds.

Instability domains of δ Scuti and Slowly Pulsating B stars : How will the CoRoT satellite help to determine the limits ?

L. Lefèvre[1], E. Michel[1], C. Aerts[2], A. Kaiser[3], C. Neiner[4], E. Poretti[5], R. Garrido[6], A. Baglin[1], M. Auvergne[1], C. Catala[1], W. Weiss[3], L. Balaguer-Nùñez[7], C. Maceroni[8], I. Ribas[9].

[1] LESIA, Observatoire de Paris, CNRS, Université Paris Diderot; 5 place Jules Janssen, 92190 Meudon, France
[2] Katholieke Universiteit Leuven, Inst. voor Sterrenkunde, B-3001 Leuven, Belgium
[3] Institut für Astronomie, Türkenschanzstrasse 17, 1180 Vienna, Austria
[4] GEPI, Observatoire de Paris, CNRS, Université Paris Diderot; 5 place Jules Janssen, 92190 Meudon, France
[5] INAF-Osservatorio Astron. di Brera, Via E. Bianchi 46,I-23807 Merate (LC), Italy
[6] Instituto de Astrofsica de Andaluca-CSIC, Apdo 3004, 18080 Granada, Spain
[7] Dept d'Astronomia i Meteorologia, Univ. de Barcelona, Av. Diagonal, 647, 08028 Barcelona, Spain
[8] INAF-Osservatorio Astr. di Roma, via Frascati 33, I-00040 Monteporzio C. Italy.
[9] IEEC, CSIC, Campus UAB, Fac. de Cincies, Torre C5, E-08193 Bellaterra, Spain.

Abstract

This work is intended to illustrate the possibilities offered by the CoRoT satellite observations to study the different instability strips, and through them, physical processes and specific features of stellar interiors.

The CoRoT space mission (Baglin A. et al. 2002), launched on December 27th 2006, has been developped and is operated by CNES, with the contribution of Austria, Belgium, Brasil, ESA, Germany and Spain. It enables us to observe oscillations from stars down to a noise level of less than a ppm, much lower than the limit usually obtained from the ground. During the nominal duration of the mission, about 6 long runs ($\sim 150d$ each) and 6 short runs ($\sim 20d$ each) will take place (CoRoT Book, 2006). Only 2 long runs and 1 short run are illustrated in this study. This means that the number of available targets will have more than tripled by the end of the mission.

These data might help testing the "purity" of the instability strips (i.e. the presence/absence of photometrically constant stars within) and lead to the discovery of new classes of pulsating stars (Degroote et al 2008). We address this problem in the frame of the B and A main sequence stars.

Individual Objects: HD 50846, HD 175869, HD 181231, HD 49330, HD 50209, HD 174844, HD 180642, HD 181440, HD 182198, HD 50230, HD 50405, HD 50844, HD 174936, HD 174966, HD 174987, HD 175542, HD 181072, HD 181555, HD 49294

Observations and variability detection

The CoRoT mission has 2 main scientific programs: stellar seismology and search for extrasolar planets. Onboard the satellite are 4 CCDs dedicated to science. Two of them are optimized for seismology (seismofield) and permit the observation of 5 stars each during a run of the satellite. The other two are optimized for the exo-planet search (exofield) and can process up to 6000 stars each. The main difference is that the CCDs for the seismofield handle stars

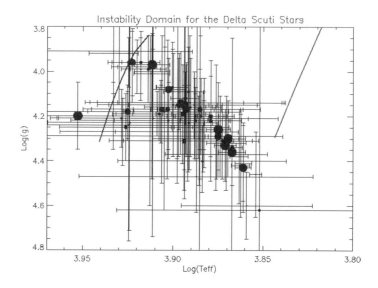

Figure 1: δ Scuti-like pulsators from the Initial Run of observations for the exofield (IR01). The diagram shows the theoretical red edge and blue edge (Dupret et al., 2004). The size of the symbols is scaled to the amplitude of the most prominent frequency and ranges from 300 to 30000 ppm (Kaiser, A., private Communication).

with magnitude $m_V \in [5.4, 9.5]$ with a 1s time-sampling, while those for the exoplanet-search are dedicated to stars with $m_V \in [10.5, 16]$ with a much lower time sampling (i.e. 512s). For the study at hand, both fields have their own importance as the seismofield offers data with a very low noise level while the large number of stars observed in the exofield allows statistical studies.

Figures 1 and 2 represent HR diagrams related to CoRoT exofield and seismofield observations respectively. In both figures the blue and red edges of the δ Scuti instability strips are plotted and figure 2 also shows the red edge of the SPB stars instability strip. Those edges are sensitive to different physical processes, which are developed below.

Figure 1 illustrates the above-mentioned great interest of the exofield for statistics. It features approximately 100 δ Scuti stars extracted from the observations of the Initial Run (IRa01) in the exofield.

Figure 2 summarizes the results of the variability analysis we conducted on 19 A and B stars observed in the CoRoT seismofield during the first runs of observations. The point size is proportional to the amplitude of the star's observed variations. Known eclipsing binaries have been plotted with empty squares, and CoRoT data, in addition to ground-based complementary observations, will enable us to improve the quality of the binarity parameters as well as discriminate between binarity-related variations and possible pulsations.

Results and Discussion

We describe here the main characteristics of the classes of pulsating stars studied here and the associated results. Table 1 summarizes the parameters and results for these stars, divided into categories each associated with their typical frequency domains.

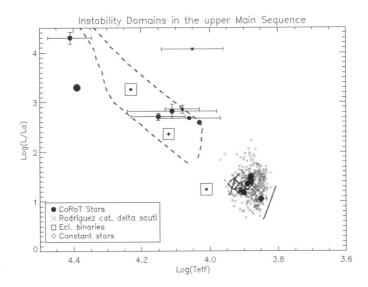

Figure 2: HR Diagram representing all the targets of interest observed from IR01 to LRa01 for the seismofield. The parameters were retrieved from various sources, i.e. Frémat et al. 2006, Poretti et al. 2007, Morel & Aerts 2007 and Miglio et al. (in preparation). The dashed lines represent the blue and red sides of the instability strip for SPB stars (Miglio et al., 2007a, 2007b) and the solid lines represent the blue and red edges of the δ Scuti strip (Dupret et al., 2004). The size of the symbol represents the amplitude of the variability of the stars. Squares represent eclipsing binary stars and diamonds are overplotted on "non-variable" stars.

B stars: SPB stars and Be stars

Slowly pulsating B stars are variable mid-B-type (B3-B8) with periods in the range of 0.5 to 5 days ($\sim$ 3-20 μHz) and g-mode pulsations are reported to be the cause of their variability. Be stars are still on the main sequence or close to it, rapidly rotating, and surrounded by a disk. Early types are close to the β Cephei part of the HR diagram, while later types have pulsational characteristics similar to that of SPB stars. The red edge of the SPB instability strip (IS) is essentially sensitive to the abundance of iron-group elements (Dziembowski et al. 1993, Miglio et al., 2007a) and while the accumulation of iron modifies it as shown in Miglio et al (2007b), a quantitative physical justification remains to be found.

All the studied stars in these classes clearly show variability in the expected frequency range. Among them are low amplitude (less than 100 ppm) pulsators that could be detected thanks to the low noise-level (estimated in the power spectrum in areas *free of signal* close to the frequency range of interest) of the seismology field (see Table 1). Note the presence of a β Cephei star in this sample (HD180642/B1.5II-III), which is also clearly pulsating in its frequency domain (up to a few 100 μHz). However, its position in figure 2 should not be trusted yet due to a log(L/L$_\odot$) value which is clearly underestimated.

A stars: δ Scuti stars

δ Scuti stars have masses between 1.5 and 2.5 M$_\odot$ and usually pulsate with periods of a few hours ($f = 50$ μHz to 600 μHz). The κ mechanism, associated with the opacity bump

Table 1: Summary of results for the seismology channel

T_{tot} (days)	HD	Sp.T. CoRoT	Log(T_{eff})	log($L/L_\odot$)	var?	Signal (ppm)	Noise (ppm)	Vsini (km.s^{-1})
Range: approx. 3 μHz to 20 μHz - Be stars								
52(ec)	50846	B5e+F0	4.23	3.26	y			135
22	175869	B8IIIe	4.08	2.86	y	100	1	140
157	181231	B5IVe	4.14	2.71	y	1400	1	169
62	49330	B0.5IVe	4.41	4.3	y	1000	1	210
62	50209	B8IVe	4.11	2.82	y	1500	1	200
Range: approx. 3 μHz to 20 μHz - B stars								
27(ec)	174844	B8	4.12	2.36	?			114.6
157	180642	B1.5II-III	4.39	3.30	y	36800	1	44
157	181440	B9III	4.05	4.07	y	70	0.5	60
96	182198	B9V	4.06	2.68	y	400	2	25
62	50230	B3	4.03	2.59	y	1000	1	23
Range: approx. 50 μHz to 600 μHz - A stars								
52(ec)	50405	A0	4.01	1.24	?			
52	50844	A2	3.88	1.4	y	15000	1	64.2
16	174936	A2	3.9	1.17	y	2000	5	169.7
27	174966	A3	3.88	1.15	y	6000	5	125.1
27(ec)	174987	A2	3.89	1.35	n ?	340	5	140
16	175542	A0	3.93	1.35	n	5	5	80
118	181072	A2	3.91	1.24	n	5	1	52.8
157	181555	A5	3.85	1.05	y	3000	1	200
62	49294	A2	3.91	1.21	y	500	1	110

of HeII, is responsible for the variability of those stars. The red edge of the δ Scuti IS can be attributed to the coupling between oscillations and convection. The position of the blue edge, however, is dependent on the abundance of Helium inside the star and the position of the resulting opacity bump.

Eight stars correspond to this class and most of them did show photometric variability. However, this small sample also revealed a few stars with no identified variability down to the 1 ppm level within the δ Scuti theoretical IS. They are plotted as diamonds in the HR diagram of Fig. 2 and ongoing research will allow better determination of their parameters. This will enable us to confirm their positions relative to the blue border of the IS. Indeed, firm values of T_{eff} and M_V, along with determination of their *vsini* and chemical abundances will help answering the questions about the occurrence of variability in the IS and physical parameters ruling it.

Conclusions

This preliminary work is intended to stress the potential of the CoRoT satellite to probe the existing limits between the different types of excitations and variations. To this purpose, the exoplanet field is of utter importance as it contains the greater numbers of stars. The seismology field, however, with very low detection limits and precise individual stellar parameters will bring valuable complementary information.

On one hand, our results show that all B-stars considered here are found variable. At this stage, variability due to e.g. ellipsoidal distortion cannot be rejected, but if confirmed, these results would suggest that the pulsation mechanisms in these stars can be apprehended with a limited amount of parameters.

On the other hand, we found A-stars that are constant down to the *ppm* level. This will help assessing which parameters, and beyond them which physical processes, are needed to understand the pulsational instability in this domain of the HR diagram.

Acknowledgments. AK and WW were supported by the Austrian Research Promotion Agency (FFG-ARL).EP acknowledges financial support from the ASI Project ASI/INAF/ I/015/07/0, WP 03170.

References

Baglin, A. et al. 2002, ESA-SP, 485, 17

CoRoT Book, 2006, ESA-SP-1306, 538P

Degroote, P., Miglio, A., Debosscher, J., et al. 2009, CoAst, 158, 167

Dupret, M.-A., Grigahcene, A., Garrido, R., et al. 2004, A&A, 414, L17

Dziembowski, W. A., Moskalik, P., & Pamyatnykh, A. A. 1993, MNRAS, 265, 588

Frémat, Y., Neiner, C., et al. 2006, A&A, 451, 1053

Miglio, A., Montalbán, J., & Dupret, M.-A. 2007a, CoAst, 151, 48

Miglio, A., Bourge, P.-O., Montalbán, J., & Dupret, M.-A. 2007b, CoAst, 150, 209

Morel, T., & Aerts, C. 2007, CoAst, 150, 201

Poretti, E., Rainer, M., Uytterhoeven, K., et al. 2007, MemSAIt, 78, 624

Comm. in Asteroseismology
Vol. 158, 2009, 38th LIAC/HELAS-ESTA/BAG, 2008
A. Noels, C. Aerts, J. Montalbán, A. Miglio and M. Briquet., eds.

The pulsations of Be stars

C. Neiner[1], A.-M. Hubert[1]

[1] GEPI, Observatoire de Paris, CNRS, Université Paris Diderot; 5 place Jules Janssen, 92190 Meudon, France

Individual Objects: ω Ori, μ Cen, v Cyg, ζ Oph

Be stars and their properties

Be stars are main sequence or slightly evolved non-supergiant, late-O, B, or early-A stars which show or have shown at least once emission in their Balmer lines due to the presence of a decretion circumstellar disk. Emission can also appear in other lines of the spectrum, in particular in red HeI lines or FeII lines. Moreover, the disk produces an infrared excess in the spectral energy distribution. About 20% of all B stars are Be stars in our galaxy, however this fraction depends on metallicity (Martayan et al. 2006, 2007). Therefore, in other environments with lower metallicity, the fraction of Be stars over B stars can be much higher. Be stars are known to be fast rotators (typically $v\sin i$=250 km.s^{-1}), nevertheless this velocity is not high enough to reach the critical limit at which the centrifugal force compensates gravity. Indeed, Be stars rotate on average at 88% of the critical angular velocity in our galaxy (Frémat et al. 2005). Thus, rotation by itself cannot explain the ejection of matter from the star which leads to the formation of the decretion disk. In addition to this disk, Be stars also exhibit polar winds.

Be stars vary on all timescales: rapid variations of the order of days are due to the rotation, pulsations, magnetic fields and wind. Variations of the order of months to decades are due to the changes in the structure of the circumstellar disk such as one-armed oscillation of the disk and sudden outbursts filling the disk with new material.

Since the discovery of Be stars over 140 years ago (Secchi 1867), the ejections of stellar material into a circumstellar disk, called the Be phenomenon, have remained a puzzle. Since rotation is not enough to eject this matter , another mechanism is needed to provide additional angular momentum. Several explanations have been proposed:

(i) Be stars could be binaries and the ejections triggered by tidal effects. If this statement is certainly true for some Be stars (e.g. KX And, Stefl 1987), no sign for binary interaction has been found in most of the studied Be stars and therefore this cannot be the explanation for all classical Be stars (see Plavec 1976).

(ii) A magnetic field could force matter to follow magnetic field lines. In the case of a simple dipole for example, matter would leave the star from the two magnetic poles and follow the field lines towards the magnetic equator. The particles coming from both hemispheres would then collide at the magnetic equator where they could stay confined if the magnetic field is strong enough. Although a possible detection of a magnetic field has been obtained for the Be star ω Ori (Neiner et al. 2003), current studies show that magnetic fields in Be stars are rather rare and weak. Therefore, even though this mechanism could explain the presence of a corotating disk for some Be stars, this currently does not seem to be the most common way of forming Be stars. Moreover, most disks around Be stars seem to be Keplerian rather than corotating (Meilland et al. 2006).

(iii) The beating of pulsation modes in Be stars could provide some angular momentum and,

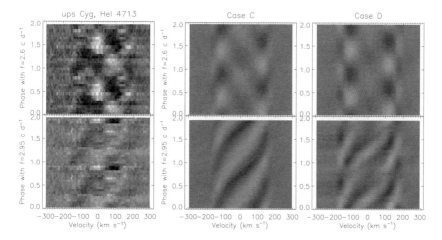

Figure 1: Observed (left) and modeled (right) with $\Omega/\Omega_c=0.8$ (case C) and 0.9 (case D) greyscale representations of the Hel 4713 line-profile variations of the Be star υ Cyg. Taken from Neiner et al. 2005.

in combination with the fast rotation, allow to reach the critical limit at which ejections can occur. Many Be stars are indeed known to pulsate (e.g. Gutiérrez-Soto et al. 2007) and Rivinius et al. (1998b) showed thanks to a spectroscopic study that the beating of pulsation modes in the Be star μ Cen seems to coincide with the ejections of matter from that star. This result, however, could not be obtained for other Be stars from the ground, probably due to the difficulty to obtain high-accuracy, long-duration, multisite spectroscopic observations. Considering the amount of pulsating Be stars already discovered, including several multimode ones (e.g. Floquet et al. 2002), this explanation of the Be phenomenon nevertheless seems the most plausible as of today.

The pulsations of Be stars and the Be phenomenon

Be stars are located in the same part of the HR diagram as β Cephei and Slowly Pulsating B (SPB) stars. It is thus not a surprise that the κ-mechanism ignates p- and/or g-mode pulsations in Be stars like in these two types of pulsators. Periods of 0.3 to 2 days have been measured in most of the studied early Be stars but in only 30% of later ones (Hubert & Floquet 1998). This could be due to the lower amplitudes of the modes expected for late type Be stars, which can be below the amplitude detectable from the ground. Amplitudes measured from the ground vary from 0.01 to 0.1 magnitude (e.g. Percy 1987). Note also that the amplitude of the spectral line profile variations decreases with increasing $v\sin i$, which makes the detection of pulsations through spectroscopy more difficult for faster rotator and higher inclination angle.

The spectroscopic study of Be stars provided various results regarding short-term periodicities depending on the studied star and used analysis technique. For example the study of μ Cen by Rivinius et al. (1998a) resulted in the detection of two groups of periods (P_1 to $P_4 \sim 0.505$ d, and $P_5 - P_6 \sim 0.280$ d) identified as retrograde g-modes with $l = |m| = 2$. This mode seems to be typical of many Be stars (Townsend 1997) although other modes are also observed. For example Neiner et al. (2005) detected a l=3, m=0 mode with P=0.38 d and a l=3, m=3 mode with P=0.33 d in the star υ Cyg (see Fig. 1). The rotation period of the star is sometimes also detected.

The results obtained from ground-based photometric studies of Be stars, however, are mostly interpreted in terms of the rotation period (e.g. Balona & James 2002) i.e. of spots at the stellar surface, but pulsation periods are also sometimes detected and identified thanks to amplitude ratio measured in various colors (e.g. Gutiérrez-Soto et al. 2007).

Recent results obtained by the MOST satellite for a few Be stars showed that many pulsation frequencies are detected in Be stars from space photometry. For example, a dozen of frequencies between 0.1 and 1 day have been detected in ζ Oph by Walker et al. (2005) including 6 frequencies also detected from the ground. The frequency peaks detected by MOST in several Be stars are mostly found in two groups. These two groups of frequencies have been modeled as prograde g-modes with $|m|=1$ and 2 (e.g. Cameron et al. 2008) separated by the rotation frequency, thanks to the models developed by Saio & Lee.

The various results (pulsation versus rotation, prograde versus retrograde modes) obtained from spectroscopy, ground-based photometry and space photometry are rather difficult to reconcile. To understand the short-term variations of Be stars and the possible role of pulsations in the Be phenomenon, it is necessary to study these variations in more details.

What does CoRoT bring to Be stars studies?

The CoRoT satellite provides us with unprecedented accuracy, long duration, highly sampled photometric data of Be stars. On average 1 bright Be star is observed in each seismology field of CoRoT in the core program, while a few tens of fainter Be stars are observed in each exoplanet field of CoRoT as part of the additional program. As of today 5 bright Be stars have already been observed and tens of fainter ones. CoRoT observing runs last from a few weeks (short runs) to 5 months (long runs). In the seismology field, the time sampling corresponds to one measurement every 32 seconds, while in the exoplanet one measurement is obtained every 512 seconds. CoRoT data of Be stars will thus allow us to detect many frequencies with a precision in amplitude down to 10^{-6} mag, to disentangle very close frequencies ($\Delta f = 0.006$ $c.d^{-1}$) which could produce beatings, and to detect low-amplitude modes ($> 10^{-6}$ mag).

The identification of pulsation modes and the disentangling of pulsations from rotation will benefit from ground-based spectroscopy as well as from modeling including the effects of rapid rotation. In particular we will be able to test whether the interpretation of the results obtained with the MOST satellite still holds with the more accurate CoRoT data. Our final goal is to study the link between pulsation beating and the Be phenomenon to decide if pulsations play a triggering role in the ejection of material from Be stars and the formation of the circumstellar disk. Preliminary results can be found in Neiner et al.(2009).

Conclusions

While it is clear that most, if not all, Be stars pulsate with p- and/or g-modes, the identification of these modes remains difficult due to the rapid rotation of these stars. The high-precision, long duration CoRoT data, in combination with state-of-the-art modeling will allow us to study these pulsations in more detail. In particular we will be able to test whether the beating of pulsation modes provides the additional angular momentum needed to eject material from the star and create a circumstellar disk. If pulsations are indeed the trigger of ejections, the long-lasting mystery of the Be phenomenon will finally be solved. Otherwise other explanations will have to be considered such as the presence of weak magnetic fields.

References

Balona, L.A., & James, D.J. 2002, MNRAS, 332, 714

Cameron, C., Saio, H., Kuschnig, R., et al. 2008, ApJ, 685, 489

Floquet, M., Neiner, C., Janot-Pacheco, E., et al. 2002, A&A, 394, 137

Frémat, Y., Zorec, J., Hubert, A.-M., & Floquet, M. 2005, A&A, 440, 305

Gutiérrez-Soto, J., Fabregat, J., Suso, J., et al. 2007, A&A, 472, 565

Hubert, A.-M. & Floquet, M. 1998, A&A, 335, 565

Martayan, C., Frémat, Y., Hubert, A.-M., et al. 2006, A&A, 452, 273

Martayan, C., Frémat, Y., Hubert, A.-M., et al. 2007, A&A, 462, 683

Meilland, A., Stee, P., Zorec, J., & Kanaan, S. 2006, A&A, 455, 953

Neiner, C., Gutiérrez-Soto, J., Floquet, M., et al. 2009, CoAst, 158, 319

Neiner, C., Floquet, M., Hubert, A.-M., et al. 2005, A&A, 437, 257

Neiner, C., Hubert, A.-M., Frémat, Y., et al. 2003, A&A, 409, 275

Percy, J.R. 1987, IAU Coll. 92 "Physics of Be stars", eds Slettebak & Snow, p. 49

Plavec, M. 1976, IAUS 70 "Be and shell stars", ed Slettebak, p. 439

Rivinius, T., Baade, D., Stefl, S., et al. 1998b, ESO workshop on "Cyclical variability in stellar winds", eds L. Kaper & A.W. Fullerton, p. 207

Rivinius, T., Baade, D., Stefl, S., et al. 1998a, A&A, 336, 177

Secchi, A. 1867, AN, 68, 63

Stefl, S. 1987, IAUS 122 "Circumstellar matter", eds Appenzeller & Jordan, p. 483

Townsend, R.H.D. 1997, MNRAS, 284, 839

Walker, G.A.H., Kuschnig, R., Matthews, J.M., et al. 2005, ApJ, 623, L145

DISCUSSION

Rauw: With respect to the explanation of the spots on Be stars as being due to magnetic fields, can you say something about the location (latitude) of the spot with respect to the disk?

Neiner: Surface spots due to a magnetic field are located at or close to the magnetic poles. For example in He-strong stars, patches of helium are found close to the magnetic poles. However, most of the time the axis of the magnetic dipole is not aligned with the rotation axis. Therefore these spots appear at a different latitude with respect to the rotation equator depending on the star. In addition, note that when the disk is confined by the magnetic field, the corotating disk is at the magnetic equator, whereas in the case of a Keplerian disk it is located at the rotation equator.

Rauw: And what about the effect of the magnetic field on the disk-like wind?

Neiner: If the magnetic field is strong enough to confine the wind, particles coming from both magnetic hemispheres and escaping at the magnetic poles follow the field lines and collide at the magnetic equator thus forming the disk. If the magnetic field is too weak, however, or if rotation dominates, as it seems to be the case for Be stars, the disk confinement cannot occur but denser clouds of material can still exist at the intersections of the magnetic and rotation equators. (See Neiner et al. 2003.)

Karoff: Is it possible to have spots and outflows on Be stars without a magnetic field and how?

Neiner: Outflows are possible without magnetic field, for example due to tidal interactions with a companion or to the collaborative work of rapid rotation with the beating of pulsations.

De Ridder: Do you believe that all Be stars pulsate?

Neiner: At least it is clear that many Be stars pulsate, and the more precise data we get, the more pulsating Be stars we discover and the more modes we detect in each Be star. However, it is not necessary to assume that all Be stars are formed in the same way. The definition of a Be star is purely phenomenological. It is very well possible that some stars become Be stars because of their pulsations, some others because of their magnetic field or their binary companion.

Puls: Two related questions: Due to the rapid rotation of the Be stars, the star is strongly distorted and shows gravity darkening (von Zeipel). i.) In how far are the oscillations affected by this strong deviation from a spherical star? ii.) In how far are these 2D effects accounted for in the spectroscopic analysis?

Neiner: Rapid rotation concentrates pulsational activity towards the stellar equator and it affects in particular g modes. When combined with gravity darkening, this equatorial concentration produces different line-profile variations for lines formed at the equator and at the pole. We use a code based on Bruce (Townsend et al. 1997) to model the pulsations observed in line profiles. This code treats these effects. Using this code allows to reproduce very well the patterns observed in the lines (see Fig. 1). In addition, when we calculate synthetic spectra to determine fundamental parameters and abundances by fitting observed averaged spectra, we use the code Fastrot (Frémat et al. 2005) which accounts for stellar geometrical distortion and gravity darkening effects on the lines. We can then determine the apparent stellar parameters and the parameters of the non-rotating counterpart.

Comm. in Asteroseismology
Vol. 158, 2009, 38th LIAC/HELAS-ESTA/BAG, 2008
A. Noels, C. Aerts, J. Montalbán, A. Miglio and M. Briquet., eds.

Photometric campaign on massive stars in the open cluster NGC 5617

F. Carrier[1], S. Saesen[1], M. Cherix[2], G. Bourban[2], G. Burki[2], J. Debosscher[1], D. Debruyne[1], P. Gruyters[1], L. M. Sarro[3], M. Spano[2], and L. Weber[2]

[1] Instituut voor Sterrenkunde, Katholieke Universiteit Leuven, Belgium
[2] Observatoire de Genève, Université de Genève, Switzerland
[3] Dpt. De Inteligencia Artificial, UNED, Madrid, Spain

Abstract

A campaign on the open cluster NGC 5617 was organized in order to characterize the pulsations and to better understand the internal structure of its stars. The variability of the cluster members was never studied before. We present the results of the observations and an up-to-date analysis of the obtained time series, especially of several SPB candidates we discovered.

Individual Objects: NGC 5617

Introduction

Since 2003, we are monitoring a set of selected southern open clusters with the aim of detecting and characterizing their variable stars. This program is conducted at Euler, the 1.2-m Swiss telescope at La Silla Observatory, Chile, with the C2 CCD camera. The large time span of the observations, typically a few years, makes it possible to study long-period variables, while repeated observations during a given run allow us to detect short-term variations as well.

The majority of the clusters are still being observed now, and the already obtained data are being reduced and analyzed. In these proceedings we present the first results of this survey by showing light curves of massive stars, like SPBs, but also δ Scuti and γ Dor stars, as well as eclipsing or ellipsoidal binaries.

Observations and reductions

Over a period of 5.5 years, we took 4400 measurements in the V-band and 750 in each of the Geneva U- and B-bands. The fluxes were extracted using a personal revised version of DAOPHOT (Stetson 1987) using combined PSF and aperture photometry. Then the effects of airmass, atmospheric extinction, ..., were corrected using multi-differential photometry. The data presented here are the V-band observations and are still not detrended. Their accuracy currently reaches 2.5 mmag for the brightest non-variable stars, over a period of more than 5 years. We plan to use the software SysRem (Tamuz et al. 2005) to correct the remaining systematic effects.

Variable stars in NGC 5617

We found a total of 218 stars displaying variability in the V-band with periods shorter than 50 days. The stars with longer periods have to be taken with caution (possible instrumental drifts) and will be analyzed later. Our automated classification software (Debosscher et al. 2007) found amongst others about 35 SPB, 30 δ Scuti and 20 γ Dor candidates, and 40 eclipsing and 15 ellipsoidal binaries, but the method does not take spectral information into account (known thanks to the membership to the cluster). The variability of these stars is thus doubtless but the classification has to be verified. Hereafter we present some of the candidate pulsators.

Two SPB candidates

In the top panels of Fig. 1, we show the amplitude spectra in different stages of prewhitening of two SPB star candidates. For one of the candidates, seven significant frequencies, with values between $0.495\,d^{-1}$ and $0.784\,d^{-1}$, were detected. Peaks near $0.58\,d^{-1}$ could belong to a quintuplet $\ell = 2$. The amplitudes range from 16.3 to 4.0 mmag. For the other candidate SPB pulsator, three modes were found: $f_1 = 0.677\,d^{-1}$, $f_2 = 0.574\,d^{-1}$, $f_3 = 0.768\,d^{-1}$. The amplitudes are 6.3, 3.6 and 2.7 mmag respectively.

A δ Scuti and γ Dor candidate

In the middle panels of Fig. 1, we plot the amplitude spectra in different stages of prewhitening of a δ Scuti and a γ Dor candidate. We determined seven significant frequencies in this δ Scuti candidate, all between 17.995 and $35.228\,d^{-1}$. The amplitudes vary from 3.2 to 0.8 mmag. For the γ Dor candidate we could identify three frequencies: $f_1 = 1.746\,d^{-1}$, $f_2 = 1.912\,d^{-1}$, $f_3 = 1.710\,d^{-1}$. Their amplitudes measure 14.0, 5.1 and 3.3 mmag.

Some eclipsing binaries

Several eclipsing binaries were identified in NGC 5617. In the lower panels of Fig. 1, we show the phase plots for two of them: one with a short period of P=0.296d and another one with a longer period of P=2.178d.

Future work

The first variability results in NGC 5617 encourage a more in-depth analysis of the observations. We will search in detail for oscillating stars and try to identify their main modes by means of the multicolour photometry. With the absolute photometry we have at our disposal, we will select some interesting oscillating cluster members and finally attempt to model them.

Acknowledgments.   F. Carrier is a Postdoctoral Fellow and S. Saesen is an Aspirant Fellow of the Fund for Scientific Research, Flanders (FWO). Part of this work was financially supported by the Swiss National Science Foundation.

References

Debosscher, J., Sarro, L. M., Aerts, C. et al. 2007, A&A, 475,1159

Stetson, P. B. 1987, PASP, 99, 191

Tamuz, O., Mazeh, T., & Zucker, S. 2005, MNRAS, 356, 1466

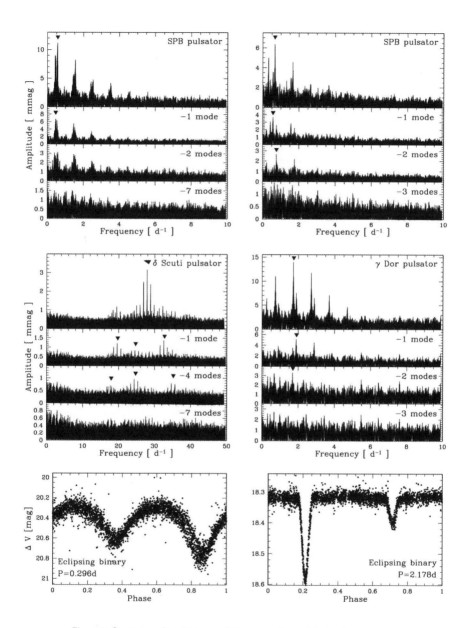

Figure 1: Some example pulsator candidates and eclipsing binaries in NGC 5617.

Comm. in Asteroseismology
Vol. 158, 2009, 38th LIAC/HELAS-ESTA/BAG, 2008
A. Noels, C. Aerts, J. Montalbán, A. Miglio and M. Briquet., eds.

Evidence for line-profile variability in the spectrum of the O supergiant HD 152249: preliminary results.

E. Gosset[1], H. Sana[2], N. Linder[1], and G. Rauw[1]

[1] Institut d'Astrophysique et de Géophysique,
Université de Liège, Allée du 6 Août 17, B-4000 Liège, Belgium
[2] European Southern Observatory, Alonso de Cordova 3107, Vitacura, Santiago 19, Chile

Abstract

Already suspected to be variable, the O9Ib((f)) supergiant HD 152249 has been the subject of a dedicated follow-up spectroscopic run. We report here on the preliminary results. This star is definitely exhibiting significant line-profile variations which are most probably a sign of the existence of non-radial pulsations. HD 152249 could thus belong to the newly identified group of pulsating OB supergiants.

Individual Objects: HD 152249

Introduction

As part of our extensive study of very massive stars and of very young open clusters, we acquired series of spectra in order to systematically classify O stars and to search for signs of binarity and/or variability among these stars.

As part of the survey of NGC 6231, we detected a few presumably single stars displaying variations suggesting an intrinsic origin (see e.g. Sana et al. 2008 and references therein). Some of these stars, our best candidates, were reobserved during a dedicated campaign. This is particularly the case for the supergiant star HD 152249 (see also Sana & Gosset, 2009). We report here on preliminary results of four consecutive nights of spectroscopic observations.

Observations and physical parameters

The star HD 152249 was one of the main targets of the four-day run that took place in June 2006. The telescope used was the ESO/MPG 2.2m telescope at La Silla equipped with the high-resolution FEROS spectrograph (resolving power 48000). Several spectra ($\sim$10) were acquired each night with a S/N ratio of 250-300. The data were reduced in a classical way using the MIDAS software as well as our own codes. The spectra were processed using an improved version of the FEROS pipeline, and were then normalized to the observed continuum. The best spectra were selected and an average spectrum was elaborated.

From the measurements of the equivalent widths of the He Iλ4471 and He IIλ4542 lines, we confirm the O9 spectral type. From the general appearance of the spectrum and from the ratio of the Si IVλ4089 and of the He Iλ4144 lines, we favour a luminosity class I. This is further confirmed by the measurements of the He Iλ4388 and He IIλ4686 lines. An emission in the N III$\lambda\lambda$4634-4641 lines implies an O9Ib((f)) type. Assuming an effective temperature of 30000 K (from the spectral type), we fitted synthetic spectra generated with the model atmosphere code TLUSTY (e.g. Hubeny & Lanz 1995) to the Hγ line profile in order to determine the $\log g$. We obtain a value around 3.0 whereas a similar fit to the He IIλ4200 line rather points out 3.2, further supporting the supergiant classification. This conclusion is again further supported by the presence of a marked P Cygni profile with strong emission for the doublet Si IV$\lambda\lambda$1394-1403 (see Fig. 4 of Walborn & Panek 1985). These are preliminary results; other models should be utilized because we found out that TLUSTY is unable to reproduce several lines in the global spectrum. An estimation of the rotational velocity on the basis of the Helium lines yields a value $v \sin i = 60$ to 85 km s^{-1}. Using the method of determination via the Fourier transform (Simón-Díaz & Herrero 2007) of the profile of O IIIλ5592 and C IVλ5801, we arrive at a value of 55-58 km s^{-1}. In any case, from the various line fits and from the derived errors, we consider 110 km s^{-1} as a strict upper limit.

Line-profile variability

In order to analyze the variability of the line profiles, we divided each individual spectrum by the average one. The variability of the He I lines is illustrated in Fig. 1 by the case of λ4471. It is immediately clear that the line profile exhibits well-marked transiting features (at the $\sim$1% level). These features travel over the line profile from blue to red: some 1.5 to 2 cycles are simultaneously visible. Along the time axis (ordinate), at a specific place in the line profile, the length of the cycle is about 0.3-0.4 days. These line-profile variations could not be due to the rotation of the star because it would necessitate either a large rotational velocity not compatible with the above derived $v \sin i$, or a very low inclination of the rotation axis that does not favour the visibility of variations due to surface inhomogeneities. The most probable hypothesis is that the line-profile variability is due to pulsations. Further detailed analyses are certainly necessary and will be conducted on all isolated lines. It could be that HD 152249 represents an extension of the pulsating B supergiant group, a new class of objects identified by Saio et al. (2006, see also Saio and Godart, 2009). In this case, this object is presently among the hottest representatives of the group. Identification of the true nature of the pulsation is awaiting further theoretical work on this particular object.

References

Godart, M., Dupret, M.-A., & Noels, A. 2009, CoAst, 158, 308

Hubeny, I., & Lanz, T. 1995, ApJ, 439, 875

Saio, H., Kuschnig, R., Gautschy, A., et al. 2006, ApJ, 650, 1111

Saio, H. 2009, CoAst, 158, 245

Sana, H., & Gosset, E. 2009, CoAst, 158, 205

Sana, H., Gosset, E., Nazé, Y., et al. 2008, MNRAS, 386, 447

Simón-Díaz, S., & Herrero, A. 2007, A&A, 468, 1063

Walborn, N.R., & Panek, R.J. 1985, ApJ, 291, 806

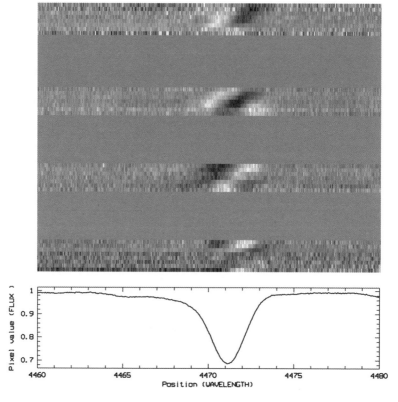

Figure 1: Variability of the He I λ4471 line in HD 152249. *Lower panel:* Spectrum as a function of wavelength; average line profile. *Upper panel:* Deviation from the average profile as a function of time. Time is linearly running from bottom to top and covers the four nights. The three internight gaps are clearly visible (uniform grey). Each horizontal line represents an individual deviation spectrum. The maximum deviations (black-white) amount to $\pm$ 1-2 %.

Comm. in Asteroseismology
Vol. 158, 2009, 38th LIAC/HELAS-ESTA/BAG, 2008
A. Noels, C. Aerts, J. Montalbán, A. Miglio and M. Briquet., eds.

Line profile variability in the massive binary HD 152219

H. Sana[1], and E. Gosset[2]

[1] European Southern Observatory, Alonso de Cordova 3107, Vitacura, Santiago 19, Chile
[2] Institut d'Astrophysique et de Géophysique, Université de Liège, Allée du 6 Août 17, B-4000 Liège, Belgium

Abstract

HD 152219 is a massive binary system with O9.5 III and B1-2 III/V components and a short orbital period of 4.2 d. In a previous work, we showed that the primary star ($M_{prim} \sim$ 21 $M_\odot$) was presenting clear line profile variabilities (LPVs) that might be caused by nonradial pulsations (NRPs). In the present work, we report on an intensive spectroscopic monitoring, that aimed at unveiling the nature of the detected LPVs. Based on this new data set, we discard the NRPs and point out the Rossiter-McLaughlin effect as being the cause of the observed LPVs. The upper limit derived on the amplitude of undetected NRPs, if any, is set at a couple of part per thousands of the continuum level.

Individual Objects: HD 152219

Introduction

With only a couple of objects known to present pulsations, asteroseismology of massive stars remains essentially a *terra incognita*. This partly results from the limited number of dedicated observational studies and from the difficulty to disentangle possible pulsations from wind effects or co-rotating structures. HD 152219 is an O9 III + B1-2 III/V short period binary ($P \sim 4.2$ d) in the core of the NGC 6231 cluster. In Sana et al. (2006, Paper I), we showed that the primary component was displaying clear line profile variability (LPV). With $\log T_{eff} = 4.504$ and $\log L/L_\odot = 5.07$, the primary star lays in the prolongation of the β Cep instability strip (e.g. Pamyatnykh 1999, Miglio et al. 2007), within the domain of instability predicted by Pamyatnykh (1999) for high-luminosity stars.

In the present paper, we briefly report on a 4-night intensive monitoring campaign (134 spectra with SNR $\sim$ 250-300) using the FEROS spectrograph at La Silla (Chile). As in Paper I, we will focus on the He II λ4686 line as it is the only well isolated, strong primary line that is uncontaminated by the secondary signature or by neighbouring lines. This strategy not only renders the analysis more straightforward, but also allows us to avoid any ill-quantified effects that are hampering spectral disentangling methods in the presence of varying spectral signatures (for an example, see Linder et al. 2008).

Analysis of the He IIλ4686 Line Profile Variability

Fig. 1 (a) displays the evolution of the He IIλ4686 line profile in the primary reference frame. The main LPV is observed at $\phi \sim 0.79 - 0.91$, which corresponds to the time of the primary eclipse, when the companion is passing in front of the primary (an effect known as the Rossiter-McLaughlin effect). Additional variations are seen during the secondary eclipse ($\phi \sim$

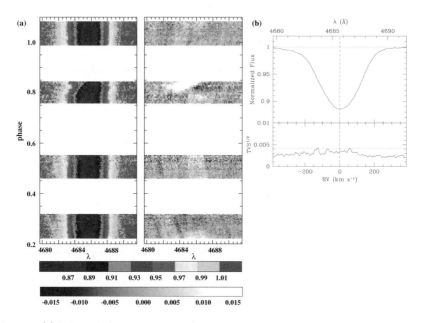

Figure 1: **(a)** *Left panel:* Color-coded image of the evolution of the He IIλ4686 profile with the phase. Spectra are displayed in the reference frame of the primary. *Right panel:* same as left panel for the difference spectra, i.e. the spectra minus the median spectrum computed over all non-eclipse phases. Below are the color codings for both panels **(b)** Averaged line profile (upper panel) and square root of the TVS (lower panel). On the upper panel, the dotted line shows the continuum level while the dashed line indicates the primary velocity frame. On the lower panel, the dotted line indicates the variability threshold corresponding to a significance level of 0.01.

0.23 − 0.37), resulting from the changing dilution of the line by the varying continuum. However, there seems to be little or no profile variations outside those phase intervals.

As a second, more quantitative step, we computed the Time Variance Spectrum (TVS, Fullerton et al. 1996) using only spectra obtained outside the eclipse phases. Fig. 1 (b) shows the averaged (outside eclipse) He II λ4686 line profile and the TVS spectrum. Using the formalism proposed by Fullerton et al. (1996), we tested our data against the null hypothesis '*not variable*'. Even adopting a rejection threshold corresponding to a loose significance level of 0.01 ($(\text{TVS})^{1/2}_{\alpha=0.01} = 3.9 \times 10^{-3}$), one cannot reject the null hypothesis. As a consequence, we consider the null hypothesis to be very likely and conclude that no significant variability is seen in the data (except for the eclipse effects).

Conclusion

From the analysis of the He II λ4686 line profile in the HD 152219 spectra collected over 4 consecutive nights, we cannot detect significant LPV beside the Rossiter-McLaughlin effect. The data displayed a SNR above 300 on average and allow us to place an upper limit for the variability level (thus for the amplitude of possible NRPs) as low as a few parts per thousand of the continuum level. The present result further suggests that the predictions for the location of the high-luminosity end of the β Cep instability strip might need to be refined.

Results from ongoing theoretical works are thus eagerly awaited. The present result however awaits confirmation by the LPV analysis of other He lines. This is left for future work.

References

Fullerton, A.W., Gies, D.R., & Bolton, C.T. 1996, ApJS, 103, 475

Linder, N., Rauw, G., Martins, F., et al. 2008, A&A, 489, 713

Miglio, A., Montalbán, J., & Dupret, M.-A. 2007, CoAst, 151, 48

Pamyatnykh, A.A. 1999, Acta Astronomica, 49, 119

Sana, H., Gosset, E., & Rauw, G. 2006, MNRAS, 371, 67 (Paper I)

Comm. in Asteroseismology
Vol. 158, 2009, 38th LIAC/HELAS-ESTA/BAG, 2008
A. Noels, C. Aerts, J. Montalbán, A. Miglio and M. Briquet., eds.

Low-amplitude variations detected by CoRoT in the late type Be star HD 175869

J. Gutiérrez-Soto[1,2,3], M. Floquet[2], C. Neiner[2], A.-M. Hubert[2], Y. Frémat[4], L. Andrade[5], B. de Batz[2], P.D. Diago[3], M. Emilio[6], J. Fabregat[2,3], W. Facanha[5], A.-L. Huat[2], E. Janot-Pacheco[5], B. Leroy[1], C. Martayan[2,4], J. Suso[3], R. Garrido[7]

[1] LESIA, Observatoire de Paris, CNRS, Université Paris Diderot; place Jules Janssen 92195 Meudon Cedex, France
[2] GEPI, Observatoire de Paris, CNRS, Université Paris Diderot; place Jules Janssen 92195 Meudon Cedex, France
[3] Observatori Astronòmic de la Universitat de València, Ed. Instituts d'Investigació, Polígon La Coma, 46980 Paterna, València, Spain
[4] Royal Observatory of Belgium, 3 Avenue Circulaire, B-1180 Brussels, Belgium
[5] Universidade de São Paulo, Instituto de Astronomia, Geofísica e Ciências Atmosféricas - IAG, Departamento de Astronomia, Rua do Matão 1226, 05508-900 São Paulo
[6] Departamento de Geociências, Universidade Estadual de Ponta Grossa, Av. General Carlos Cavalcanti, 4748 Ponta Grossa, PR 84.030-900, Brazil
[7] Instituto de Astrofísica de Andalucía, Granada, Spain

Abstract

We present the analysis of the CoRoT data of the B8IIIe star HD 175869 observed during a short run (27.3 days). The light curve shows low-amplitude variations of the order of 0.3 mmag. A frequency within the range determined for the rotational frequency and its 5 harmonics are detected. Other significant frequencies with a low amplitude of a few ppm are also found. The analysis of line profiles from ground-based spectroscopic data does not show any variation.

Individual Objects: HD 175869

Introduction

Be stars are non-supergiant B stars that show or have shown emission in Balmer lines, due to the presence of a circumstellar disk. Short-term variability is present in these stars due to nonradial pulsations or/and rotational modulation (see Neiner et al. 2009).

HD 175869 is a B8IIIe star of magnitude V=5.56 which has been observed by CoRoT in the seismo field during the first short run in the center direction (27.3 days, from 11 April to 8 May 2007). The fundamental parameters of this star have been determined from high-resolution spectra obtained during the preparation of the CoRoT mission and collected in the GAUDI database (Frémat et al. 2006): T_{eff}=12000 ± 600 K, log g = 3.43 ± 0.12, $v \sin i$=171 ± 10 km s^{-1}, i=47 ± 10 deg. Similar values within errors have been obtained with the new spectroscopic data (see below). The rotational frequency of the star is between 0.54 and 0.68 c d^{-1}. Note that the star is in the last stage of the main sequence.

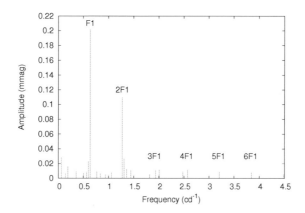

Figure 1: Power spectrum of the frequencies detected in the star with the main frequency at F1=0.63 c d^{-1}. The low-amplitude frequencies are probably due to the presence of nonradial pulsations.

Results

The CoRoT light curve shows variations with a peak to peak amplitude of 0.3 mmag. A frequency search provides 22 significant frequencies with semi-amplitudes from 0.2 to a few 0.001 mmag. The power spectrum is shown in Fig. 1. Most of the variations can be re-produced using the main frequency F1=0.63 c d^{-1} and its 5 first harmonics. However, the maxima and minima of the sinusoidal signal are modulated by a long-term trend and the contribution of the low-amplitude frequencies close to F1 and 2F1. The phase diagram with the frequency F1 clearly shows a double-wave, with different maxima and minima. A spectro-scopic study of the star has also been performed using 100 spectra obtained at the DDO and 18 NARVAL spectra. The line profiles do not show any coherent variation. The frequencies observed in the CoRoT data are not detected in spectroscopy either. In addition, the star does not show any clear Zeeman signature in the NARVAL spectro-polarimetry.

Discussion and conclusions

The high accuracy of the CoRoT data (several ppm) allows us to detect very low-amplitude frequencies in the studied Be star. In particular, a main frequency F1=0.63 c d^{-1} with a semi-amplitude of 0.2 mmag is detected in the CoRoT light curve. This frequency is within the range determined for the rotational frequency and shows a double-wave phase diagram. Therefore, this frequency could be interpreted in terms of rotational modulation (Balona 1990). However, no sign of magnetic cloud, starspot or activity has been detected in the analysis of the spectro-polarimetric data. Another explanation is that this frequency could be due to pulsations. Saio et al. (2007) found similar results in the MOST data of a late-type Be star and interpreted it as prograde g-mode. The other low frequencies could also be interpreted in terms of pulsation. The identifications of the pulsation modes will require models taking into account the specificity of Be stars, in particular fast rotation.

Acknowledgments. We wish to thank the CoRoT team for the acquisition and reduction of the CoRoT data. This research is also based on data obtained at the David Dunlap Observatory (University of Toronto) and at the Télescope Bernard Lyot (Pic du Midi).

References

Balona, L. 1990, MNRAS, 245, 92

Frémat, Y., et al. 2006, A&A 451, 1053

Neiner, C., Gutiérrez-Soto, J., Floquet, M., et al. 2009, CoAst, 158, 319

Saio, H., Cameron, C., Kuschnig, R., et al. 2007, ApJ, 654, 544

Comm. in Asteroseismology
Vol. 158, 2009, 38th LIAC/HELAS-ESTA/BAG, 2008
A. Noels, C. Aerts, J. Montalbán, A. Miglio and M. Briquet., eds.

Application of the TiSAFT code (Time Series Analysis with Fisher's Test)

A.-L. Huat[1], B. Leroy[2], and P. D. Diago[3]

[1] GEPI, Observatoire de Paris, CNRS, Université Paris Diderot ;
5 Place Jules Janssen, 92190 Meudon, France
[2] LESIA, Observatoire de Paris, CNRS, UPMC, Université Paris Diderot ;
5 Place Jules Janssen, 92190 Meudon, France
[3] Observatori Astronòmic de la Universitat de València, Edifici Instituts d'Invertigació,
Polígon La Coma, 46980 Patema, València

Abstract

A new code for extracting frequencies, amplitudes and phases in a time series has been developed for the CoRoT data analysis: TiSAFT (Time Series Analysis with Fisher's Test of significance). We provide an application of the TiSAFT code on a simulated β Cephei light curve and on CoRoT data of a Be star, as well as a comparison with other codes already used in the CoRoT Be team.

Individual Objects: CoRoT ID 0102791482

Introduction

CoRoT data will provide us with many new results on Be stars, thanks to their high precision and long duration. In order to extract reliable information, we need to use different codes and compare their results. However, codes already used in the CoRoT Be team only use empirical rejection criteria for the frequency search. Therefore we introduced a statistical method to compute a more physical significance level in TiSAFT.

TiSAFT code: Time Series Analysis with Fisher's Test of significance

The TiSAFT code uses a statistical method to compute the significance of frequencies obtained from seismic data. This statistical test is based on Fisher's test of significance (for more information, see Fisher 1929).

We check whether each peak in the power spectrum is due to noise or to signal by performing a comparison between the power of the peak and a fixed threshold of significance. When the computed value of significance of the peak is above this threshold, the peak is considered as significant, otherwise it is rejected and considered as due to noise.

Test of the TiSAFT code on simulated data

A light curve of the β Cephei HD180642 has been simulated for the CoRoT Hare & Hound 5 exercice (De Ridder). The light curve has a total duration of 150 days and is regularly sampled with $s = 32$ seconds. It contains 51 frequencies. A prewhitening of the powerful radial mode has been performed before the test because it introduces a bias in the estimate of the significance value.

To perform this test, we used three other codes to compare the results: Period04 (Lenz & Breger 2004), Pasper and Clean-NG (developed in the CoRoT Be team). Period04 and Pasper both use a rejection criterion which is the Signal-to-Noise ratio, here $SNR > 4$. Clean-NG has a stopping criterion based on the relative rate, r, of increase of the total power in the clean components from one iteration to the next one. There is no rule for getting the value of this criterion. Therefore we choose it so as to find a number of frequencies close to the input number of frequencies, i.e. $r = 5.10^{-6}$. TiSAFT uses a probability of error as a criterion (the significance of the test). We fixed this probability to $P = 0.1$.

We summarize hereafter the results of the four codes used:

Code	Rejection	Nbr freq found
TiSAFT	$P = 0.1$	47
Pasper	$S/N > 4$	25
Period04	$S/N > 4$	> 75
Clean-NG	$r = 5.10^{-6}$	44

For each code, the errors on the value of the frequencies are very low ($\pm 0.02\%$) while the errors on the amplitudes show an upward shift by 10% (see Fig. 1). The reason of this systematic shift is not yet understood, however note that each code uses a different algorithm to compute the amplitudes. The phase determination is less precise ($\pm 35\%$) since it strongly depends on the accuracy on the frequencies.

Pasper and Period04 seem to have a different definition of the noise since they give a quite different number of frequencies. We cannot judge Clean-NG on the number of frequencies because we tuned its criterion. Thus, TiSAFT is the only code that gives a number of frequencies near the input one.

Test of the TiSAFT code on CoRoT data of a Be star

TiSAFT code has also been tested on the CoRoT Be star 0102791482 from the exoplanet field of the initial run. We also used Period04, Pasper and Clean-NG to compare the results with each other. The four codes give approximatively the same amount of frequencies. The values found for the frequencies and the amplitudes are very similar from one code to another. However, note that following the results for the simulated data, there might be a systematic amplitude shift for all codes. Moreover, the phases are largely spread over $[0; 2\pi]$ for most of the frequencies. Phases are indeed the most difficult parameter to determine.

Code	Rejection	Nbr freq found
TiSAFT	$P = 0.1$	37
Pasper	$S/N > 4$	40
Period04	$S/N > 4$	47
Clean-NG	$r = 5.10^{-4}$	43

Conclusion

Simulated light curve: The statistical criterium used by Fisher's test of significance is very reliable since TiSAFT finds the closest number of frequencies to the input. In order to increase the accuracy on the phases, the frequency determination process needs to be improved. This work showed that the usual methods provide amplitudes with a systematic bias. This shift will be investigated further by testing the codes on other simulated datasets.

CoRoT data of an early Be star: All the codes used during this study give almost the same

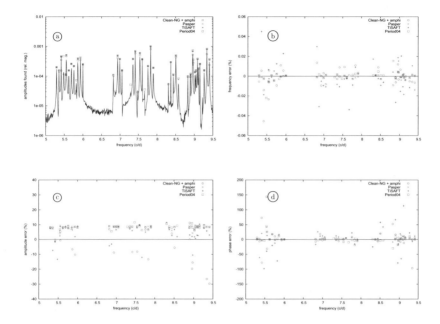

Figure 1: Results of TiSAFT, Pasper, Clean-NG and Period04 codes on the simulated light curve of HD180642. a: frequencies and amplitudes found; b: error on frequencies; c: error on amplitudes; d: error on phases.

results for the frequencies and amplitudes. However, phases are difficult to determine. The frequency accuracy and the resolution of the spectra given by the time series duration are the main parameters influencing the phases in TiSAFT.

Acknowledgments. We wish to thank the CoRoT team, in particular the CoRoT Hare & Hound team.

References

De Ridder, J., CoRoT Hare and Hound exercise 5
      http://www.ias.u-psud.fr/virgo/html/corot/datagroup/fifth_hh.html

Fisher, R. A. 1929, Proc. Roy. Soc. London Ser., 45, 234

Lenz, P., & Breger, M. 2005, CoAst, 146, 53

Comm. in Asteroseismology
Vol. 158, 2009, 38th LIAC/HELAS-ESTA/BAG, 2008
A. Noels, C. Aerts, J. Montalbán, A. Miglio and M. Briquet., eds.

Rapid-photometry of Wolf-Rayet Stars: a search for strange-mode pulsations

A.-N. Chené[1], O. Schnurr[2], A. F. J. Moffat[3], and N. St-Louis[3]

[1] Herzberg Institute of Astrophysics
5071, West Saanich Road, Victoria (BC), V9E 2E7, Canada
[2] Department of Physics & Astronomy
University of Sheffield, Hicks Building, Hounsfield Rd, Sheffield, S3 7RH, UK.
[3] Département de Physique
Université de Montréal, C.P. 6128, Succ. Centre-Ville, Montréal (Qc) H3C 3J7, Canada.

Abstract

Theoretical work suggests that strange-mode pulsations (SMPs) are present in the envelope of hot and luminous stars with a large luminosity-to-mass ratio, where the thermal timescale is short compared to the dynamical timescale, and where radiation pressure dominates (Glatzel et al. 1993). The most violent SMPs are expected in classical Wolf-Rayet (WR) stars, i.e the bare, compact helium-burning cores of evolved massive stars (Glatzel et al. 1999), where SMPs manifest themselves in cyclic photometric variability with periods ranging from minutes to hours. However, these variabilities are expected to be epoch-dependent. Here we report on our attempts to detect SMPs in several WR stars using rapid, high-precision photometry.

Individual Objects: BAT99 47, WR 2

Rapid Photometry in the Optical

Conventional optical cameras feature CCDs that normally have read-out times ranging from a few tens of seconds to minutes to ensure minimum read-out noise. This severely limits the possible duty cycle for rapid-photometry applications. Windowing-down of the detector is a possible solution, but the cost is a highly limited field of view (FOV) which makes it extremely difficult to frame suitable comparison stars for differential photometry. The problem becomes very acute if very bright (and thus very rare) stars are targeted to ensure high S/N within very short exposure times. During semesters 2003B and 2004A, the Acquisition Camera (AcqCam) was attached to Gemini South's Cassegrain focus. With its 1k x 1k, thermo-electrically cooled CDD, AcqCam offers the capability for rapid (6.3 sec) read-out of a full 2' x 2' frame (0.12" per pixel). To provide both a sufficiently large number of comparison stars in the relatively small field, we targeted BAT99-47 (WN3b), a hot, hydrogen-free WR star in the LMC. Over a period of 8 hours, we obtained a short (2 sec exposure time) broadband-V image every 8.3 seconds. Despite the fact that BAT99-47 turned out to be considerably fainter than the V = 14.1 mag listed in the literature (cf. BAT99), and crowded with a (fainter) star (cf. Figure 1a), a photometric precision of $\sigma \sim 3$ mmag for the WR star was achieved for each data point using DAOPHOT, a package for stellar photometry designed to deal with crowded fields. Preliminary results (cf. Figure 1b) indicate that while no periodic, short-time scale variability seems to be present at a level of 3 mmag (except for some small uncorrected local atmospheric variations), there is a slow variation in the flux within 8 hours. The analysis of

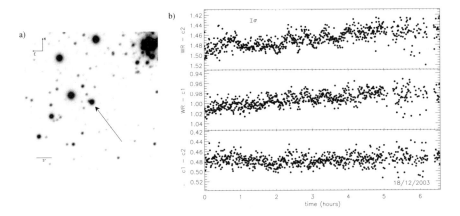

Figure 1: a) A 40"x40"-wide zoom of the field around BAT99 47 (pointed with the arrow) observed with AcqCam. A faint star can be seen at the east, very close to BAT99 47. b) Light-curve of BAT99 47 relative to two different comparison stars, compared to the difference between the two comparison stars.

this phenomenon is ongoing, but it could be that the timescale of the variability of BAT99-47 is longer than the duration of our run.

The Near-Infrared Cameras

If fast-readout, low-noise optical CCDs are not (yet) widely available, near-infrared (NIR) detectors have very short read-out times by design, and wide-field NIR cameras (e.g. WIRCAM, WFCAM,...) are readily available at all major observatories. Due to interstellar extinction, many Galactic OB and WR stars are optically faint, but very bright in the NIR, thereby limiting the sky-brightness issues that normally relate to NIR imaging. Moreover, the fact that massive stars reside in clusters makes for an excellent availablity of many equally bright comparison stars. Thus, rapid, high-precision differential photometry should be relatively easy with conventional and existing instruments. As a first test, we have used SIMON (Spectromètre Infrarouge de Montréal), a NIR spectro-imager that provides a 8' x 8' FOV at the Observatoire du mont Mégantic (OMM) (Québec). We have observed WR 2 (WN2), the hottest Galactic WR star known, and obtained a full 2k x 2k frame in J band every 5.4 sec during 4 partial nights of 3 hours each. Differential photometry yielded a precision of $\sigma \sim 10$ mmag for the preliminary light-curve which covers 3.2 hours (cf. Figure 2a). WR 2 shows a marginal, sinusoidal curve with an amplitude of ~ 20 mmag (2 σ) and a possible frequency of ~ 0.55 cycle per hour (P = 1.8 hours) that is not present in the comparison stars (see the periodogram in Figure 2b). The same behaviour is observed in data that were obtained 3 days later. A more detailed analysis is in progress.

Conclusions

To search for SMPs (or any other type of short-term variability) in OB and WR stars, rapid, high-precision, differential photometry is required. Moderately wide-field, full-frame imaging dramatically increases the availability of suitable comparison stars. While optical CCD cameras are known to deliver very high photometric accuracy (a few mmags), initial tests indicate that conventional NIR cameras can be used with excellent results; these tests indicate that

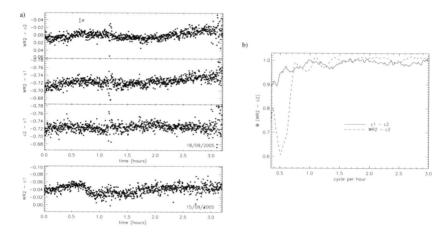

Figure 2: a) WR 2 light-curve versus two comparison stars and between two comparison stars on 18/09/2005 and WR2-c1 on 15/09/2005. b) Periodogram calculated using PDM on the two light-curves c1 - c2 (solid line) and WR 2 - c2 (dashed line) observed on 18/09/2005.

even under suboptimal observing conditions, a 10 mmag photometric precision can easily be reached. Better observational conditions and more sophisticated data reduction and analysis techniques might yield even higher accuracy. Thus, rapid NIR photometry using existing designs on 2 to 4m-class telescopes is a highly viable approach, in particular if the variability is expected to reach amplitudes of few tens of mmags. Preliminary results of our ongoing rapid-photometry campaign indicate that in none of the two observed, hot WN stars, BAT99-47 in the LMC and WR 2 in the Milky Way, is a short-period (few minutes) cyclic variability present. However, a period of 1.8 hours may have been found in the WR 2 light-curve.

References

Breysacher, J., Azzopardi, M., & Testor, G. 1999, A&A, 137, 117 (BAT99)

Glatzel, W., Kiriakidis, M., Chernigovskij, S., & Fricke, K. J. 1999, MNRAS, 303, 116

Glatzel, W., Kiriakidis, M., & Fricke, K. J. 1993, MNRAS, 262L, 7

Comm. in Asteroseismology
Vol. 158, 2009, 38th LIAC/HELAS-ESTA/BAG, 2008
A. Noels, C. Aerts, J. Montalbán, A. Miglio and M. Briquet., eds.

SPB guide star photometry with MOST

D. Gruber[1], R. Kuschnig[1], M. Gruberbauer[1], M. Hareter[1], W. W. Weiss[1], and J. M. Matthews[2]

[1] Institut für Astronomie, Türkenschanzstrasse 17, A-1180 Vienna, Austria
[2] Dept. of Physics and Astronomy, University of British Columbia, 6224 Agricultural Road, Vancouver, Canada

Abstract

MOST, the first Canadian space telescope (launched in June 2003), was designed primarily to perform high-precision photometry of bright stars ($V \leq 6$) nearly continuously over long time spans of up to 60 days. In the past five years, the capabilities of MOST have been extended and enhanced to allow photometry of stars as faint as $V \approx 13$. MOST has monitored more than 1100 stars, most of which are guide stars in the fields containing Primary Science Targets. Many variable stars have been discovered in this large guide star sample, spread across the HR Diagram, including more than 30 SPB (Slowly Pulsating B) stars. For this class of stars, MOST offers excellent time coverage, precision and frequency resolution.

SPB star obervations with MOST

In the last 5 years, MOST observed and discovered approximately 30 SPB stars. The results of HD 163830 were already published, taking advantage of the fact that independent spectroscopy for this object was obtained from the ground (Aerts et al., 2006).

Another example, shown in Fig.1, was continuously observed for $\approx$ 20 days. It is a confirmed cluster member of NGC 2244 and classified as a B9V star ($V = 10.9$ mag). Unfortunately, for this object no fundamental parameters are yet determined.

Outlook

The main problem in analyzing these stars is the lack of knowledge of the fundamental parameters for most of these relatively faint stars. Therefore ground-based follow-up spectroscopy is needed to determine T_{eff} and $\log g$ of these stars to constrain their location in evolutionary model grids and hence in pulsational model grids to fit the observed g-mode frequencies to structural models.

Acknowledgments. This project is funded by the Canadian Space Agency (CSA) and the Natural Sciences and Engineering Research Council (NSERC) of Canada. Support is also provided by the University of Vienna and the Austrian Science Promotion Agency (FFG-ARL).

References

Aerts, C., De Cat, P., Kuschnig, R., et al. 2006, ApJ, 642, 165

Matthews, J.M. 2004, AAS, 467, 1353

Waelkens, C. 1991, A&A, 467, 453

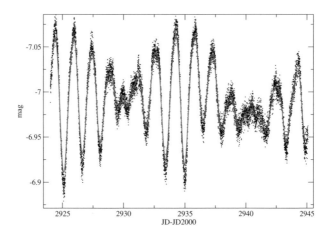

Figure 1: Light curve of the newly discovered SPB star in NGC 2244.

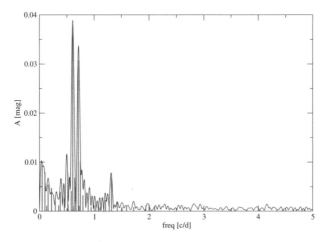

Figure 2: Amplitude spectrum of the frequencies detected in the star.

Session 5
What can asteroseismology do to solve the problems ?

Comm. in Asteroseismology
Vol. 158, 2009, 38th LIAC/HELAS-ESTA/BAG, 2008
A. Noels, C. Aerts, J. Montalbán, A. Miglio and M. Briquet., eds.

Seismic diagnostics of rotation for massive stars

M.J. Goupil[1], S. Talon[2]

[1] Observatoire de Paris
LESIA, UMR 8109, 5 place Jules Janssen, 92190 Meudon principal cedex - France
[2] Réseau québécois de calcul de haute performance, C.P. 6128, succ. centre-ville, Montréal, Québec,
Canada H3C 3J7

Abstract

We recall the main seismic diagnostics for probing stellar rotation and what they can teach
us when they are used for massive stars such as β Cephei stars.
Individual Objects: θ Ophiuchi, HD129929, ν Eri, 12 Lac

Introduction

An important issue in stellar physics is to determine the rotation profile inside stars. This
knowledge would help constrain the respective importance of various processes in the transport
of angular momentum. For instance, a total absence of radial structure could be achieved by
magnetic fields, either via the Spruit-Taylor dynamo or an internal fossil field. On the other
hand, very large radial gradients on the Main Sequence (MS) would point towards the role
of μ gradient (i.e. gradient of mean molecular weight) on the meridional circulation, while
an inverse gradient of the rotation rate Ω (i.e. Ω decreasing toward the interior) would favor
wave induced transport.

Seismic studies of the rotation nowadays aim at identifying regions of uniform and differential rotations (i.e. the depth and latitude dependences). The knowledge of the ratio
$\Omega_{core}/\Omega_{env}$ for instance already brings some constraints on transport of angular momentum
from the central to the surface layers. Another important issue is whether it is possible to
disentangle the effects of overshooting and rotation induced mixing in central regions and the
extension of the convective core.

Prior to any seismic analysis, one has to establish whether the star is a fast or a slow
rotator. Indeed, this determines the way the theoretical frequencies of rotating models can
be computed.

The ratio of the rotation rate to the centrifugal and Coriolis accelerations which are
quantified with $\epsilon = \Omega^2 R/(GM/R^2)$ and Ω/ω where Ω is the star's rotation rate, ω, the
wave frequency and other symbols have their usual meaning, determine whether perturbation
techniques can be used or not, as these are valid only provided ϵ and Ω/ω are small enough.
The main advantage of perturbation methods lies in their relative simplicity and the fact
that they are less time-consuming. However, they have a limited range of validity. Results
discussed here are obtained using perturbation methods only. Whether a star is a fast or a
slow rotator is a relative notion and the questions to be addressed then are: what is the limit of
validity of perturbation methods (PM) in terms of rotation rate? What are the consequences
of overusing PM? Are they severe? Some answers can be found in Reese et al. (2009) and
in Ouazzani et al. (2009).

Several seismic diagnostics for probing rotation exist; these are rotational splittings, their asymmetric pattern in a Fourier spectrum and the impact of rotation on axisymmetric modes. As pulsating massive stars, we consider here β Cephei stars, that-is B type stars on the main sequence with masses larger than 5-7 $M_\odot$. β Cephei stars are pulsating with a few modes around the fundamental radial mode that-is with low radial order, low degree p and g modes with periods around 3-8 h; identified modes most often are p_1, p_2, g_1 modes which are often of mixed p and g nature. Among these stars, one encounters both slow rotators (less than 50 km/s) and rapid rotators (larger than 250 km/s). Four β Cephei stars have been the subject of seismic analyses which yielded information about internal rotation, namely V386 Cen (or HD129929), ν Eri 12 Lac and θ Oph. For reviews, see Stankov & Handler (2005), Pigulski (2007), Aerts (2008) and references therein.

Rotational splittings

Rotational splitting is defined as $S = (\nu_{\ell,n,m} - \nu_{\ell,n,0})/m$ where $\nu_{\ell,n,m}$ is the mode frequency. One also defines it as $S = (\nu_{\ell,n,m} - \nu_{\ell,n,-m})/(2m) = \nu_{\ell,n,m} - \nu_{\ell,n,m-1}$. These various definitions are equivalent only at first perturbation order in the rotation rate Ω; the last two are used when only a few components are available. At first perturbation order in Ω, only the Coriolis acceleration plays a role; for a shellular rotation $\Omega(r)$, S becomes m independent and one has: $S = \int_0^R K_{n,\ell} \frac{\Omega(r)}{2\pi} dr$. The corresponding analytical expression for the kernels $K_{n,\ell}$ can be found in Schou et al. (1994a,b), Pijpers (1997) and involves the eigenfunction for the nonrotating stellar model. For a uniform rotation, one is left with $S = \Omega \, \mathcal{K}$ where S is measured and $\mathcal{K}$ is assumed to be known from a nonrotating stellar model such that Ω can be inferred.

The β Cepheid HD129929 is a 9 $M_\odot$ MS star for which a $\ell = 1$, p_1 triplet has been detected and identified as well as 2 successive components of the $\ell = 2$, g_1 mode. From the triplet and assuming a solid body rotation, one deduces Ω and $v_{rot} = 3.61$ km/s but from the other mode, one obtains $v_{rot} = 4.21$ km/s, clearly indicating a nonuniform rotation (Dupret et al., 2004). Assuming a uniform rotation for the convective core with angular velocity $\Omega = \Omega_c$ and a uniform rotation $\Omega = \Omega_e$ for the envelope of the star, the splittings then obey $S = \Omega_c \beta_c + \Omega_e \beta_e$ where β_j is the integral for the core or the envelope. It is found that $\beta_c << \beta_e$ that-is actually the detected modes do not efficiently probe the convective core. This can be seen with the associated rotational kernels (Fig.7 Dupret et al. 2004). Therefore Ω_c is taken as the rotation rate of the radiative region in the μ gradient region above the convective core. A rotation gradient in the envelope with $\Omega_c/\Omega_e = 3.6$ is obtained. The seismic analysis also favors a core overshooting distance of ~ 0.1 pressure scale height (H_p) rather than 0 while an overshoot of 0.2 H_p is rejected.

The β Cepheid θ Ophiuchi is also a MS 9 $M_\odot$ star and is known to pulsate with at least 7 frequencies which are identified as the fundamental radial $\ell = 0, p1$; one triplet $\ell = 1, p1$ and 3 components $m = -1, 1, 2$ of a quintuplet $\ell = 2, g_1$. The seismic analysis performed by Briquet et al. (2007) shows a case similar to HD129929, although for θ Ophiuchi, data are compatible with a uniform or a quite slowly varying rotation of the envelope. The convective core overshoot distance is found to be $(0.44 \pm 0.07) \, H_p$. The star ν Eridani oscillates with 3 triplets $\ell = 1$ (g_1, p_1, p_2), one radial mode p_1 and one $\ell = 2$ component. The seismic studies for this star indicate that the detected modes can probe the rotation of the core, which is rotating faster than the envelope (Pamyatnykh et al. 2004, PD04; Ausseloos et al. 2004). The quality of the seismic data enabled Dziembowski & Pamyatnykh (2008, DP08) to assume a linear gradient as a transition (in the μ gradient zone) between the uniform fast rotation of the core and the uniform slow rotation of the envelope above the μ gradient region with a ratio $\Omega_c/\Omega_e = 5.3 - 5.8$. On the other hand, model fitting yields an extension of the mixed central region of 0.1-0.2 H_p above the Schwarzschild radius.

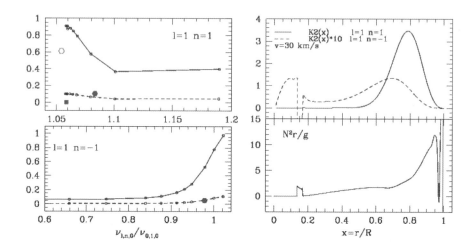

Figure 1: **left:** Scaled asymmetries $R\ 10^3$ for $\ell = 1\ n = 1$ (top) and $n = -1$ (bottom) modes in function of the $m = 0$ frequency scaled by the radial fundamental mode frequency. The open dot (resp. full dot, full square) represents the observed asymmetry for θ Oph, (resp. for V386 Cen, ν Eri). The solid (resp. dashed) line corresponds to $v = 30$ km/s (resp. 10 km/s) $8.2M_\odot$ models. The central hydrogen content X_c is decreasing toward the right. **top right:** Kernels $K_2(x)$ for splitting asymmetries of $\ell = 1, n = 1$ (p1) mode (solid line) and $\ell = 1, n = -1$ (g1) mode (dashed line) for model with $Xc = 0.35$. The abscissae is the normalized radius. **bottom right** Run of the normalized Brünt-Väissälä profile $N^2 r/g$ for the corresponding model with r/R.

These studies lead to the conclusion that a few rotationally split modes are enough to obtain important information about internal rotation of β Cephei stars if the modes are identified, enough precise measurements are obtained and if the age of the star is such that excited modes have mixed p, g characteristics.

Splitting asymmetries

Splitting asymmetry for acoustic modes is mainly due to the oblateness of the star caused by the centrifugal force although for low radial order modes, the Coriolis contribution is also significant. The asymmetry of each multiplet of modes (ℓ, n) is defined as $A = \nu_{\ell,n,0} - 1/2(\nu_{\ell,n,m} + \nu_{\ell,n,-m})$. Fig.1 represents the normalized splitting asymmetries $R \equiv A/\nu_{0,1,0}$ for the $\ell = 1$, p1 and g1 modes in function of the scaled frequency $y = \nu_{\ell,n,0}/\nu_{0,1,0}$ where $\nu_{0,1,0}$ is the frequency of the radial fundamental mode. R is plotted for θ Ophiuchi, HD129929 and ν Eri. The same quantities for 8.2 $M_\odot$ nonrotating MS models are also represented. The models have been computed with CESAM code (Morel, 1997) assuming standard physics (Goupil 2008a, Lebreton et al. 2008) including a core overshooting distance of 0.1 H_p and an initial hydrogen abundance $X = 0.71$ and metal abundance $Z = 0.014$. The selected models have decreasing X_c from 0.5 to 0.2. The frequencies have been computed using a second order perturbation method and an adiabatic oscillation code adapted from the Warsaw' s oscillation code (Daszyńska-Daszkiewicz et al. 2002). For each model, two sets of frequencies are computed assuming a uniform rotation corresponding to $v = 30$ km/s and $v = 10$ km/s respectively. These sequences of models do not represent true evolutionary sequences as in

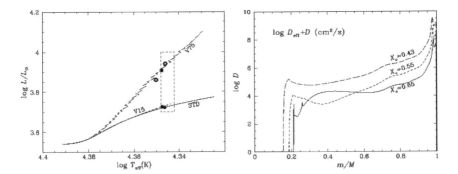

Figure 2: **left:** HR diagram: the squared box corresponds to the star θ Ophiuchi. Full dots represent the approximate location of HD129929 and ν Eri. The open dots represent selected models. The evolutionary track for model V15 (resp. STD) is represented by a dashed (resp. solid) curve. The highest luminosity evolutionary track corresponds to model V75. Wiggles along this last track are due to regions close to semiconvective instability which are not smoothed out as no diffusion is included. They are of no consequence for the present study. **right:** Run of the rotationally induced turbulent coefficient with relative shell mass at 3 different evolutionary stages- labelled with the central hydrogen content X_c- leading to the stellar model V15 ($X_c = 0.43$).

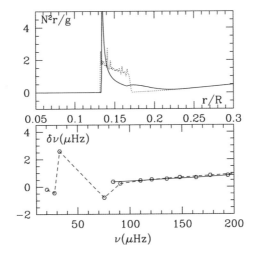

Figure 3: **top** Zoom of Brünt-Väissälä frequency profile in the vicinity of the edge of convective core and normalised rotation profile for model V15 (solid line) and STD(dashed line). **Bottom** Differences between frequencies computed from model STD and model V15 for $\ell = 0$ (solid line) and $\ell = 1$ (dashed line) $m = 0$ modes in μHz.

realistic conditions, the rotation changes with time and likely is nonuniform. They however illustrate the evolution of the asymmetry when a mode changes its nature during evolution, from pure p mode to mixed p and g mode for instance. Indeed pure g modes have small asymmetries compared with pure p modes because they have much smaller amplitude in the outer envelope where distorsion has its most significant effect. This is illustrated in Fig.1. In a perturbation description, one finds that R is a second order effect proportional to Ω^2 (Dziembowski & Goode 1992; Goupil et al. 2000; Goupil, 2008b and references therein). The variation of R with the scaled frequency y (i.e. with aging models) is similar for the $v = 30$ and $v = 10$ km/s sequences of models but R is roughly 9 times (ie ratio of Ω^2) larger for $v = 30$ km/s models than $v = 10$ km/s models. For pure p modes, the asymmetry amounts to $R \sim 0.8 \ 10^{-3}$ whereas for pure g modes it almost vanishes. R for $\ell = 1, n = 1$ decreases for older models (larger y). The reverse happens for the $\ell = 1, n = -1$ mode. The reason is that for young models, $\ell = 1, n = 1$ and $n = -1$ modes are pure p and g modes respectively. When the model is more evolved, these 2 modes experience an avoided crossing and exchange their nature. From a perturbative approach, one derives: $A = \nu_{\ell,n,0} \int_0^1 \hat{\Omega}^2(x) \ K_2(x) \ dx$ where $\hat{\Omega}^2 = \Omega^2/(GM/R^3)$ and $K_2(x)$ depends on the centrifugal perturbation part of pressure and density as well as the differential rotation $\Omega(r)$ and the mode eigenfunction. Fig.1 shows $K_2(x)$ in function of the normalized radius $x = r/R$ for $\ell = 1, n = 1$ (p1) and $\ell = 1, n = -1$ (g1) modes for the $v = 30$ km/s, 8.2 $M_\odot$ model $X_c = 0.5$. The inner layers contribute to the asymmetry of $\ell = 1, g1$ multiplet in contrast with the $\ell = 1, p1$ multiplet for which the kernel K_2 is concentrated toward the surface layers. The asymmetry of the $\ell = 1, g1$ multiplet is sensitive to the inner maximum of the Brünt Väissälä, arising from the μ gradient, which contributes negatively to K_2. As the negative contribution is very localized, it decreases the asymmetry only slightly compared to a pure p mode for a uniform rotation. However, one can expect a larger decrease in case of a rotation faster in the inner regions than the surface. Observed asymmetries deduced from $\ell = 2$ modes seem to disagree for θ Ophiuchi (Briquet et al. 2007) and ν Eridani for $\ell = 1, p2$ (Dziembowski & Jerzykiewicz, 1999). Is the disagreement real? The question has some relevance as the asymmetry values are only marginally above the observation uncertainties. Or is it possible that the observed frequencies do not belong to the same multiplet as suggested by DP08 for ν Eri?

Axisymmetric modes: distorsion and rotationally induced mixing

Distorsion has strong effects on all modes including the axisymmetric modes. This has been extensively discussed in past publications (for a review, see Goupil, 2008b). In a perturbation approach, mixed modes for instance with very close frequencies must be corrected for degeneracy effects. Oblateness also causes radiative desequilibrium which generates large-scale motions (meridional circulation), differential rotation and consequently shear turbulence. All this concurs to affect the rotation profile. It also causes mixing which affects the prior evolution of the observed star and therefore its structure. This equilibrium structure modifications, compared to a nonrotating star, affect all modes, including the $m = 0$ axisymmetric modes. These structure changes must be computed by coupling both evolutions of the angular momentum and the chemicals. Evolution of angular momentum is governed by a time and 4th order differential partial equation (see for details S. Talon, 2009). The chemical species must satisfy the following equation:

$$\rho \frac{dc}{dt} = \rho c_{nuc} + \frac{1}{r^2} \frac{\partial}{\partial r} \left[r^2 \rho V_{ip} c \right] + \frac{1}{r^2} \frac{\partial}{\partial r} \left[\rho r^2 D_t \frac{\partial c}{\partial r} \right] \tag{1}$$

where the first term is the effect of nuclear transformation, the second term is due to atomic diffusion. The diffusion coefficient $D_t = (D_{\text{eff}} + D_v)$ in this last contribution expresses effect of extra mixing in radiative region to occur whenever meridional circulation and turbulence coexist.

Uniform and constant diffusion coefficient D_t: Montalbán et al. (2008) and Miglio et al.(2008) have investigated the effect of turbulent mixing on g mode frequency spectrum and the ability of such modes to probe the size of stellar convective cores. They assumed a constant in time and uniform in space global diffusion coefficient D_t. The constant value for D_t is chosen to correspond to the value near the convective core given by Geneva models including rotationally induced mixing. This is valid for g modes which do have most of their amplitude there (see Miglio et al. 2009). Fig. 2 of Montalbán et al. (2008) shows the Brünt-Väisälä frequency (N) profile for a model with extra element mixing. The model is a mid main sequence 10 $M_\odot$ with $D_t = 7 \ 10^4 \ cm^2/s$ chosen to correspond to a rotational velocity $v = 50$ km/s. The Brünt-Väisälä frequency of the model with 'extra' turbulent mixing behaves more smoothly in the μ gradient region above the convective core than for the model computed with no 'extra' mixing but with an overshoot distance of 0.1 H_p. With the adopted value of D_t, the rotation profile has a core to envelope ratio of 1.6. The differences between the two profiles arising at the edge of the convective core cause significant changes on frequencies of g modes and mixed modes. The frequency separations $\Delta_{n,\ell} = \nu_{\ell,n,0} - \nu_{\ell,n-1,0}$ differ by a few μHz for radial order $n = -1$ and $n = -2$, $\ell = 2$ modes between the model with overshoot 0.1 H_p and the model with extra turbulent mixing (Fig.5 of Montalbán et al. 2008). At higher frequencies, as mixed modes do not exist, no differences in $\Delta_{n,\ell}$ are seen when adding extra turbulent mixing or not.

Rotationally induced diffusion coefficient: Evolution including the coupling between rotationally induced mixing and momentum transport as described by Talon (2009) [1] and computed with the Toulouse-Geneva evolutionary code [2] leads to models which show a smoother Brünt-Väisälä profile at the edge of the convective core. To illustrate the impact of such a smoothing on the oscillation frequencies, the models presented here have been evolved from premain sequence up to a central hydrogen content $Xc = 0.3$. A 8.5 $M_\odot$ mass has been chosen so that the models evolve through the HR diagram location where the star θ Ophiuchi is expected (Fig.2). Model V75 was computed assuming $V = 75$ km/s on the pms and crosses the error box in a HR diagram with a surface velocity of 167 km/s, a rather extreme case. With such an extreme rotation, stars as luminous as HD129929 for instance would have a much smaller mass than deduced assuming no rotation. A model V15 with no overshoot initiated with a rotational velocity $V = 15$ km/s on the pms reaches a surface velocity of 48.2 km/s when crossing the θ Ophiuchi error box at $\log L/L_\odot = 3.73$, $T_{eff} = 4.35$ with an age of 19.65 Myr. Neither overshoot, nor rotationally induced mixing is included in the stellar model STD which is evolved to an age of 19.86 Myr where $\log L/L_\odot = 3.72$, $T_{eff} = 4.34$. The diffusion coefficient, D_t depends on the meridional circulation velocity and the local turbulence strength, it varies with depth and evolves with time as illustrated in Fig.2. The D_t profile is represented for 3 models with ages 0.5 Myr, 1 Myr and 1.5 Myr built assuming an initial 15 km/s velocity on the pms. From the pms, the rotation evolves from uniform to strongly differential rotation which causes a relaxation toward a stationary profile which persists with only an adjustments due to expansion and contraction with evolution.

Effect of turbulent mixing on the structure is significant at the edge of the convective core as emphasized in Fig.3 where we compare the N^2 profile in the vicinity of the edge of convective core for model V15 and STD.

Modes p_1, p_2, g_1 for these models have amplitudes near the edge of the convective core. As Fig.3 shows, this can result in significant frequency differences for the same mode easily detectable with CoRoT observations. The frequencies of these modes are quite sensitive to

---

[1] Note that the impact of internal gravity waves in the mixing of massive stars, where waves are excited by the convective core, has not been properly described yet and is ignored here.

[2] The rotational evolution of the star begins from solid body when the core is still radiative, shortly after the star leaves the Hayashi track. All transport processes are considered from there on.

the detail of the Brünt-Väissälä profile in this region. This means that some care must be taken when computing these frequencies and drawing conclusions. The frequencies of these modes are indeed sensitive not only to the physics but unfortunately also to the numerics which can be quite inaccurate in this region of the star.

References

Aerts, C. 2008, IAUS, 250, 237

Ausseloos, M., Scuflaire, R., Thoul, A., et al. 2004, MNRAS, 355, 352

Briquet, M., Morel, T., Thoul, A., et al. 2007, MNRAS, 381, 1482

Daszyńska-Daszkiewicz, J., Dziembowski, W. A., Pamyatnykh, A. A., & Goupil, M.-J. 2002, A&A, 392, 151

Dupret, M.-A., Thoul, A., Scuflaire, R., et al. 2004, A&A, 415, 251

Dziembowski, W. A., & Goode, P. R. 1992, ApJ, 394, 670

Dziembowski, W. A., & Jerzykiewicz, M. 1999, A&A, 341, 480

Dziembowski, W. A., & Pamyatnykh, A. A. 2008, MNRAS, 385, 2061

Goupil, M.J., Dziembowski, W. A., Pamyatnykh, A. A., et al. 2000, ASPC, 210, 267

Goupil, M.J. 2008a, Ap&SS, 316, 251

Goupil, M.J. 2008b, in Lecture Notes in Physics, Springer

Lebreton, Y., Montalbán, J., Christensen-Dalsgaard, J., et al. 2008, Ap&SS, 316, 187

Miglio, A., Montalbán, J., Noels, A., et al. 2008, MNRAS, 386, 1487

Montalbán, J., Miglio, A., Eggenberger, P., et al. 2008, AN, 329, 535

Miglio, A., Montalbán, J., Eggenberger, P., et al. 2009, CoAst, 158, 233

Morel, P. 1997, A&AS, 124, 597

Ouazzani, R.-M., Goupil, M.-J., Dupret, M.-A., et al. 2009, CoAst, 158, 283

Pamyatnykh, A. A., Handler, G., & Dziembowski, W. A. 2004, MNRAS, 350, 1022

Pigulski, A. 2007, CoAst, 150, 159

Pijpers, F. P. 1997, MNRAS, 326, 1235

Reese, D.R., MacGregor, K.B., Jackson, S., et al. 2009, CoAst, 158, 264

Schou, J., Christensen-Dalsgaard, J., & Thompson, M.J. 1994a, ApJ, 433, 389

Schou, J., & Brown, T.M. 1994b, ApJ, 434, 378

Stankov, A., & Handler, G. 2005, ApJS, 158, 193

Comm. in Asteroseismology
Vol. 158, 2009, 38th LIAC/HELAS-ESTA/BAG, 2008
A. Noels, C. Aerts, J. Montalbán, A. Miglio and M. Briquet., eds.

Driving mechanism in massive B-type pulsators

W.A. Dziembowski

Warsaw University Observatory, Al. Ujazdowskie 4, 00-478 and
Copernicus Astronomical Center, Bartycka 18, 00-716 Warsaw, Poland

Abstract

After a historical introduction, I present the current status of our understanding of the mechanism responsible for pulsation in β Cephei and SPB stars.

Introduction

Variability of the certain β Cephei stars has been known since the beginning of previous century but serious efforts to explain its origin were undertaken much later. To my knowledge Paul Ledoux was the first theorist who showed an interest in these objects and the first who invoked nonradial oscillations to explain pulsation in any stars. In his pioneering work, Ledoux (1951) used data on spectral line profile changes to reveal the nature of modes responsible for the variability of prototype stars with two close frequencies. More than ten years later, Chandrasekhar & Lebovitz (1962) proposed a different interpretation of the two frequencies. These two papers had a lasting impact on studies of nonradial oscillations in stars. However, the matter of driving was only briefly touched upon. The search for the driving mechanism began a few years later. There have been a number of proposals suggesting the deep interior as the site of the driving effect. As it turned out, this was not correct.

The first step towards a correct answer was made by Stellingwerf (1978). He noted that a slight bump in opacity connected with the HeII ionization edge at $T = 1.5 \times 10^5$K produces a substantial driving effect in β Cep models. The effect was not big enough to cause an instability of any mode in any of the models. However, Stellingwerf suggested that an improvement in the opacity calculations might lead to an enhancement of the bump and consequently to instability of modes corresponding to β Cep pulsations. Subsequently, Simon (1982) pointed out that augmenting the heavy element opacities by the 2–3 factor would resolve the period ratio discrepancy in classical Cepheid, which has been another outstanding puzzle in stellar pulsation theory, in addition to significant enhancement of the bump postulated by Stellingwerf. Simon's plea for re-examination of the heavy element contribution to stellar opacity has been an inspiration for the OPAL (Iglesias et al. 1987) and the OP (Seaton 1993) projects. Inclusion of hitherto neglected transitions in heavy element ions resulted in a large (up to factor 3) increase of opacities in the temperature range near $T \approx 2 \times 10^5$K leading to a pronounced bump, which is often referred to as the Fe-bump, because the transitions within the iron M-shell are the primary contributors.

Not long after the OPAL opacities became available for stellar modeling, first papers demonstrating that there are unstable modes in β Cep stars with periods consistent with observations were published (Cox et al. 1992; Kiriakidis et al. 1992; Moskalik & Dziembowski 1992). However, as pointed out in the last paper, the explanation of pulsation in lower luminosity β Cep required a metallicity parameter $Z \geq 0.03$, which appeared too large even then.

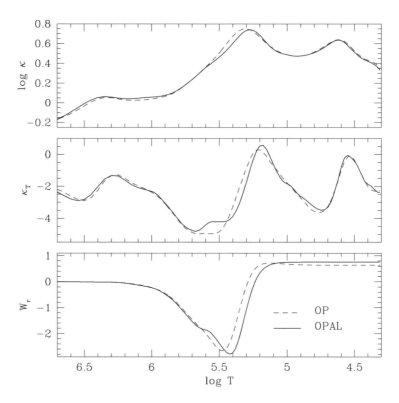

Figure 1: Panels from top to bottom show Rosseland mean opacity, κ, its logarithmic temperature derivative, κ_T, and the cumulative work integral for the fundamental radial mode, W_r, plotted against temperature in two stellar models. The models are characterized by the same parameters: $M = 9.63 M_\odot$,
$\log T_{\text{eff}} = 4.336$, $\log L = 3.891$ $X = 0.7$, $Z = 0.0185$, and the same heavy element mix (Asplund et al.
2005), but use different opacity data. Results obtained with the OPAL data (Iglesias & Rogers 1996) are
indicated with solid lines and those obtained with the OP data (Seaton 2005) are indicated with dashed
lines.

An improvement in atomic physics introduced in the subsequent release of the OPAL opacities
(Rogers & Iglesias, 1992) ruled out this discrepancy. In the first models employing the new
opacity data (Dziembowski & Pamyatnykh 1993; Gautschy & Saio 1993; Dziembowski et al.
1993) the instability was found already at $Z = 0.02$. Moreover, in addition to low-order p-
and g-mode instability responsible for β Cep pulsation, an instability of high-order g-modes in
a detached lower frequency range was found. In the β Cep domains instability was found only
at high angular degrees ($\ell \geq 6$). However, for lower luminosity stars the instability extended
to more easily detectable low degree modes and this provided a natural explanation of the
origin of pulsation in Slowly Pulsating B (SPB) stars. This new type of variable stars was
defined only two years earlier by Waelkens (1991).

Opacity mechanism in B stars

After 1993 not much was added to our basic understanding on how the opacity mechanism in massive B stars works. The plots in Figure 1 are very similar to those shown in Fig.1 of Dziembowski & Pamyatnykh (1993). The two envelope models selected for the plots correspond to an object in the mid of the β Cep domain in the H-R diagram and they differ only in the opacity data. Here, the newest versions of the data from the OPAL and OP projects were adopted. Shown in the upper panel, the run of the opacity coefficient, κ, reveals differences between the two data. The largest difference is seen at the inner side of the Fe-bump, which is centered at $\log T \approx 5.3$. Centered at $\log T \approx 4.6$, the bump caused by the HeII ionization plays no role in driving B star pulsation. The bump at $\log T \approx 6.4$, which is associated mainly with L-shell transitions and ultimate ionization of C, O, and Ne may be active in still hotter stars.

To understand how pulsation is driven, more helpful than the plot of opacity itself is the plot of its logarithmic temperature derivative at constant density, κ_T, which is depicted in the mid plot. The close connection of κ_T with mode excitation is seen when its run is compared with the run of the normalized cumulative work integral, W_r, which is shown in the bottom panel. The latter describes the pulsation energy gain or loss by a mode per unit of time between the center and the distance r. An expression for W_r in terms of eigenfunction describing the Lagrangian perturbation of temperature and total flux, respectively, δT and δL, may be written as follows,

$$W_r = -\frac{1}{L} \int_0^r dr \oint dt \Re \left[\left(\frac{\delta T}{T} \right)^* \frac{d\delta L}{dr} \right] = \int_0^r dr \left| \frac{\delta T}{T} \right|^2 \frac{d\kappa_T}{dr} + ...$$

where in the second equality, I wrote explicitly only the term giving rise to the opacity effect. This is just one of several terms arising from δL but it is the one that matters here. We may see in Figure 1 that in the driving zone, where W_r increases too and the opposite is true in the damping zone. It is the slope of κ_T and not its value which is really relevant. In our models there are three driving slopes, but the only one active occurs in the thin layer extending $\log T \approx 5.5$ to 5.2 and is associated with the Fe-bump. The remaining two are inactive for different reasons. In the layer of the deep bump, the pulsation amplitude is very low while in the HeII-bump zone the thermal relaxation time, τ_r, is much shorter than the pulsation period, Π, so that the δL gradient cannot be maintained. All the damping arises in the layer of decreasing κ_T below the Fe-bump. In the considered cases, it is overcompensated by the driving above which renders the mode unstable.

The conditions for the mode instability are the same as for all opacity-driven pulsation. Within the zone of rising κ_T, the mode amplitude must be large and slowly varying with distance, so that the opacity perturbation dominates in the perturbed flux, and the thermal relaxation time is not significantly shorter than the pulsation period ($\tau_r \gtrsim \Pi$). Outside such a zone, the amplitude must be low or we must have $\tau_r \ll \Pi$. In B stars we encounter the opacity mechanism in the cleanest form. There is a convective layer in the Fe-bump zone of the early B-type stars but by far most of the flux is carried by radiation. The adiabatic temperature gradient changes within this zone but the variations are small, thus there is no significant role of the γ-mechanism.

In the two models considered, only one radial mode is unstable. At somewhat higher effective temperature, the first overtone is unstable, in addition to the fundamental. In our two models, the number of unstable nonradial modes is huge and the same is true for nearly all models of SPB and β Cep stars. The change in mode geometry does not change conditions for instability. The occurrence of the two instability ranges may be easily understood by considering changes in the shape of the radial eigenfunction describing Lagrangian perturbation of pressure, $y_p(r) \propto \delta p/p$, with period, Π, at specified degree, ℓ. For greater generality, it is

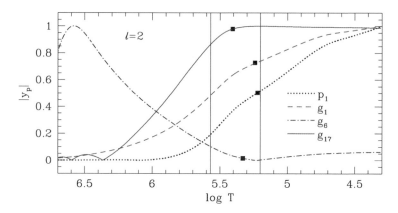

Figure 2: Absolute value of the eigenfunction describing Lagrangian pressure perturbation for selected quadrupole modes in the model calculated with the OP opacity data. The model parameters are given in the caption of Figure 1. Modes p_1 ($\Pi = 0.115$d) and g_6 (0.529 d) are stable while modes g_1 (0.175 d) and g_{17} (1.376 d) are unstable. Dots in the curves indicate places where $\tau_r = \Pi$. The two vertical lines mark the boundaries of the driving slope associated with the Fe-bump.

better to consider changes with the dimensionless frequency,

$$\sigma \equiv \frac{2\pi}{\Pi} \sqrt{\frac{R^3}{GM}}.$$

Let us focus on the model employing the OP data. This is one of the seismic models considered for the β Cep star ν Eri (Dziembowski & Pamyatnykh 2008). Behavior of y_p in the outer layers for four selected quadrupole modes is shown in Figure 2. Mode p_1 has the period similar to first radial overtone and, likewise, is stable. As long as $\ell \ll \sigma^2$, which is true in this case, the ℓ value has little effect on y_p in the outer layers. Mode g_1 has the shape of $y_p(r)$ and a period suitable for driving. Unstable are also g_2 and g_3 modes. Periods of these three modes are in the range of 0.14 to 0.28 d, which is typical for β Cep stars. At $\ell = 1$ and ($3 \leq \ell \leq 8$) there are two or three unstable modes. Then the number decreases to 1 and at $\ell = 11$ the instability disappears.

In the intermediate frequency range ($\sigma \sim \sqrt{\ell}$), the absolute maximum of $|y_p|$ occurs below the Fe-bump. The mode g_6 in Figure 2 is an example. This mode is stable, though there is a significant destabilizing contribution arising at the driving slope of the deep bump but it compensates less than 60% of the damping occurring below. The layers above $\log T = 6.3$ do not contribute significantly to the total work integral. The $\ell = 2$ g-modes between $n = 5$ and $n = 13$ are stable. At still lower frequencies, the maximum is again close to the surface and a new instability range may appear. All the $\ell = 2$ modes in the range $13 \leq n \leq 20$ are unstable. We see in Figure 2 that the g_{17} mode has the shape y_p perfect for driving. Similar shapes are found for other degrees if $0.06 \lesssim \sigma^2/\ell \lesssim 0.3$, which implies shorter periods and higher radial orders at higher degrees. In our model, there are no unstable modes of this type at $\ell = 1$ because the $\tau_r \gtrsim \Pi$ condition is not well satisfied along the driving slope. The number of unstable modes increases up to $\ell = 24$ and the instability disappears at $\ell = 35$. With the m-dependence included, there are about 28 000 unstable high-order g-modes, which is nearly two orders of magnitude more than the low-order modes in this β Cep star model.

In the model calculated with the OPAL opacity data, the slow mode instability begins only at $\ell = 4$. This shift is connected with the upward shift of the Fe-bump (see Figure 1), hence lower τ_r within its range. Modes of lower ℓs with suitable $y_p(r)$ have a period too long to satisfy the $\tau_r \gtrsim \Pi$ condition. To find instability with the OPAL data, we have to go to stars with lower L and/or T_{eff} where the Fe-bump occurs at higher density, hence at higher τ_r. This explains, as Pamyatnykh (1999) first noted, a significant difference between blue boundaries of the instability domains of low degree modes in the H-R diagrams calculated with OPAL and OP data. The latter place the boundaries, both of SPB and β Cep at a higher T_{eff}. Miglio et al. (2007), who used the new OP data (Seaton 2005) found differences In log T_{eff} reaching up to 0.05. The two domains partially overlap.

Questions that remain to be answered

Are the current opacity calculations adequate?

The theory should account for the excitation of all detected modes in the star but still it does not in all cases. Particularly challenging are hybrid objects, where both high and low frequency modes are found. One such object is ν Eri. The difficulty with explaining excitation of modes at both ends of the observed frequency range were discussed by Pamyatnykh et al. (2004), who suggested that the iron abundance in the driving layer is significantly enhanced due to selective radiation pressure. The enhancement required to destabilize high frequency modes was somewhat less than factor 4 and somewhat higher for the low frequency mode. The authors based their proposal on the Charpinet et al. (1996) solution of the driving problem for sdB pulsators. Unfortunately, they ignored the work of Seaton (1999) which states that in massive B stars levitation leads to the enhancement of the iron abundance not only in the bump zone but also in the photosphere, and thus cannot be hidden.

We revisited the problem of mode excitation in ν Eri and in another hybrid pulsator, 12 Lac, in our recent paper (Dziembowski & Pamyatnykh 2008). Since there is no spectroscopic evidence for chemical anomalies in any of these objects, we argued that levitation must be offset by a macroscopic mixing. We showed that even in the very slowly rotating ν Eri, mixing in outer layers by meridional circulation may be fast enough. With the OP opacities, there are unstable high-order g-modes $\ell = 2$ and their periods are in the 1.1 -1.6 d range in the ν Eri model. The observed periods are 1.6 and 2.3d. This might suggest that it is only a matter of further improvement in the opacity calculations to get the agreement. This is possible. However, in the case of 12 Lac, which is a brighter object, the explanation of the long period (2.8 d) requires a much larger opacity modification. Moreover, the use of the OP data did not help a bit in solving the difficulty with mode driving at the short period end. These discrepancies may justify a new plea to atomic physicists for revisiting opacity calculations.

How does rotation affect driving?

The effect of the Coriolis force becomes significant as soon as the spin parameter, $s \equiv 2\Pi/\Pi_{rot}$, approaches one and, for the high-order g-modes, this happens well before the rotation rate approaches the maximum value. Modes cannot be described in spherical harmonics but within the *traditional approximation*, separation of the radial and angular dependencies in the pulsation amplitudes is still possible in terms of the *Hough functions*. Then, the Coriolis force effect is reduced to the replacement $\ell(\ell+1) \rightarrow \lambda(s)$. The essence of the driving effect is not changed. The range of models having unstable low degree modes is somewhat increased (Townsend 2005). However, this approximation may be inadequate (Lee & Saio 1989). There are intriguing discoveries of a large number of modes in four Be stars with the *MOST* satellite and discrepant conclusions from model calculations regarding stability of these modes. (Cameron et al. 2008, and references therein). Accurate modeling of oscillations in these extreme rotators lies still ahead of us. This is important because the detected modes may yield us the clue to understanding the Be star phenomenon.

What is the role of high-degree modes?

A vast majority of the unstable modes are of high degree $\ell > 4$ and, thus, cannot be easily detected if their intrinsic pulsation amplitudes are similar to those of those low degrees. Smolec and Moskalik (2007) proposed that such modes play a role in collective saturation of the driving. This was their explanation of why the amplitude they determined by nonlinear modeling of radial pulsation in β Cep stars was much higher than observed in any star of this type. They noted, however, that the postulated high-ℓ modes may be difficult to hide, as they should contribute to spectral line broadening. According to my crude estimate, if saturation is mostly due to excitation of g-modes of high orders and degrees, then the r.m.s. velocity in the atmosphere should be between 50 and 100 km/s. This seems unacceptably high and, thus, we must conclude that the instability is not saturated. In such a case the terminal state of pulsation must be determined by a resonant excitation of damped modes. If all unstable modes had the same chances to grow, then the occurrence of detectable pulsation would have be regarded a miracle. The modes with $\ell \leq 2$ constitute only about 0.2 percent of the total population. Apparently, the coupling is more effective for high-degree modes. Questions why it is so and what is the contribution of the invisible modes to spectral line broadening are awaiting answers.

Acknowledgments.   This paper was supported by the Polish MNiSW Grant No. 1 P03D 011 28.

References

Asplund, M., Grevesse, N., & Sauval, A.J. 2005, ASP Conf. Ser., Vol. 336, p. 25

Cameron, C., Saio, H., Kuschnig, R., et al. 2008, 685, 489

Chandrasekhar, S., & Lebovitz, N. 1996, ApJ, 136, 1105

Charpinet, S., Fontaine, G., Brassard, P., & Dorman, B. 1996, ApJ, 471, L103

Cox, A. N., Morgan, S. M., Rogers, & F. J., Iglesias, C. A. 1992, ApJ, 393, 272

Dziembowski, W. A., & Pamyatnykh, A. A. 1993, MNRAS, 262, 204

Dziembowski, W. A., & Pamyatnykh, A. A. 2008, MNRAS, 385, 206

Dziembowski, W. A., Moskalik, P., & Pamyatnykh, A. A. 1993, MNRAS, 265, 588

Gautschy, A., & Saio, H. 1993, MNRAS, 262, 213

Iglesias, C. A., Rogers, F. J., & Wilson, B. G. 1987, ApJ, 322, L24

Iglesias, C. A., Rogers, F. J., & Wilson, B. G. 1992, ApJ, 397, 717

Iglesias, C. A., & Rogers F. J. 1996, ApJ, 464, 943

Kiriakidis, M., El Eid, M. F., & Glatzel, W. 1992, MNRAS, 255, 1

Ledoux, P. 1951, ApJ, 114, 373

Lee, U., & Saio, H. 1989, MNRAS, 237, 875

Miglio, A., Montalbán, J., & Dupret, M.-A. 2007, MNRAS, 37, L21

Pamyatnykh, A. A. 1999, Acta Astron., 49, 119

Moskalik, P., & Dziembowski, W. A. 1992, A&A, 256, L5

Pamyatnykh, A. A., Handler, G., & Dziembowski, W. A. 2004, MNRAS, 350, 1022

Seaton, M. 1993, ASP Conf. Ser., Vol. 40, p. 222

Seaton, M. 1999, MNRAS, 307, 1008

Seaton, M. 2005, MNRAS, 362, L1

Simon, N. R. 1982, ApJ, 260, L87

Smolec, R., & Moskalik, P. 2007, MNRAS, 277, 645

Stellingwer, R. F. 1978, AJ, 83, 1184

Townsend, R.H.D. 2005, MNRAS, 360, 465

Waelkens, C. 1991, A&A, 246, 453

Comm. in Asteroseismology
Vol. 158, 2009, 38th LIAC/HELAS-ESTA/BAG, 2008
A. Noels, C. Aerts, J. Montalbán, A. Miglio and M. Briquet., eds.

Discriminating between overshooting and rotational mixing in massive stars: any help from asteroseismology?

A. Miglio, J. Montalbán, P. Eggenberger and A. Noels

Institut d'Astrophysique et de Géophysique,
Université de Liège, Allée du 6 Août 17 - B 4000 Liège - Belgique

Abstract

Chemical turbulent mixing induced by rotation can affect the internal distribution of μ near the energy-generating core of main-sequence stars, having an effect on the evolutionary tracks similar to that of overshooting. However, this mixing also leads to a smoother chemical composition profile near the edge of the convective core, which is reflected in the behavior of the buoyancy frequency and, therefore, in the frequencies of gravity modes. We show that for rotational velocities typical of main-sequence B-type pulsating stars, the signature of a rotationally induced mixing significantly perturbs the spectrum of gravity modes and mixed modes, and can be distinguished from that of overshooting. The cases of high-order gravity modes in Slowly Pulsating B stars and of low-order g modes and mixed modes in β Cephei stars are discussed.

Introduction

Asteroseismology of main-sequence B-type stars is now providing us with constraints on the structure and the internal rotation rate of massive stars (see e.g. the recent review by Aerts 2008). In particular, evidence for non-rigid rotation and for extra mixing near the core was found in several β Cep pulsators (see e.g. Aerts et al. 2003, Mazumdar et al. 2006, Briquet et al. 2008, Dziembowski & Pamyatnykh 2008). In this context, an even deeper insight into the internal structure of B stars could be obtained if the seismic probe provided by the observed oscillation modes is fine enough to differentiate the near-core extra mixing due to overshoot from that induced by rotation.

To study the effects of different extra-mixing processes on the frequency spectra of SPB and β Cep pulsators we focus, respectively, on models of 6 and 10 M$_\odot$. We consider the effect of overshooting and of turbulent mixing near the core that could be induced by the effects of rotation on the transport of chemicals. The models with overshooting were computed with CLES (Code Liégeois d'Evolution Stellaire, Scuflaire et al. 2008). In this code the thickness of the overshooting layer Λ_{OV} is parameterized in terms of the local pressure scale height H_p: $\Lambda_{OV} = \alpha_{OV} \times (\min(r_{cc}, H_p(r_{cc}))$, where r_{cc} is the radius of the convective core given by the Schwarzschild criterion and α_{OV} is a free parameter. In the overshooting region mixing is assumed to be instantaneous and the temperature stratification radiative.

As a first approach, the turbulent mixing has been modelled in CLES by a diffusion process with a parametric turbulent diffusion coefficient D_T that is uniform inside the star and independent of age. We computed also models using the Geneva code that includes the treatment of rotation and related transport processes as described in Eggenberger et al. 2008. The comparison between the two series of models shows that the chemical composition profiles in

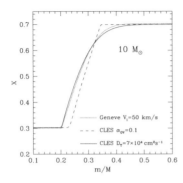
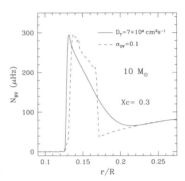

Figure 1: Left panel: Hydrogen mass fraction profile in the central regions of 10 $M_\odot$ models with $X_C \simeq$ 0.3. The different lines correspond to models computed with CLES, with overshooting (dashed line), and turbulent diffusion coefficient D_T (solid line), and with the Geneva code with an initial rotational velocity of 50 km s^{-1} (dotted line). Right panel: Brunt-Väisälä frequency profile in the central regions of the 10 $M_\odot$ CLES models shown in the left panel.

the central regions provided by a uniform and time independent diffusion coefficient represent a first approximation, at least for massive stars, of the effect that would be produced by such a rotationally induced chemical transport (see Fig. 1). The values of the parameter D_T were chosen in order to be close to the value of the chemical diffusion coefficient near the core provided by the Geneva models (see Montalbán et al. 2008 for more details on this parametrization).

SPBs and β Cep pulsators are considered as slow or moderate rotators. SPBs show a typical rotational velocity of 25 km s^{-1} (Briquet et al. 2007), whereas the range of projected rotational velocity in β Cep stars extends from 0 to 300 km s^{-1} with an average of 100 km s^{-1} (Stankov & Handler 2005). The Geneva code calculations for 10 $M_\odot$ models provide values of the chemical diffusion coefficient near the convective core with $X_C = 0.3$, of the order of 5×10^4 cm^{2}s^{-1} for a rotational velocity on the zero-age main sequence (hereafter named initial rotational velocity, V_i) of 20 km s^{-1}, 7×10^4 cm^{2}s^{-1} for V_i=50 km s^{-1}, and 1.6 $\times$ 10^5 cm^{2}s^{-1} for V_i=100 km s^{-1}. On the other hand, the effect of an initial rotational velocity of 25km s^{-1} on the central hydrogen distribution of a 6 $M_\odot$ model is well mimicked by a $D_T \sim 5000$ cm^{2}s^{-1}. The results presented in this paper mainly concern the parametric models with $D_T = 5 \times 10^3$ cm^{2}s^{-1} for SPB models and $D_T = 7 \times 10^4$ cm^{2}s^{-1} for β Cep ones, and they will be compared with overshooting models closely located in the HR diagram, that means α_{OV}=0 for the SPB case and 0.1 for the β Cep one.

As shown in Fig. 1, models located at the same position in the HR diagram, but computed with overshooting or with turbulent mixing, show a significantly different chemical composition profile near the core (see Fig. 1 left panel) and consequently a different behaviour of the sharp feature in the Brunt-Väisälä frequency (N) located in the μ-gradient region (Fig. 1 right panel).

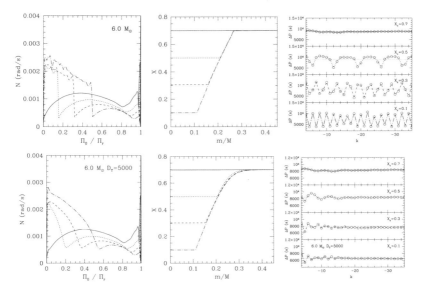

Figure 2: Behaviour of the Brunt-Väisäla frequency (left panels), of the hydrogen abundance profile (middle panels) and of the $\ell = 1$ g-mode period spacing in models of 6 $M_\odot$ (right panels). We consider several models along the main sequence with decreasing central hydrogen abundance ($X_c = 0.7$, 0.5, 0.3 and 0.1). In the right panels the uppermost model has $X_c = 0.7$ and the lowermost $X_c = 0.1$. Upper panels refer to models computed with no extra mixing, whereas lower panels in models of 6.0 $M_\odot$ computed with a turbulent diffusion coefficient $D_T = 5000$ cm^2 s^{-1}.

High-order g modes in Slowly Pulsating B stars

As presented in Miglio et al. (2008a) (hereafter Paper I), the periods of high-order gravity modes in main-sequence stars can be related, by means of analytical expressions, to the detailed characteristics of the μ-gradient region that develops near the energy-generating core, and thus to the mixing processes that affect the behaviour of μ in the central regions. The period spacing (ΔP) of high-order g modes can be described as a superposition of a constant term predicted by the first order approximation by Tassoul (1980) and periodic components directly related to the location and sharpness of ∇_μ.

We recall that in the asymptotic approximation presented in Tassoul (1980) the periods of high-order gravity modes are given by $P_{kl} = \pi^2 \Pi_0 L^{-1}(2k + cte)$, where $L = [\ell(\ell + 1)]^{1/2}$ (with ℓ the mode degree), k the order of the mode and $\Pi_0^{-1} = \int_{x_0}^{1} \frac{N}{x'} dx'$, where x' is the normalized radius and x_0 corresponds to the boundary of the convective core.

In main-sequence stars ∇_μ depends on the size of the convective core and on extra-mixing processes (e.g. overshooting, diffusion, turbulent mixing) that may alter the chemical composition profile in the central region of the star. We here briefly describe the effect of rotationally induced mixing by considering models of a 6 $M_\odot$ star computed including turbulent diffusion near the core (described by a turbulent-diffusion coefficient $D_T = 5 \times 10^3$ cm^2 s^{-1}), which mimics the effects of an initial rotational velocity of 25km s^{-1}. In Fig. 2 we compare the period spacing of such models (lower-right panel) with models that share the same location in the HR diagram, but are computed without any extra-mixing (upper-right

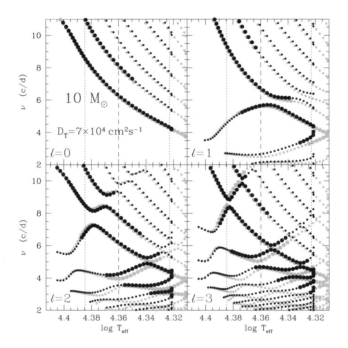

Figure 3: Frequencies of pulsation modes with angular degree $\ell = 0 - 3$ as a function of $\log T_{\text{eff}}$ for main-sequence models of a 10 $M_\odot$ star. Gray dots correspond to the frequencies of models computed with an overshooting parameter $\alpha_{\text{OV}} = 0.1$, whereas black dots are the frequencies of models computed with a turbulent diffusion coefficient $D_{\text{T}} = 7 \times 10^4$ cm s^{-2}. Excited modes are represented by thicker symbols. The vertical lines indicate the effective temperature of models with a hydrogen mass fraction at the center of the order of 0.5, 0.3 and 0.1 (left to right).

panel). We notice that turbulent mixing has a substantial effect on the period spacing: the amplitude of the periodic components in ΔP becomes a decreasing function of the radial order k. This behaviour can be easily explained by the analytical approximation in Paper I, provided that the smoother μ profile (and thus N, see Fig. 2) is taken into account in the analysis.

Low-order g modes and mixed modes in β Cephei stars

As recalled in the previous section, changes of μ induced by the turbulent mixing acting near the core lead to significant variations in the properties of the spectrum of high-order g modes. Although the approximation presented in Paper I and used to relate the sharpness of the "bump" in the Brunt-Väisälä frequency to the properties of g modes is no longer valid for low-order modes, the analytical description of gravity modes mentioned there is still able to qualitatively relate some properties of low-order g-mode spectra to the characteristics of the μ-gradient region near the core (see Miglio et al. 2008b).

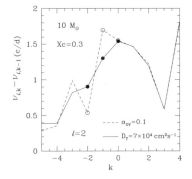

Figure 4: Frequency difference (in c/d) between modes with consecutive radial order (k) and degree $\ell = 2$ for the 10 M$_\odot$ CLES models in Fig. 1.

During the main sequence, the combined action of nuclear reactions and convective mixing leads to a chemical composition gradient at the boundary of the convective core, to a decrease of Π_0, and thus to an increase of the frequencies of gravity modes. The latter interact with pressure modes of similar frequency and affect the properties of non-radial oscillations by the so-called avoided-crossing phenomenon. The modes undergoing an avoided crossing (mixed modes) are therefore sensitive probes of the core structure of the star.

As presented in Montalbán et al. (2008), we describe the effects of extra-mixing on the oscillation spectrum of a typical β Cep star by comparing the properties of low-order g and p modes in models computed with overshooting and with chemical turbulent diffusion. Since the parameters α_{OV} and D_T were chosen to lead to similar evolutionary tracks on the HR diagram, such comparison allows us to remove the differences in the frequencies due to a different stellar radius. In Fig. 3 we plot the oscillation frequencies for 10 M$_\odot$ models along the main-sequence phase. As expected, the differences between frequencies of overshooting models and turbulent mixing ones is very small for pressure modes, while significant differences appear for gravity and mixed modes. These differences increase with the radial order of g modes and, for a given frequency, with the angular degree of the modes. In Fig. 3 we have also bold-marked the frequencies of modes that the non-adiabatic oscillation code MAD (Dupret et al. 2003) predicts to be excited. Here we show only the case with $D_T = 7 \times 10^4$ cm^{2}s^{-1} but computations with lower or higher efficiency of the turbulent mixing show that the differences between overshooting and turbulent-mixing models increase with the value of D_T.

Most relevant from an asteroseismic point of view is the change in the distance between consecutive frequencies of g-modes or modes of mixed p-g character ($\Delta\nu$). In Fig. 4 we plot these differences for the $\ell = 2$ modes of the 10 M$_\odot$ models with an effective temperature $T_{eff} \simeq 22550$ K ($X_C \simeq 0.3$). The dots indicate differences computed between pairs of excited modes. Therefore the difference of $\Delta\nu$ that we can expect between models with sharp ∇_μ (overshooting models for instance) and models with a chemical composition gradient smoothed by the effect of, for instance, a slow rotation ($V_{rot} \sim 50$ km/s) is of the order of 0.4 c/d ($\sim 5\,\mu$Hz), much larger than the precision of present and forthcoming observations.

Conclusions

We presented how the frequencies of gravity modes and mixed modes in β Cep and SPB stars depend on the detailed characteristics of the μ-gradient region that develops near the energy generating core, and thus on the mixing processes that can affect the behaviour of μ in the central regions. In particular we have shown that for rotational velocities typical of main-sequence B-type pulsating stars, the signature of a rotationally induced mixing significantly perturbs the spectrum of gravity modes and mixed modes, and can be distinguished from that of overshooting.

Such a sensitivity of g-mode oscillation frequencies to near-core mixing can provide an additional constraint to the modelling of massive stars with rotation, especially when coupled with seismic inferences on the internal rotational profile (see e.g. the review by Goupil & Talon 2009), and with constraints on chemical enrichments in the photosphere (e.g. Maeder et al. 2008 and Morel 2008). Further investigations are however needed to assess under which observational conditions such information can be recovered from the oscillation frequencies, given realistic uncertainties on stellar global parameters, errors on oscillation frequencies and, in the case of SPBs, given the severe influence of rotation on the spectrum of g modes.

Acknowledgments. A.M. and J.M. acknowledge financial support from the Prodex-ESA Contract Prodex 8 COROT (C90199). A.M. is a *Chargé de Recherches* of the FRS-FNRS. P.E. is thankful to the Swiss National Science Foundation for support.

References

Aerts, C. 2008, IAUS, 250, 237

Aerts, C., Thoul, A., Daszynska, J., et al. 2003, Science, 300, 1926

Briquet, M., Hubrig, S., De Cat, P., et al. 2007, A&A, 466, 269

Briquet, M., Morel, T., Thoul, A., et al. 2007, MNRAS, 381, 1428

Dupret, M.-A., De Ridder, J., De Cat, P., et al. 2003, A&A, 398, 677

Dziembowski, W., & Pamyatnykh, A. 2008, MNRAS, 385, 206

Eggenberger, P., Meynet, G., Maeder, A., et al. 2008, Ap&SS, 316, 43

Goupil, M.J., & Talon., S. 2008, CoAst, 158, 220

Maeder, A., Meynet, G., Ekstrom, S., & Georgy, C. 2009, CoAst, 158, 72

Mazumdar, A., Briquet, M., Desmet, M., & Aerts, C. 2006, A&A, 459, 589

Miglio, A., Montalbán, J., Noels, A., & Eggenberger, P. 2008a, MNRAS, 386, 1487

Miglio, A., Montalbán, J., Eggenberger, P., & Noels, A. 2008b, AN, 329, 529

Montalbán, J., Miglio, A., Eggenberger, P., & Noels, A. 2008, AN, 329, 535

Morel, T. 2009, CoAst, 158, 122

Scuflaire, R., Montalbán, J., Théado, S., et al. 2008, ApSS, 316, 149

Stankov, A., & Handler, G. 2005, ApJS, 158, 193

Tassoul, M. 1980, ApJS, 43, 469

Comm. in Asteroseismology
Vol. 158, 2009, 38th LIAC/HELAS-ESTA/BAG, 2008
A. Noels, C. Aerts, J. Montalbán, A. Miglio and M. Briquet., eds.

Signature of main sequence internal structure in post-main sequence stars

M.-A. Dupret[1], M. Godart[2], A. Noels[2], and Y. Lebreton[3,4]

[1] Observatoire de Paris, LESIA, CNRS UMR 8109, 92195 Meudon, France
[2] Institut d'Astrophysique et de Géophysique, University of Liège, Belgium
[3] Observatoire de Paris, GEPI, CNRS UMR 8111, 92195 Meudon, France
[4] Université de Rennes 1, 35042, Rennes, France

Abstract

Post Main Sequence (post MS) stars keep a trace of their past main sequence history. Typically, the presence or not of an Intermediate Convective Zone (ICZ) above the H-burning shell of massive stars critically depends on the details of the Main Sequence (MS) phase modeling (convection criterion, mass loss, ...). We show here how the excitation of g-modes in blue supergiant stars is closely related to this ICZ, allowing to constrain different associated physical processes.

Introduction

Based on Hipparcos data, about 30 periodically variable supergiant stars have been detected and analysed (Waelkens et al. 1998; Aerts et al. 1999; Mathias et al. 2001; Lefever et al. 2007). From observations by the MOST satellite, Saio et al. (2006) (S06) also reported the discovery of p and g-mode pulsations (about 48 frequencies) in the B supergiant star HD 163899. The presence of g-mode pulsations in a supergiant star is quite challenging but a theoretical explanation has been first proposed by S06. After a review of some important characteristics of the deep interior of these stars, we show how such seismic observations allow to constrain these regions.

Internal structure of blue supergiants and relation with MS phase

B supergiants are post MS massive stars that have a hydrogen burning shell and have (or not have) begun core helium burning. This contracting high density core is surrounded by an expanding diluted envelope. The characteristics of the deep regions around the H-burning shell strongly depend on the details of the MS phase. In particular, depending on this past history, an ICZ can be present just above the H-burning shell. During the MS phase of massive stars, the radiative and the adiabatic temperature gradients decrease in the central regions: the first because of the decreasing opacity due to decreasing X, and the second because of the increasing radiative pressure. As a consequence, the convective core either grows or slowly recedes and a region where $\nabla_{rad} \sim \nabla_{ad}$ can develop above it. Afterwards during the post MS phase, hydrogen burning takes place in a shell around the core; so L/m increases and thus ∇_{rad} increases. As a result, the region in which the temperature gradients are close to each other easily becomes convective since it just needs a slight increase in the radiative temperature gradient, and an ICZ is formed.

However, the appearance of this ICZ critically depends on other physical processes. If a large enough mass loss rate is taken into account during the MS phase, the central temperature increases less quickly. Hence, the adiabatic gradient no longer decreases significantly, the convective core recedes more quickly and does not leave behind a region where $\nabla_{rad} \sim \nabla_{ad}$ (Chiosi & Maeder 1986). Therefore, with significant mass loss, no ICZ appears during the post MS phase (Godart et al. 2009). Overshooting during MS can also prevent the formation of an ICZ during the post MS phase. Finally, depending on the criterion used to define the boundaries of the convective zones: Schwarzschild $\nabla_{rad} = \nabla_{ad}$ or Ledoux $\nabla_{rad} = \nabla_{ad} + \beta/(4 - 3\beta)\nabla_\mu$, ICZ of very different kinds can appear (Lebreton et al. 2009).

Non-radial modes in blue supergiants

The structure of a supergiant star is characterized by a high density contrast between the small size contracting core and the low density expanding envelope. This particular structure has important consequences on the physics of non-radial oscillations. In the deep very dense radiative regions, the Brunt-Väisälä (BV) and Lamb frequencies, respectively N and L_ℓ take huge values compared to those reached in the envelope. The direct consequence is that modes of different types (p- and g-modes) in the envelope have all a g-mode character in the deep interior where $\sigma^2 < N^2$, L_ℓ^2.

A first consequence of the large buoyancy restoring force in the core is that the eigenfunctions present a huge number of nodes there: $n \approx \sqrt{\ell(\ell+1)} \int N/r \, dr \, P / (2\pi^2)$, where P is the pulsation period. For a β Cep type mode with period around the fundamental radial one, $n \approx 100$ and for a typical SPB type mode with larger period, $n \approx 1000$! This implies that the frequency spectrum is extremely dense, the number of non-radial modes being multiplied by about 50 compared to what is obtained in an envelope model without the core. This is the first bad news since it seems that there is no hope to identify the individual non-radial modes and use their frequencies to constrain the internal physics.

The second bad news concerns the energetic aspects of the oscillations. In a radiative g-mode cavity with short wavelength oscillations, radiative damping always occurs. Later we will give a simple expression for the mechanical energy lost per length unit by the mode during each pulsation cycle, the dominating term is of the order of:

$$dW_{rad}/dr \approx \frac{L}{\sigma \, d\ln T/dr} \frac{\delta T}{T} \frac{d^2(\delta T/T)}{dr^2} = \ell(\ell+1) \frac{N^2}{\sigma^3} \frac{T^4}{\kappa\rho} \left|\frac{\delta T}{T}\right|^2 \frac{16\pi ac}{3}. \qquad (1)$$

This radiative damping of non-radial modes is thus very large due to the huge values of N^2 in the core, and it seems that there is no hope to observe them in post MS stars.

However, in the reasoning above we have assumed that the eigenfunction modulus $|\delta T/T|^2$ is not negligible, and this is not always the case due to the presence of an ICZ. In a convection zone $N^2 < 0$ and g-modes are evanescent (left panel of Fig. 1). Therefore, the ICZ can act as a potential barrier: some modes can cross it, others are reflected. This is illustrated in Fig. 1 (right panel). For the mode that crosses the convective barrier (grey), the amplitudes of short wavelength oscillations are significant in the radiative core and strong radiative damping ensues (bottom left panel of Fig. 2): this mode cannot be observed. But for another mode with close frequency, the amplitudes are small in the radiative core (black in the right panel of Fig. 1). Hence, the radiative damping remains small compared to the κ-mechanism occurring in the iron opacity bump near the surface, as shown in top left panel of Fig 2; this mode is unstable and could be observed. So this is a first good news: in the dense spectrum of non-radial modes, some reflect on the convective shell and can be excited and observed. It is useful to note that, if only reflected modes are considered, the mode propagation cavity is similar to a MS B star: because of the reflection, the radiative core can be somehow forgotten

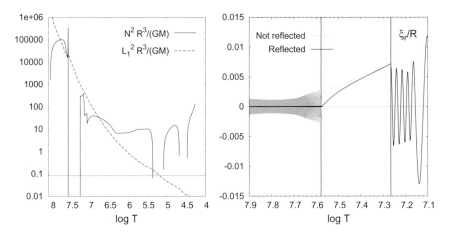

Figure 1: Left: Propagation diagram showing the dimensionless squared BV and Lamb frequencies, in a model with an ICZ and $M = 13M_\odot$, $T_{eff} = 18325\,K$, $\log(L/L_\odot) = 4.5$ and $\log g = 3.06$. Right: Radial displacement ξ_r/R for a reflected and a non reflected $\ell = 1$ mode with dimensionless frequency $\omega = \sigma\sqrt{R^3/GM} = 0.29$.

and the bottom of the cavity is a convective region. As a consequence, the frequencies of reflected modes behave as they do in a MS star. This is illustrated in the right panel of Fig. 2 where we give the typical evolution of the reflected $\ell = 1$ frequencies during the post MS phase of $13M_\odot$ models. So this is the second good news: when restricting to the reflected modes, the frequency pattern is sparse enough and asteroseismology of blue supergiants becomes possible. The third good news is of course that such non-radial modes are observed, as mentioned above.

Physical constraints

As seismic interpretation of non-radial oscillations in Blue supergiants is possible, this opens the way to a probe of their very deep layers and the physical processes affecting them. As we have seen, the key point is the presence of an ICZ, so we can hope to constrain the physical processes related to its presence. First, we have mentioned above that significant mass loss can prevent the appearing of such ICZ. Godart et al. (2009) show how the bottom boundary of the instability strip is displaced towards higher luminosities (lower gravities) when significant mass loss is included. The observed instability strip of Slowly Pulsating B supergiants (SPBsg) gives thus an upper limit on the mass loss rate in this mass range. Overshooting can also prevent the formation of an ICZ during the post MS phase and extends the MS phase. Concerning for example the SPBsg HD 163899, we could imagine either that it is a MS star with large overshooting, or that it could be a post MS star with moderate overshooting during this phase. Godart et al. (2009) show that none of these two possibilities agree with the seismic observations; this gives limits on overshooting during MS and post MS phase.

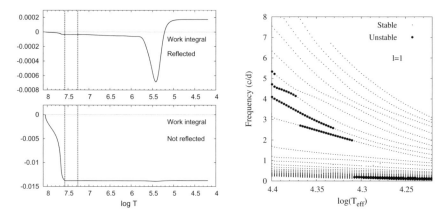

Figure 2: Left: Work integral for a reflected mode (top) and nonreflected mode (bottom), same modes and model as in Fig. 1. Right: Evolution of the frequencies of $\ell = 1$ reflected modes as a function of log T_{eff}, for models with $M = 13 M_\odot$.

Numerical recipe

It is not easy to solve the non-radial oscillation equations in post MS stars accurately and therefore it is useful to mention numerical recipes adequate in this context. First, it is possible to avoid the computation of the full dense spectrum of non-radial modes by preselecting the modes that are reflected at the bottom of the ICZ (and potentially excited). This is achieved by imposing a rigid boundary condition at the base of the ICZ: $\xi_r = 0$. This condition automatically leads to small amplitudes of all eigenfunctions in the radiative core, as we have verified by numerous full numerical computations. Moreover, we can justify this choice by considering the simple case of two cavities with constant wavenumbers k_1 and k_2 separated by an evanescent region of size h and zero wavenumber. Analytical solutions are easily obtained in this case, showing that in the limiting case where $k_1 >> h^{-1}$, the smallest amplitudes in cavity 1 are obtained for frequencies corresponding to a rigid boundary at the base of the evanescent region. Artificially imposing a boundary condition inside the star does not allow to get the real frequencies of the full star. But as the frequency spectrum of the full star is very dense, there always exists a mode with frequency close to the one corresponding to rigid reflexion and the error introduced by this approach is very small.

Once the non-adiabatic pulsation equations are solved in the envelope, we cannot completely disregard the radiative core because reflexion is only partial. In the radiative core, we have seen that the eigenfunctions have a huge number of nodes (up to $n \approx 1000$!). This makes the task of solving accurately the differential problem very difficult in this region. But at the same time, two approximations are valid with high accuracy in this region. First, the internal energy of these deep layers is very high, which justifies the so-called quasi-adiabatic approximation. This approximation can be presented in different ways (Dziembowski 1977, Van Hoolst et al. 1998, Unno et al. 1989), but leading always to the same final result. We can also present it as follows. First, the adiabatic problem is solved to get the eigenfunctions ξ_r, δP, ... Next, these adiabatic eigenfunctions are used to determine δL from the perturbed diffusion equation:

$$\frac{\delta L}{L} = 4 \frac{\xi_r}{r} + 3 \frac{\delta T}{T} - \frac{\delta \kappa}{\kappa} - \ell(\ell+1) \frac{\xi_h}{r} + \frac{d \, \delta T/dr}{d T/dr} . \tag{2}$$

Finally, with all these ingredients the work integral can be computed and divided by the inertia to get the damping rate of the modes, which gives in a pure radiative zone without nuclear reactions:

$$
\eta = \frac{\int_0^M \frac{\delta T}{T} \left(\frac{\partial \delta L}{\partial m} - \frac{\ell(\ell+1) L}{4\pi \rho r^3} \frac{T'}{dT/d \ln r} \right) dm}{2\, \sigma^2 \int_0^M |\vec{\xi}|^2 \, dm} , \tag{3}
$$

where σ (resp. η) are the real (resp. imaginary) parts of the angular frequency (time-dependence: $\exp(i\sigma t - \eta t)$). The quasi-adiabatic treatment is used for the integration in the radiative core ($r < r_0$) and the full non-adiabatic eigenfunctions are used for the envelope ($r \geq r_0$). Moreover, another important simplification is to use the asymptotic theory, which applies perfectly in the radiative core where the wavelength of the eigenfunctions is by far smaller than the scale heights of different equilibrium quantities. A full non-adiabatic asymptotic treatment was derived by Dziembowski (1977). Here we use instead the standard adiabatic asymptotic theory, which gives the following expressions for the radial (ξ_r) and transversal (ξ_h) components of the displacement far from the edges of a g-mode cavity ($\sigma^2 << L_\ell^2$, N^2):

$$
\frac{\xi_r}{r} = K \frac{[\ell(\ell+1)]^{1/4}}{\sqrt{\sigma\, r^5 \rho\, N}} \sin \left[\int_{r_0}^r k_r \, dr \right] \tag{4}
$$

$$
\frac{\xi_h}{r} = \frac{P'}{\sigma^2 r^2 \rho} = \frac{K}{[\ell(\ell+1)]^{1/4}} \sqrt{\frac{N}{\sigma^3 r^5 \rho}} \cos \left[\int_{r_0}^r k_r \, dr \right] , \tag{5}
$$

where the local radial wavenumber is given by $k_r = \sqrt{\ell(\ell+1)}\, N / (\sigma\, r)$. The K constant is obtained by applying the continuity of δP at the bottom of the ICZ (r_0), which is equivalent to the continuity of P' because of our rigid boundary condition. We note that in this asymptotic limit, we also have $|\delta P/P| >> |P'/P|$ so that $\delta P/P \simeq (d \ln P/dr)\, \xi_r$ and near incompressibility: $|\delta\rho/\rho| << \ell(\ell+1)\, |\xi_h/r|$, so that $d\xi_r/dr \simeq \ell(\ell+1)\, \xi_h/r$.

Substituting Eqs. 4 and 5 in Eq. 2 and keeping only the dominating terms in the asymptotic limit gives:

$$
\frac{\delta L}{L} \simeq \frac{d\, (\delta T/T)}{d \ln T} - \ell(\ell+1) \frac{\xi_h}{r} \simeq \ell(\ell+1) \left(\frac{\nabla_{ad}}{\nabla} - 1 \right) \frac{\xi_h}{r} . \tag{6}
$$

We obtain then from the equation of energy conservation and neglecting the transversal component of the divergence of the flux:

$$
\begin{aligned}
i\sigma T \delta s &\simeq -\frac{d\delta L}{dm} \\
&\simeq K \frac{[\ell(\ell+1)]^{5/4}}{4\pi} L \left(\frac{\nabla_{ad}}{\nabla} - 1 \right) \sqrt{\frac{N^3}{\rho^3 \sigma^5 r^{11}}} \sin \left[\int_{r_0}^r k_r \, dr \right] \\
&\simeq \frac{\ell(\ell+1)}{4\pi} \left(\frac{\nabla_{ad}}{\nabla} - 1 \right) \frac{N^2 L}{\sigma^2 r^3 \rho} \frac{\xi_r}{r} .
\end{aligned} \tag{7}
$$

The contribution of the radiative core to the numerator of Eq. (3) is thus simply given by:

$$
\begin{aligned}
\int_0^{m_0} \frac{\delta T}{T} \frac{\partial \delta L}{\partial m} \, dm &\simeq \frac{\ell(\ell+1)}{\sigma^2} \int_0^{r_0} \frac{\rho g}{P} \left(\frac{\nabla_{ad}}{\nabla} - 1 \right) \nabla_{ad} N^2 L \left(\frac{\xi_r}{r} \right)^2 dr \\
&\simeq \frac{K^2 \, [\ell(\ell+1)]^{3/2}}{2\sigma^3} \int_0^{r_0} \left(\frac{\nabla_{ad}}{\nabla} - 1 \right) \frac{\nabla_{ad} \, N\, g\, L}{P\, r^5} \, dr ,
\end{aligned} \tag{8}
$$

and for the denominator:

$$2\sigma^2 \int_0^{m_0} |\vec{\xi}|^2 \, dm \simeq 4\pi K^2 \int_0^{r_0} k_r \, dr \, . \tag{9}$$

These equations are perfectly compatible with those given in Dziembowski et al. (2001). Finally, it is important to emphasize that using Lagrangian or Eulerian formalisms can lead to different numerical results. In our first computations, we used a Lagrangian formalism, but it did not lead to the appropriate evanescent behaviour of the eigenfunctions in the ICZ; instead they showed a large wavelength oscillation. This comes from the fact that no control of the BV frequency is possible in a Lagrangian formalism. Because of numerical truncation errors, the finite difference scheme does not know the real value of N^2, nor its sign, when it is slightly negative like it is in an ICZ. With an Eulerian formalism instead, N^2 appears explicitly in the movement equation, and the correct value can be attributed to it. As was already pointed out by Dziembowski years ago, we thus emphasize that it is much better to use an Eulerian formalism for the finite difference scheme inside a deep convection zone.

Conclusions

Significant radiative damping affects most non-radial modes of the very dense spectrum of B supergiants. But if an ICZ is present above the H-burning shell, some modes can reflect on it and are not damped. Asteroseismology of B supergiants is thus possible, allowing to constrain different physical aspects of the MS and post MS history of these stars (mass loss, convection, ...).

References

Aerts, C., De Cat, P., Peeters, E., et al. 1999, A&A, 343, 872

Chiosi, C., & Maeder, A. 1986, ARA&A, 24, 329

Dziembowski, W. A. 1977, Acta Astron., 27, 95

Dziembowski, W. A., Gough, D. O., Houdek, G., & Sienkiewicz, R. 2001, MNRAS, 328, 601

Lefever, K., Puls, J., & Aerts, C. 2007, A&A, 463, 1093

Mathias, P., Aerts, C., Briquet, M., et al. 2001, A&A, 379, 905

Saio, H., Kuschnig, R., Gautschy, A., et al. 2006, ApJ, 650, 1111

Unno, W., et al. 1989, University of Tokyo Press, 2nd ed.

Van Hoolst, T., Dziembowski, W. A., & Kawaler, S. D. 1998, MNRAS, 297, 536

Waelkens, C., Aerts, C., Kestens, E., et al. 1998, A&A, 330, 215

DISCUSSION

Dziembowski: While interpreting data on supergiant pulsators one should take into account the fact that the post-MS phase is very short in comparison with the core helium burning phase.

Dupret: I agree. The question is whether these stars are helium burning or not, as most of them have too small $\log g$ values to be on the main sequence. Concerning the possible excitation of g-modes, it depends on the presence or not of an intermediate convective zone but not so much on a possible He-burning convective core as it is too deep to change significantly the radiative damping.

Jerzykiewicz: If the MOST SPB variable is really a post-MS star, the periods will increase so fast as the star evolves through the Hertzsprung gap that the effect should be observable in a few years.

Dupret: That is a good idea, but I fear it would be difficult from an observational point of view. The spectrum of g-modes is dense and we have to be sure that we follow the time variation of the frequency of the *same* mode, and with enough precision.

Comm. in Asteroseismology
Vol. 158, 2009, 38th LIAC/HELAS-ESTA/BAG, 2008
A. Noels, C. Aerts, J. Montalbán, A. Miglio and M. Briquet., eds.

Strange modes

H. Saio[1]

[1] Astronomical Institute, Graduate School of Science, Tohoku University, Sendai, Japan

Abstract

Strange modes appear when the pulsation energy is strongly confined in a region where the radiation pressure is much larger than the gas pressure. Strange-mode instability, which occurs in an extremely nonadiabatic environment, generates a pair of a strongly excited mode and a strongly damped mode. This is caused by a significant phase difference between pressure and density maximum. The strange-mode instability seems responsible for complex light variations of luminous H-deficient stars such as R CrB stars and extreme-helium B stars. It is also suggested that strange modes are involved in pulsations in Wolf-Rayet stars, Luminous Blue Variables, and F-supergiant variables.

Historical Accounts

Some 30 years ago Wood (1976) found that nonadiabatic pulsations of luminous helium stars have properties very different from adiabatic ones. The period of a mode varies as a function of the effective temperature differently and looses correspondence with that of an adiabatic mode. Generally, periods increase with a decrease in the effective temperature for a fixed luminosity. For $\log T_{\mathrm{eff}} <\sim 3.9$, however, the period of at least one nonadiabatic mode starts decreasing and gets strongly damped. For this type of mode the word "strange mode" was used for the first time by Cox et al. (1980). The strongly damped strange modes are associated with damping thermal waves, which were investigated in detail by Saio et al. (1984). We do not discuss strange modes of this type anymore in this paper.

Wood (1976) also found that a nonadiabatic mode which is close to the n th adiabatic mode in low effective temperature models, moves toward the $n+1$ st adiabatic mode in high effective temperature models. This transition is accompanied with a blueward extension of instability range on the HR diagram. Wood (1976) noted that instability occurs for $3.93 \leq \log T_{\mathrm{eff}} \leq 4.12$ and the width of this region increases with the luminosity to mass ratio; L/M. The phenomena are related to another type of strange modes discovered by Gautschy & Glatzel (1990). They have shown that strange modes exist even in the approximation in which the thermal time is artificially set to zero (NAR approximation). This indicates that these strange modes have nothing to do with thermal waves. Furthermore, Gautschy & Glatzel found that strange modes become unstable even under the NAR approximation, indicating that the instability is not related with the kappa-mechanism. Glatzel & Gautschy (1992) found that strange modes and the strange-mode instability also occur in *nonradial* pulsations with $\ell <\sim 100$.

What are strange modes?

Strange modes are easily noticeable in a modal diagram in which pulsation frequency [normal-ized by a quantity proportional to $\sqrt{GM/R^3}$] is plotted as a function of a stellar parameter (e.g., mass, effective temperature, etc.). The normalized frequency of an ordinary mode tends to vary only weakly in a modal diagram, while a strange mode varies differently so that it sometimes crosses an ordinary mode (Fig. 1).

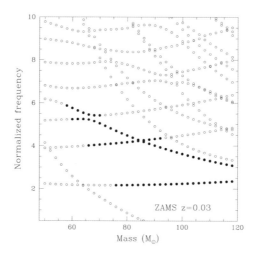

Figure 1: A modal diagram for radial pulsations of very massive zams models. Filled circles are for excited modes and open circles for damped modes.

The peculiar behavior of the frequency of a strange mode is caused by the phenomenon that the pulsation energy is trapped in a relatively narrow cavity which is often associated with density inversion produced by an opacity peak. Since the relative position of the cavity within the stellar envelope changes with a stellar parameter, the strange modes behave differently from ordinary modes in the modal diagram.

Since the density inversion is stronger when the luminosity to mass ratio L/M is larger, strange modes appear in models with $L/M > \sim 10^4 L_\odot/M_\odot$. Strange mode-like behavior sometimes occurs even for adiabatic pulsations (Kiriakidis et al. 1993; Gautschy 1993; Saio et al. 1998).

Excitation of strange modes

Strange modes are excited in two ways; enhanced kappa-mechanism and strange-mode in-stability. The growth time tends to be very short; typically a length of a few periods. To understand the excitation of strange modes we consider linear radial pulsations under the plane parallel approximation. Linearized energy-conservation equation may be written as

$$i\sigma \frac{C_p T m}{F} \frac{\Delta s}{C_p} = -\frac{d}{dq} \frac{\Delta F}{F}, \qquad (1)$$

where σ is the angular frequency of pulsation [the temporal variation of pulsation is written as $\exp(i\sigma t)$], C_p the specific heat with constant pressure, T the temperature, m the envelope mass within a column with unit cross section, q is the fractional mass measured from the bottom, F radiative flux, s is entropy, and Δ means the Lagrangian perturbation to the next variable.

The quantity in the left-hand side of Eq. (1),

$$f \equiv |\sigma| \frac{C_p Tm}{F} \frac{\Delta s}{C_p} \sim \frac{\text{thermal time}}{\text{pulsation period}},$$

represents the ratio of the thermal timescale of the envelope to the pulsation period. For weakly nonadiabatic pulsators as Classical Cepheids, $f \gg 1$, and (slow) damping or driving of pulsation at each layer in the envelope is determined by the phase relation between Δs and $d\Delta F/dq$ (Eq. 1). Since the amplitude of a strange mode tends to be trapped around a peak of opacity where kappa-mechanism driving occurs, strange modes are sometimes excited mainly by the enhanced kappa-mechanism (e.g., excited strange modes in Fig. 1).

As the thermal timescale ($\propto M/L$) and hence the ratio f decreases, the degree of nonadiabaticity of pulsations increases. As the limit of extreme nonadiabaticity Gautschy & Glatzel (1990) introduced NAR (nonadiabatic reversible) approximation, in which the left-hand side of equation (1) is set to be zero; i.e., $f = 0$. Under the NAR approximation equation (1) reduces to

$$\frac{\Delta F}{F} = 0 = -\frac{\kappa_T}{4} \frac{\Delta P_R}{P_R} - \kappa_\rho \frac{\Delta\rho}{\rho} - \frac{c}{\kappa Fm} \frac{d\Delta P_R}{dq}, \qquad (2)$$

where κ is the Rosseland-mean opacity per unit mass, $\kappa_T \equiv (\partial \ln \kappa/\partial \ln T)_\rho$, $\kappa_\rho \equiv (\partial \ln \kappa/\partial \ln \rho)_T$, P_R is the radiation pressure, ρ the density, and c the speed of light. The expression for ΔF in equation (2) was derived from the diffusion equation for the radiation flux in the plane-parallel approximation, where the radiation pressure ($P_R = aT^4/3$) is used instead of the temperature. Gautschy & Glatzel (1990) have shown that the pure strange-mode instability occurs under the NAR approximation. The origin of strange-mode instability is discussed by Glatzel (1994), Papaloizou et al. (1997), and Saio et al. (1998).

Origin of the strange-mode instability

The stability of pulsations under the NAR approximation is very different from that of weakly nonadiabatic pulsations. The κ-mechanism (which works by storing and releasing thermal energy) never works in the NAR approximation where thermal time is set to be zero. Linear pulsation equations with the NAR approximation depend on pulsation frequency σ only in the form of σ^2 and all the coefficients are purely real. Therefore, if a complex σ^2 is an eigenvalue its complex conjugate is also an eigenvalue. The eigen-frequency under the NAR approximation is purely real or complex and complex-conjugate pair; the latter corresponds to the strange-mode instability. In a modal diagram the occurrence of the strange-mode instability appears as two real frequencies collapse to form a complex and complex-conjugate pair.

Equation (2) may be re-written as

$$\frac{d(Q\Delta P_R)}{dq} = -\kappa_\rho \kappa F \frac{mQ}{c} \frac{\Delta\rho}{\rho} \quad \text{with} \quad Q \equiv \exp\left[\int^q \frac{\kappa_T \kappa Fm}{4cP_R} dq\right]. \qquad (3)$$

If the radiation pressure is much larger than the gas pressure, (which seems to be a necessary condition for the appearance of the strange mode instability,) $\Delta P \approx \Delta P_R$ and then equation (3) gives the relation between pressure and density perturbations, ΔP and $\Delta\rho$. The relation, in contrast to the adiabatic relation, is not algebraic and allows a phase difference between ΔP and $\Delta\rho$, and hence can cause oscillatory instability (or overstability).

If we assume a plane wave form $\exp(ikq)$ for ΔP and assume $|k| \gg |d \ln Q/dq|$, we obtain from equation (3)

$$\Delta P = i \frac{\kappa_\rho \kappa}{k} \frac{Fm}{c} \frac{\Delta \rho}{\rho}. \tag{4}$$

This equation indicates that depending on the sign of k (propagation direction) the phase of ΔP can delay or precede the phase of $\Delta \rho$, corresponding to amplitude growing and decaying, which just corresponds to a complex eigenfrequency and its complex conjugate for the strange modes at the occurrence of the instability.

Thus we understand that the strange-mode instability occurs when the pulsation is confined in the outer layers where the ratio of thermal timescale to pulsation period is sufficiently small and the radiation pressure is much larger than the gas pressure. Since these conditions are met at an opacity peak in a stellar envelope if L/M is sufficiently large ($>\sim 10^4$), the instability boundary in the HR diagram is nearly horizontal. We also note that Eqs. (3)(4) indicate that a finite κ_ρ (not κ_T) is essential for the strange-mode instability to occur, which is quite different from the case of the κ-mechanism excitation.

Strange modes in various stars

Very massive stars

Figure 2 shows instability range in the HR diagram for radial pulsations. In the range $L/M <\sim$ 10^4 pulsations are mainly excited by kappa-mechanisms and the stability boundaries are nearly vertical; the blue part (β Cep instability region) associated with the Fe opacity bump at $T \approx 1.5 \times 10^5$K and the redder part (Cepheid instability strip) with the opacity peak at the HeII ionization. The two parts merge in the upper part of the HR diagram ($L/M > 10^4$) where strange modes are excited by the enhanced kappa-mechanism and the strange-mode instability.

Glatzel & Kiriakidis (1993a) reported on the presence of strange modes in massive ($>\sim$ $60M_\odot$) main-sequence stars. These strange modes are trapped around the density minimum produced by the Fe opacity bump. The effect of the density minimum is so strong that the trapping and strange-mode-like behavior appear even in adiabatic pulsations (Kiriakidis et al. 1993; Saio et al. 1998). For the excitation of the strange mode the contribution from the ϵ-mechanism is negligible, although the ϵ-mechanism was considered, before the appearance of the OPAL opacity (Rogers & Iglesias 1992), to be responsible for destabilizing very massive stars and hence for determining the maximum mass of the stars.

The boundary of the strong excitation in post main-sequence models was found by Glatzel & Kiriakidis (1993b) and Kiriakidis et al (1993) to be similar to the Humphreys-Davidson limit (Humphreys & Davidson 1979). This suggests that strange modes might be related to the cause of the Luminous Blue Variables.

Glatzel & Mehren (1996) found that the normalized frequencies of *nonradial* strange modes with $\ell >\sim 100$ in post main-sequence massive stars become independent of the effective temperature. This means the frequency is perfectly normalized by $\sqrt{GM/R^3}$.

Strange modes in Wolf-Rayet stars

The presence of strange modes in models of Wolf-Rayet (WR) stars was found by Glatzel et al. (1993) and the properties were studied systematically by Kiriakidis et al (1996). Since the luminosity of a helium star is much more luminous than the main-sequence star with the same mass, the L/M ratio is much larger. They found that pulsations are so nonadiabatic that the properties are similar to those under the NAR approximation and hence strange-mode instability works for WR stars. Nonradial strange modes in WR-star models have a peculiar

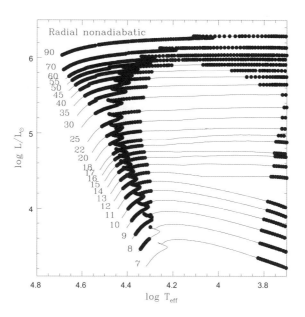

Figure 2: The range where radial pulsations are excited along evolutionary tracks for $7 \leq M/M_\odot \leq 90$ are shown by black dots. Less luminous and cooler region corresponds to the Cepheid instability strip (no red-boundary appears because the convection effect is neglected).

property (Glatzel & Kaltschmidt 2002); the frequencies of the strange modes (, which appear for $\ell <\sim 200$) are almost independent of ℓ. The property is not well understood.

Recently, the MOST satellite detected a light variation with a period of 9.8 h in the WR star WR 123 (Lefèvre et al. 2005). Dorfi et al. (2006) have shown that it is consistent with a strange mode pulsation of a massive ($22 \sim 27 M_\odot$) WR star model. However, MOST found no coherent oscillations in the WR star WR 111 (Moffat et al. 2008). It is not certain yet how ubiquitous pulsations are in WR stars, although theory predicts that all WR stars should pulsate.

Strange modes in R CrB and Extreme Helium B stars

Strange modes were first found in less massive but luminous helium star models for R CrB and Extreme Helium B stars (Wood 1976; Gautschy & Glatzel 1990). Those earlier studies were based on the LANL opacities and the modal diagrams were relatively simple. Using OPAL (or OP) opacity tables makes modal diagrams complex, and excites much more pulsation modes especially in hotter models (e.g., Saio 1995, 2008). Excitation of a large number of modes probably corresponds to the very complex light curves of luminous extreme helium stars (e.g., Kilkenny et al. 1999). The instability boundary for helium star models is similar to that shown in Fig. 2 if the vertical axis is converted to $\log(L/M)$.

Strange modes in post-AGB stars

Some F-supergiants identified as post-AGB stars are known to show irregular light variations which are thought due to pulsations. Theoretical works by Zalewski (1992), Gautschy (1993), and Aikawa (1993) indicate strange modes are involved.

Summary

Strange modes are defined as pulsation modes whose frequencies vary differently from ordinary modes in a modal diagram. The property comes from the phenomenon that the pulsation energy of a strange mode is confined in a narrow zone in the envelope where the radiation pressure is much larger than the gas pressure. Strange modes are strongly excited by enhanced kappa-mechanism or by the strange-mode instability; growth times are as short as a few pulsation periods. The strange mode instability is a dynamical-type instability which occurs in the limit of short thermal time. Excited strange modes appear in various stellar types with large luminosity to mass ratios ($L/M >\sim 10^4$). It is suspected that the strange modes might play an important role in enhancing mass-loss from these stars. To clarify the role, accurate nonlinear analyses for strange modes are needed (Glatzel 2009).

Acknowledgments. I am grateful to Wolfgang Glatzel for his hospitality and discussions at Institute for Astrophysics Göttingen, where I completed the manuscript.

References

Aikawa, T. 1993, MNRAS, 262, 893

Cox, J. P., King, D. S., Cox, A.N., et al. 1980, Space Sci. Rev., 27, 529

Dorfi, E. A., Gautschy, A., & Saio, H. 2006, A&A, 453, L35

Gautschy, A., & Glatzel, W. 1990, MNRAS, 245, 597

Gautschy, A. 1993, MNRAS, 265, 340

Glatzel, W. 1994, MNRAS, 271, 66

Glatzel, W., & Gautschy, A. 1992, MNRAS, 256, 209

Glatzel, W., & Kaltschmidt, H. O. 2002 MNRAS, 337, 743

Glatzel, W., & Kiriakidis, M. 1993a, MNRAS, 262, 85

Glatzel, W., & Kiriakidis, M. 1993b, MNRAS, 263, 375

Glatzel, W., Kiriakidis, M., & Fricke, K. J. 1993, MNRAS, 262, L7

Glatzel, W., & Mehren, S. 1996, MNRAS, 282, 1470

Glatzel, W. 2009, CoAst, 158, 252

Humphreys, R. M., & Davidson, K. 1979, AJ, 232, 409

Kilkenny, D., Lawson, W. A., Marang, F., et al. 1999, MNRAS, 305, 103

Kiriakidis, M., Fricke, K. J., & Glatzel, W. 1993, MNRAS, 264 50

Kiriakidis, M., Glatzel, W., & Fricke, K. J. 1996, MNRAS, 281, 406

Lefèvre, L., Marchenko, S. V., Moffat, A. F. J., et al. 2005, ApJ, 634, L109

Moffat, A. F. J., Marchenko, S. V., Zhilyaev, B. E., et al. 2008, ApJ, 679, L45

Papaloizou, J. C. B., Alberts, F., Pringle, J. E., & Savonije, G. J. 1997, MNRAS, 284, 821

Rogers, F. J., & Iglesias, C. A. 1992, ApJS, 79, 507

Saio, H. 1995, MNRAS, 277, 1393

Saio, H. 2008, ASPC, 391, 69

Saio, H., Baker, N. H., & Gautschy, A. 1998, MNRAS, 294, 622

Saio, H., Wheeler, J. C., & Cox, J. P. 1984, ApJ, 281, 318

Wood, P. R. 1976, MNRAS, 174, 531
Zalewski, J. 1992, PASJ, 44, 27

DISCUSSION

Scuflaire: When a model is close to dynamical instability, the dynamical and thermal time scales may be of the same order of magnitude. In this case, it is no longer possible to make a distinction between a dynamical mode and a secular mode, neither is it possible to describe an instability as dynamical, vibrational or secular (thermal).

Comm. in Asteroseismology
Vol. 158, 2009, 38th LIAC/HELAS-ESTA/BAG, 2008
A. Noels, C. Aerts, J. Montalbán, A. Miglio and M. Briquet., eds.

Nonlinear strange-mode pulsations

W. Glatzel

Institut für Astrophysik
Universität Göttingen, Friedrich - Hund - Platz 1, D-37073 Göttingen, Germany

Abstract

The final result of strange-mode instabilities in models of massive stars is determined by following their evolution into the nonlinear regime. In general the instability leads to finite amplitude pulsations which inflate the stellar envelope considerably implying an increase of pulsation periods. Thus periods derived from a linear analysis may differ significantly from periods obtained in the nonlinear regime. The relevance of this effect for the interpretation of the variability observed in the Wolf-Rayet star WR123 is discussed. Associated with strange-mode pulsations are velocity amplitudes of the order of 100 km/sec and a mean photospheric acoustic energy flux which is comparable to the kinetic energy flux observed in winds of massive stars.

Final result of strange-mode instabilities

Basic procedure, numerical treatment, requirements

To determine the final consequence of strange-mode instabilities their evolution into the nonlinear regime has to be followed by direct numerical simulation. It requires the solution of the conservation equations for mass, momentum and energy together with a prescription for energy transport. Below the photosphere and as long as convection can be considered to be negligible the latter is given by the equation for radiation diffusion. To close the system of equations it has to be supplemented by an equation of state and suitable boundary conditions at the photosphere and at the bottom of the envelope. (For a study of strange-mode instabilities it is sufficient to consider stellar envelopes .)

The first step of the analysis consists of constructing hydrostatic initial models with prescribed chemical composition, mass, luminosity and effective temperature. If convection is present, it is treated according to standard mixing length theory. These hydrostatic models are tested for linear nonadiabatic stability and the occurrence of strange-mode instabilities. Assuming spherical geometry for selected unstable models their evolution in time is then followed from hydrostatic equilibrium through the linear phase of exponential growth into the nonlinear regime using the method described in Grott et al. (2005). Both in the linear and nonlinear analysis convection is treated in the frozen - in approximation. The physical instability is required to emerge in hydrostatic equilibrium from numerical noise without any additional external perturbation. The linear regime of the instability then allows the code to be validated with the independently derived results of the linear stability analysis.

Concerning the numerical simulations on which the results presented here are based, we have to add some cautionary remarks. Since a sophisticated treatment of radiation transport is not feasible, we have disregarded the optically thin regime and restricted our analysis to subphotospheric levels. Therefore the outer boundary of the models does not correspond to the physical boundary of the system with the consequence that the outer boundary conditions for the models are ambiguous. Whether the results are sensitive to this ambiguity, requires careful study. For the results discussed here, we have required the gradient of the compression and the storage of heat at the outer boundary to vanish. These conditions appear to allow outgoing shocks to pass the boundary without reflection. For more details we refer to Grott et al. (2005). strange-mode instabilities are not restricted to spherically symmetric perturbations. Whether the geometry has a significant influence, needs to be studied, as soon as multidimensional simulations become feasible. The occurrence of shock waves requires the introduction of artificial viscosity. Contrary to many physical quantities which are not severely affected, the acoustic luminosity generated by the strange-mode driven pulsations and being of particular interest here is sensitive to the treatment of numerical viscosity.

With respect to a meaningful determination of the acoustic luminosity associated with strange-mode driven pulsations the energy balance of the system requires special consideration. It is dominated by the three terms representing the gravitational and thermal energies together with the time integral of the (thermal) luminosity. Their sum is smaller than the constituents by approximately two orders of magnitude and comparable with the kinetic energy. Finally, the remaining and smallest term in the energy balance is the time integral of the acoustic luminosity. Since phases with incoming and outgoing mechanical energy flux alternate during a pulsation cycle, this function is not monotonic. However, on the average it increases with time indicating a mean mechanical energy flux from the stellar envelope into the star's atmosphere. Its mean slope corresponds to the mean mechanical (acoustic) luminosity generated by the strange-mode driven pulsations.

Considering the various terms in the energy balance and their order of magnitude we deduce that for a meaningful correct determination of the mean acoustic luminosity the relative error in the energy balance has to be smaller than 10^{-4}. This condition can never be satisfied by ordinary numerical methods and requires the use of sophisticated (even locally) fully conservative schemes. The results presented here are based on such algorithms (for details see Grott et al., 2005).

Even if we can guarantee that the energy balance of the system is correct with a precision that allows for a meaningful determination of the acoustic luminosity, the latter still depends on the numerical viscosity necessary to handle shock waves. Viscosity has a dissipative effect and transfers mechanical energy into thermal energy. Therefore the acoustic luminosity increases with decreasing artificial viscosity and the values determined have to be regarded as lower limits of the correct mechanical luminosity.

General properties of strange-mode pulsations (WR models)

The general properties of finite amplitude strange-mode pulsations will be discussed here for the paradigm of models for Wolf-Rayet stars which exhibit the strongest strange-mode instabilities identified so far. As an example, the evolution in time of the photospheric velocity of an envelope model with the chemical composition $(X, Y, Z) = (0, 0.98, 0.02)$, the mass $M = 17.05 M_\odot$, the initial luminosity $L = 3.67 \cdot 10^5 L_\odot$ and the initial effective temperature $T_{eff} = 60000 K$ is shown in Fig. 1. This model has several unstable modes which emerge from numerical noise ($\approx 10^{-6} cm/sec$) with correct periods and growth rates in the linear regime of exponential growth and saturate at an amplitude of $\approx 4 \cdot 10^7 cm/sec$ in the nonlinear regime. Thus the final result of strange-mode instabilities in Wolf-Rayet stars are finite amplitude pulsations with velocity amplitudes of the order of $\mathcal{O} \sim 100$ km/sec, which is a significant fraction (up to 50%) of the escape speed of the object. Although they do exhibit dominant periods, we note that the final pulsations are in general not strictly periodic. Multi-periodicity with indications for chaos are common. From Fig. 1 we deduce a dominant final nonlinear period close to 10 hr, which is much longer than the periods of the unstable modes of the model determined by the linear analysis (around 0.6 hr) and seen in the linear regime in

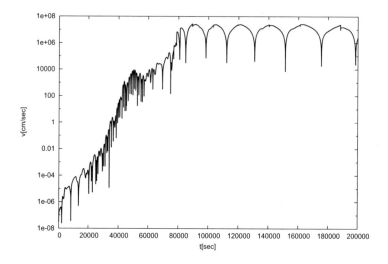

Figure 1: Modulus of the photospheric velocity as a function of time for a Wolf - Rayet model with $(X, Y, Z) = (0, 0.98, 0.02)$; $M = 17.05 M_{\odot}$; $L_{initial} = 3.67 \cdot 10^5 L_{\odot}$; $T_{eff,initial} = 60kK$

Fig.1. Thus nonlinear and linear periods may differ substantially and observed periods should be compared with nonlinear rather than with linear periods of a model. With respect to WR123 (see next section) we note that strange-mode driven pulsations with periods around 10 hr can easily be obtained even with hydrogen depleted models ($X = 0$).

The increase of pulsation periods in the nonlinear regime is caused by a substantial increase of the mean stellar radius (see Fig. 2) and − associated with it − a decrease of the mean effective temperature. (Due to the small heat capacity of the envelope the mean luminosity remains constant.)

The inflation of the star is a consequence of consecutive shock waves generated by strange-mode instabilities, and pushing the stellar matter to higher levels. For illustration, snapshots of the density stratification are shown in Fig.3. It reveals a sequence of almost isothermal shocks travelling from the instability region towards the photosphere.

The inflation of the stellar envelope by strange-mode driven shock waves leads to a decrease of its mean effective temperature. Largely independent of the initial effective temperature and the mass (or the luminosity) of the model a unique final mean effective temperature around 30kK is reached in most cases. (The mean luminosity remains constant.) That the corresponding position in the HRD is remarkably close to the observed position of many Wolf-Rayet stars, is noteworthy.

The selection of a unique final mean effective temperature around 30kK may be caused by the dependence on temperature of the opacity. This idea is supported by an (empirical) coincidence of the final mean effective temperatures reached in the simulations with a temperature strip around 30 kK defined by the descending branch of the opacity from its maximum due to the second ionization of helium towards lower temperatures. It might suggest that in

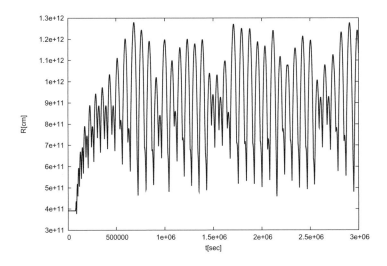

Figure 2: Stellar radius as a function of time for a Wolf-Rayet model with $(X, Y, Z) = (0, 0.98, 0.02)$; $M = 17.05 M_\odot$; $L_{initial} = 3.67 \cdot 10^5 L_\odot$; $T_{eff,initial} = 60kK$

general final mean effective temperatures only lie in temperature intervals where the opacity increases with temperature. Should the hypothesis be correct, we would expect two additional 'attracting' temperature intervals around 15 kK and above 100 kK, where final mean effective temperatures could be found. In fact, for suitably chosen initial models, final mean effective temperatures in these intervals were reached.

Associated with the finite amplitude pulsations is a transfer of acoustic energy to the atmosphere of the star. Being a nonmonotonic function of time (see Fig. 4) its mean slope corresponds to the mean mechanical luminosity of the star (see previous section). From Fig. 4 we deduce $L_{acoustic} \sim 10^{36} ergs/sec$ for its value. Identfying this mechanical luminosity with the flux of kinetic energy in a stationary stellar wind (using the relation $L_{acoustic} \sim \frac{1}{2} \dot{M} v_{escape}^2$) we derive a mass loss rate of $\dot{M} \sim 10^{-6} M_\odot/yr$. Thus, concerning the energetics strange-mode driven pulsations in Wolf-Rayet stars could easily drive the winds observed in these objects. With respect to the violent variability (even deep below the photosphere) of the Wolf-Rayet model envelopes presented involving highly supersonic flows and strong shocks we may ask, whether stationary wind models are meaningful approaches for the description of such objects.

WR123: Interpretation of the observation

Using the MOST satellite photometric monitoring of the Wolf-Rayet star WR123 has revealed a light variation of this object with a period of 9.8 hr (Lefévre et al., 2005). Strange-mode periods for He-ZAMS models, however, lie below 1 hr (see Glatzel et al., 1993, and Kiriakidis et al., 1996). Therefore Lefévre et al. (2005) have excluded strange-mode instabilities and

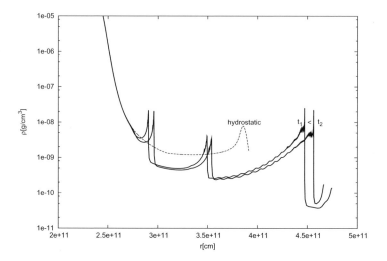

Figure 3: Snapshots of the density stratification in the nonlinear regime of strange-mode driven pulsations of a Wolf-Rayet model with $(X, Y, Z) = (0, 0.98, 0.02)$; $M = 17.05 M_\odot$; $L_{initial} = 3.67 \cdot 10^5 L_\odot$; $T_{eff,initial} = 60kK$

strange-mode pulsations as the origin of the observed light variation in WR123.

Subsequently, Dorfi et al. (2006) argued that the observed period should be compared to strange-mode periods of a model with a radius comparable to the observed radius of 15.3 $R_\odot$ for WR123 (Crowther 1997) rather than to the strange-modes periods of He-ZAMS models with radii smaller than 1 $R_\odot$. According to the period density relation the period of a given mode should be proportional to $R^{3/2}$, where R denotes the stellar radius. Correcting the strange-mode periods of the He-ZAMS models for the radius according to this relation would shift them into a range compatible with the observed period of WR123. The authors support their argument by presenting a variety of RHD models with pulsation periods close to that observed in WR123 and conclude that its variability could well be attributed to a strange-mode instability. However, the investigations of Dorfi et al. (2006) were based on hydrogen-rich envelopes with X=0.35, whereas WR123 seems to be largely depleted of hydrogen ($X < 0.005$; see Crowther et al., 1995).

Townsend & MacDonald (2006) have studied the excitation of low degree ($l = 1, 2$) g-modes in models for Wolf-Rayet stars by the κ- mechanism due to an opacity bump around $\log T \approx 6.25$. They find the periods of unstable modes to depend on the mass fraction of hydrogen, where only those of hydrogen - depleted models are consistent with the observed period in WR123. Since both the chemical composition and the observed periods of WR123 are compatible with g-modes excited by the κ- mechanism, they suggest the latter to be the source of the observed light variation in WR123. The explanation of Dorfi et al. (2006) in terms of strange-mode driven pulsations is rejected, since for their models they have adopted a for WR123 unrealistically high mass fraction of hydrogen of $X = 0.35$.

In the previous Section we have shown that even models containing no hydrogen ($X = 0$)

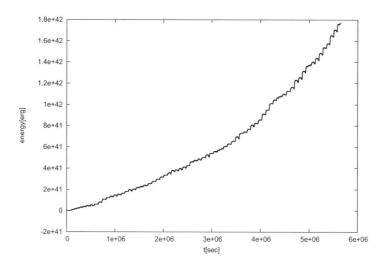

Figure 4: Acoustic energy transferred to the atmosphere by pulsations as a function of time for a Wolf - Rayet model with $(X, Y, Z) = (0, 0.98, 0.02)$; $M = 17.05 M_\odot$; $L_{initial} = 3.67 \cdot 10^5 L_\odot$; $T_{eff,initial} = 60kK$

exhibit strange-mode driven pulsations with (nonlinear) periods around 10 hr comparable to the observed period of WR123. Thus the observations may be interpreted in terms of strange-mode driven pulsations even on the basis of models containing no hydrogen ($X = 0$). The explanation is based on proper numerical simulations of the evolution of strange-mode instabilities into the nonlinear regime. Consecutive shock waves generated by strange-mode instabilities inflate the stellar envelope considerably thus increasing the pulsation periods and shifting them into a range compatible with the observations of WR123. According to this nonlinear effect, the interpretation of observed periods in terms of linear strange-mode periods is not correct here and should in general be done with caution. Thus a mass fraction of hydrogen of $X = 0.35$ adopted by Dorfi et al. (2006) is not necessary to provide periods close to that observed in WR123 and the argument by Townsend & MacDonald (2006) against strange-mode driven pulsations in WR123 does not hold. That the growth rates of the g-modes considered by Townsend & MacDonald (2006) are smaller than the typical growth rates of strange modes by at least three orders of magnitude, seems to be noteworthy.

Acknowledgments. Contributions to the subject by S. Chernigovski, M. Grott, S. Kühnrich and S. Wende are gratefully acknowledged.

References

Crowther, P.A. 1997, in Luminous Blue Variables: Massive Stars in Transition, ed. A. Nota & H.J.G.L.M. Lamers (ASP), 120, 51

Crowther, P.A., Hillier, D.J., & Smith, L.J. 1995, A&A, 293, 403

Dorfi, E.A., Gautschy, A., & Saio, H. 2006, A&A, 453, L35

Glatzel, W., Kiriakidis, M., & Fricke, K.J. 1993, MNRAS, 262, L7

Grott, M., Chernigovski, S., & Glatzel, W. 2005, MNRAS, 360, 1532

Kiriakidis, M., Glatzel, W., & Fricke, K.J. 1996, MNRAS, 281, 406

Lefèvre, L., Marchenko, S.V., Moffat, A.F.J., et al. 2005, ApJ, 634, L109

Townsend, R.H.D., & MacDonald, J. 2006, MNRAS, 368, L57

DISCUSSION

Noels: I would like to propose the following origin of strange modes. In massive, highly luminous stars, the thermal timescale becomes of the same order as the dynamical timescale. In that case, the secular spectrum cannot be separated from the "normal" spectrum. So strange modes would be secular modes appearing in the same frequency range than the dynamical modes.

Comm. in Asteroseismology
Vol. 158, 2009, 38th LIAC/HELAS-ESTA/BAG, 2008
A. Noels, C. Aerts, J. Montalbán, A. Miglio and M. Briquet., eds.

Pulsation and convection in Luminous Blue Variable stars

A. N. Cox and J. A. Guzik

Los Alamos National Laboratory, Los Alamos, NM 87545, USA

Abstract

Evolution, linear pulsation studies, and hydrodynamic calculations at Los Alamos over the last 12 years will be reviewed to discuss mechanisms that can cause at least the observed milder, S Doradus type, outbursts. Eddington found long ago that where the radiation luminosity exceeds the Eddington luminosity, the internal pressure gradient is steeper than can be constrained by the local gravity. Then local outward motions can happen and, when the super-Eddington luminosities occur in thick layers, hydrodynamic outbursts can occur. Large luminosities are usually accommodated by stellar models having a significant convection zone to carry the large luminosity, but if the convection turns on and off during pulsations, and is not able to adapt rapidly enough relative to the natural pulsation period of the model, significant outward motions can occur. Then both radial and nonradial outflow occur. We believe that super-Eddington luminosities and time-dependent convection are important mechanisms for at least low mass loss outbursts.

Luminous Blue Variables

The brightest stars in galaxies are often variable in light because they display self-excited radial and maybe even nonradial pulsations. On the main sequence where large masses become stars, burning hydrogen to helium, it has been long known that if the stellar mass is about 80 or more times the solar mass, the central convective core burning can vary with the pulsation period, and the pulsations may grow to a disruptive amplitude. A recent study of this can be found in Cahn Cox & Ostlie (1987). When the star produces considerable central helium, the luminosity source for the star shifts to a convective hydrogen burning shell outside the original convection core. The core gradually becomes inert until much later in the evolution when shrinkage of the core produces high enough temperatures to burn helium to carbon and oxygen.

Actually, it may be that, depending on the composition and therefore the internal opacity, stars of such large mass may have a luminosity above the Eddington limit, and therefore no hydrostatic model can be constructed for them anyway.

These stars immediately begin to expand from their main sequence radius to show cooler surface temperatures. With a small depletion of the central hydrogen, they no longer display that hydrogen-burning epsilon effect pulsation driving. They still are driven into pulsation, though, by another well-known kappa effect driving near 250,000 K caused by a large opacity from iron atoms ionized down to their M atomic shell. This large opacity is produced by the very large number of same-atomic shell electron transitions, giving much more mean opacity than thought before about 1990.

The blue and red edges on the H-R diagram for this iron kappa effect pulsation discussed in this paper has been investigated by Soukup Cox Guzik & Morgan (1994). It may overlap with the main sequence region epsilon effect instability strip.

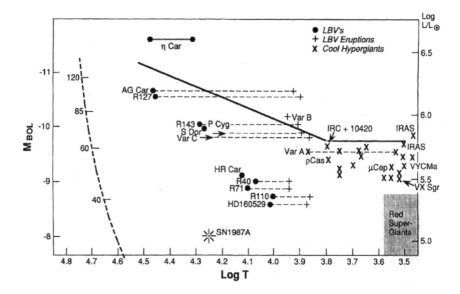

Figure 1: Many luminous blue variables are plotted on the H-R diagram at their normal and outburst positions. The pre-supernova 1987A position is also plotted to show that the evolved massive stars are more luminous than this highly evolved pre-supernova.

Figure 1 shows the Hertzsprung-Russell diagram for many luminous blue supergiants and the Humphreys-Davidson line drawn there. This H-D line marks the coolest of the observed luminous blue variables on the H-R diagram, and it may actually be the red edge of the iron kappa effect pulsation strip for the radial and nonradial pulsation modes.

Our first LBV work originated at a Hawaii meeting producing Cox et al. (1997) and Guzik et al. (1997).

Blue Variable Star Model

For the study reported here, we use a model developed by Soukup that originally had a mass of 35 solar mass, but after considerable mass loss from its hot stellar wind, the current mass is only 22 solar mass. The luminosity for this model is close to that observed for HR Car in our galaxy, about 0.37 million solar luminosities, and the surface effective temperature is 21,040 K. To get good spatial resolution, the model has 1400 mass shells. Such a young star should have carbon, oxygen, and iron abundance enhancements over that for our older Sun, but the recent Asplund et al. solar element abundances Asplund et al. (2005) and Livermore OPAL opacities are used anyway for this exploratory case. The Soukup model using the solar composition has an opacity structure given in Figure 2. About half the mass deep in this model, the local hydrogen has been converted to almost pure helium and just a bit deeper is a convective helium core, already half burned out to carbon and oxygen.

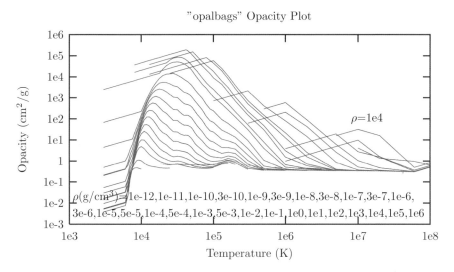

Figure 2: The opacity structure of the Soukup model using the solar composition. This displays the entire opacity table, but the opacity in this model is given by the line at the bottom of the figure. The opacity behavior shows that it increases with temperature near 200,000 K, and this results in the well-known kappa effect pulsational driving.

Linear Pulsation Solution

Construction of a massive star model may not be successful if the radiative luminosity exceeds the Eddington limit. This luminosity limit is reached when the temperature gradient needed to carry this Eddington luminosity results in a (radiative) pressure gradient steeper than what gravity can balance. Then the model cannot have any hydrostatic equilibrium structure.

The only pulsationally unstable mode found for the Soukup model is the radial fundamental mode at a period of 3.4 days and a growth rate of 7.2e-2 per period (see Fig. 3). Searches were made for the first and second radial overtones and the l=1 and l=2 low order p, f, and g modes. It would be expected that with more short spatial displacements, as for these latter modes, radiative luminosity can flow away from layers that are driving the pulsations instead of converting luminosity excursions to motions.

The usual procedure for this LBV study is to find the most pulsationally unstable mode from the linear analysis, and to insert this motion structure into the hydrodynamic code to follow its growth. Since this code is only 1D, only radial modes can be inserted. Experience has shown that the mode to follow is the fundamental with no motion nodes, or the first radial overtone.

Nonlinear Pulsation Solution

It is found that the luminosity gets very large periodically, and the temperature gradient becomes superadiabatic. In this case convection turns on to start carrying this luminosity, but it can adapt slowly compared to the pulsation period. Then if the radiative part of the luminosity is super-Eddington, it can cause outward motions in the entire mass shell that is super-Eddington (see Fig. 4). The hydrodynamic calculations demonstrate this, and after many pulsation cycles, an outburst with mass loss occurs.

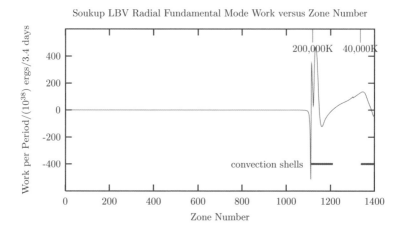

Figure 3: The work per zone done on and by each of the 1400 mass shells each period.

The actual equations for how the convective luminosity behaves with time are given by Cox (1990). In many layers of the stellar model, the sluggish start for the convection does not allow convective elements to move across very much of their mixing length each hydrodynamic time step, and convection often never turns on to its full strength before the luminosity starts its cyclical decrease to require only a sub-adiabatic temperature gradient.

Conclusions

Hydrodynamic calculations for many different models have been done by our Los Alamos team, and they have been reported in the last few years. The Eddington luminosity at any position in our stellar model depends inversely on the opacity, and for larger heavy element (larger Z including iron) abundance, it is easier to get super-Eddington luminosities. The fourth figure shows this outburst (a positive velocity as seen from the star, but a negative radial velocity as seen from the earth.

The outbursts we get in the computer start right away as there are just a few pulsation cycles, of about a million seconds each for this fundamental mode, before the rapid growth. Real massive stars, though, have outbursts, presumably like we get, but then after any outburst, it may take many years before the star can reassemble itself as we have for our model,

Unfortunately treatment of the actual mass loss is not yet done here as it is for nova and supernova cases. The amount of mass involved in this surface eruption is typically only 0.0001 of the total model mass. But as the mass loss continues, original deeper layers now become the surface layers, and the entire phenomenon can occur again. Only when levels so deep that they have experienced hydrogen and helium burning, and associated convection mixing, will the general behavior be different. At that time also the total stellar mass will be significantly less so that maybe the Eddington luminosity can no longer be reached. Then the outward pushing is not available anymore. Each luminous blue variable might have thousands of outbursts in its life.

Acknowledgments. Our thanks to Jay Onifer at the University of Florida who worked extensively on luminous blue variables when he was a postdoc with Joyce Guzik at Los Alamos.

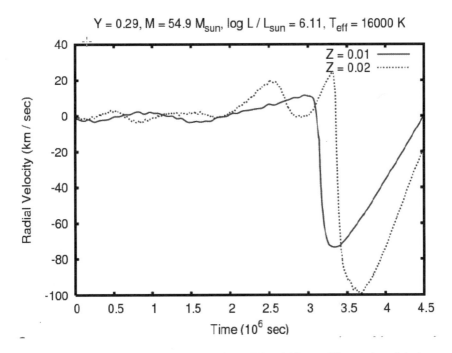

Figure 4: The radial velocity of the outer surface of the LBV model for two different values of the heavy element abundance.

References

Asplund, M., Grevesse, N., & Sauval, A.J. 2005, ASP Conf. Ser., 336, 25

Cahn, J.H, Cox, A.N., & Ostlie, D.A. 1987, Lecture Notes in Physics, 274, 51

Cox, A.N., 1990, Annals of the New York Academy of Sciences, 617, 66

Cox, A.N., Guzik, J.A., & Soukup, M.S. 1997, ASP Conf. Ser., 120, 133

Guzik, J.A., Cox, A.N., Despain, K.M., & Soukup, M.S. 1997, ASP Conf. Ser., 120, 138

Soukup, M.S., Cox, A.N., Guzik, J.A., & Morgan. S.M. 1994, AAS, 184, 3101

Comm. in Asteroseismology
Vol. 158, 2009, 38th LIAC/HELAS-ESTA/BAG, 2008
A. Noels, C. Aerts, J. Montalbán, A. Miglio and M. Briquet., eds.

Comparison of pulsation modes in rapidly rotating polytropic and SCF models

D. Reese[1], K. B. MacGregor[2], S. Jackson[2], A. Skumanich[2], and T. S. Metcalfe[2]

[1] Department of Applied Mathematics, University of Sheffield, Hicks Building, Hounsfield Road, Sheffield
S3 7RH, UK
[2] High Altitude Observatory, National Center for Atmospheric Research, Boulder, CO 80307, USA

Abstract

In this talk, I will show numerical calculations of pulsation modes in rapidly differentially rotating stellar models based on the self-consistent field (SCF) method (e.g. MacGregor et al. 2007). The pulsation calculations are based on the numerical method presented in Lignières et al. (2006) and Reese et al. (2006). I will compare these results with previous calculations based on simpler polytropic models, and discuss how the structure of the frequency spectrum is affected by differential rotation and other stellar parameters. In particular, differential rotation can lead to a breakdown of regularities in the frequency spectrum in favour of a more chaotic behaviour.

Introduction

A number of studies have shown the important role played by stellar rotation in gravity (e.g. Townsend 2005), r (e.g. Lee 2006), gravito-inertial (e.g. Dintrans & Rieutord 2000) and acoustic modes (Reese 2008 and references therein). In particular, it was shown, using polytropic models, that rapid rotation and stellar deformation have a major impact on the geometric structure, frequency organisation and ray dynamics of acoustic modes. The following questions naturally arise: Do similar results apply for realistic models of rapidly rotating stars? What happens when the rotation profile is differential?

In the present paper, we aim to address these questions by applying the numerical method for calculating pulsation modes developed in Lignières et al. (2006) and Reese et al. (2006) to the more realistic models of rapidly rotating stars based on the SCF method (Jackson et al. 2004, 2005, MacGregor et al. 2007). The first section briefly describes the SCF models as well as the numerical method for calculating pulsation modes. The next section deals with pulsations in uniformly or nearly uniformly rotating models. The following section looks at what happens when the rotation profile becomes strongly differential.

Description of the models and the numerical method

The self-consistent field (SCF) method is an iterative procedure which produces rapidly rotating stellar models. The basic idea is to alternate between solving Poisson's equation, and solving the rest of the stellar equilibrium equations. Currently, models based on the SCF method are homogeneous ZAMS models with a conservative, i.e. cylindrical, rotation profile:

$$\Omega(s) = \frac{\eta \Omega_{cr}}{1 + \left(\frac{\alpha s}{R_{eq}}\right)^2} \tag{1}$$

where s is the distance to the rotation axis, R_{eq} the equatorial radius, Ω_{cr} the equatorial critical rotation rate, and η and α two parameters which control how fast and how differential the rotation profile is. The quantity Ω_{cr} is given by $\sqrt{g_{eq}/R_{eq}}$, where g_{eq} is the gravity at the equator, excluding the centrifugal force. As a result of the rotation profile being conservative, thermodynamic quantities remain constant on level surfaces.

It is important to note that other works have used different rotation profiles. For instance, Zahn's shellular rotation profile (Zahn, 1992) has been used in a series of papers which model the effects of rotation on stellar evolution and which have been backed by a number of observations (Meynet & Maeder 2005 and references therein). More recently, Rieutord (2006) came up with 2D rotation profiles based on baroclinic effects. The pulsation modes of such models will need to be studied for future asteroseismological investigations.

The numerical method for calculating pulsation modes makes use of an adapted set of coordinates which follows the shape of the star. Then the set of pulsation equations are projected onto the spherical harmonic basis. From there, it is possible to discretise in the radial direction using either finite-differences or Chebyshev polynomials. A more detailed description of this method, its accuracy, the pulsation equations and the SCF models can be found in Reese et al. (2008b) and references therein.

Uniform or nearly uniform rotation profile

Mode classification

Lignières & Georgeot (2008) have previously shown that for rotating polytropic models, acoustic modes fall into three categories: island modes, chaotic modes and whispering gallery modes. These are the rotating counterparts to low, intermediate and high degree modes, respectively. We have found that similar results also apply to pulsation modes in SCF models with uniform or mildly differential rotation (at least up to $\alpha = 0.4$, which, based on Eq. (1), gives an equatorial rotation rate which is 84% of the polar rotation rate). Figure 1 compares pulsation modes from both types of models. As can be seen in the figure, corresponding modes have an analogous structure.

Island modes

In the polytropic case, it is possible to introduce a new set of quantum numbers $(\tilde{n}, \tilde{\ell}, m)$ for the island modes, based on their geometry (e.g. Reese 2008). These quantum numbers then intervene in a new asymptotic formula which describes the frequency organisation of island modes (Lignières & Georgeot 2008, Reese et al. 2008a):

$$\omega = \tilde{n}\Delta_{\tilde{n}} + \tilde{\ell}\Delta_{\tilde{\ell}} + |m|\Delta_{\tilde{m}} - m\Omega + \tilde{\alpha} \tag{2}$$

where $\Delta_{\tilde{n}}$, $\Delta_{\tilde{\ell}}$, $\Delta_{\tilde{m}}$ and $\tilde{\alpha}$ are four parameters which depend on stellar structure and Ω the rotation rate. The same formula also applies for pulsation modes in SCF models with uniform or mildly differential rotation[1]. Table 1 gives the values of these parameters for selected SCF models as well as for a polytropic model. Due to difficulties in mode identification, the parameters given for the most rapidly rotating models ($\eta = 0.9$) must be taken with caution. As was noted in Reese et al. (2008a) for polytropic models, the ratio $\Delta_{\tilde{\ell}}/\Delta_{\tilde{n}}$ decreases for increasing rotation rates, although in this case other factors like stellar mass also play a role.

In the last column, the average deviation between the asymptotic and numerical frequencies is also included. Although the asymptotic formula captures the basic structure of the frequency spectrum (at least for low values of m), these differences are significant. Also, the value of Ω obtained for the differentially rotating 25 $M_{\odot}$, $\eta = 0.6$, $\alpha = 0.2$ model is

---

[1]When the rotation profile is differential, we treat Ω as a fifth parameter in Eq. (2).

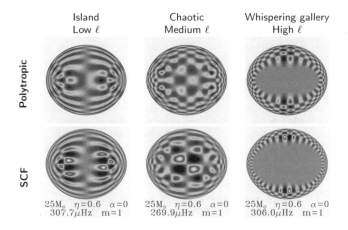

Figure 1: A comparison between pulsation modes in polytropic models and in SCF models. The plots in this and the other figures represent a meridional cross-section of the Eulerian pressure perturbation divided by the square root of the equilibrium density. The same three categories apply in both cases as can be seen by the analogous geometric structure.

$\dfrac{M}{M_\odot}$	η	α	$\dfrac{\Delta_{\tilde{n}}}{(\mu Hz)}$	$\dfrac{\Delta_{\tilde{\ell}}}{\Delta_{\tilde{n}}}$	$\dfrac{\Delta_{\tilde{m}}}{\Delta_{\tilde{n}}}$	$\dfrac{\tilde{\alpha}}{\Delta_{\tilde{n}}}$	$\dfrac{\Omega}{\Delta_{\tilde{n}}}$	$\dfrac{\langle \delta\omega^2\rangle^{1/2}}{\Delta_{\tilde{n}}}$
poly*	0.6	0.0	36.7	0.660	0.063	2.901	0.838	0.047
1.7	0.7	0.0	37.2	0.77	0.04	3.46	0.98	0.025
1.8	0.9	0.0	33.5	0.42	0.01	2.85	1.17	0.038
25.0	0.6	0.0	15.3	0.80	0.04	3.31	0.97	0.038
25.0	0.6	0.2	15.1	0.85	0.04	3.60	0.94†	0.044
25.0	0.6	0.4	15.5	0.89	0.12	3.37	0.92†	0.066
25.0	0.9	0.0	12.4	0.70	0.002	3.35	1.38	0.034

* polytropic model with $N = 3$, $M = 1.7 M_\odot$ and $R_{eq} = 1.84 R_\odot$
† in this case, Ω is treated like a fifth parameter in Eq. (2)

Table 1: Values of the different parameters from Eq. (2) for selected SCF models as well as for a polytropic model (first line). The last column contains the average deviation between asymptotic frequencies based on Eq. (2) and the numerical frequencies. The lower mass models resemble HD 181555 (observed by CoRoT) and the other models are similar to ζ Oph (observed by MOST).

marginally below the range covered by the rotation profile. Both of these point towards the need for a more accurate formula as will be discussed in Reese et al. (2008c).

Highly differential rotation

When the rotation profile becomes highly differential, the stellar structure becomes very deformed and the polar regions can, in some cases, become concave. This naturally affects the structure and organisation of pulsation modes. Figure 2 shows a chaotic and a whispering gallery mode in models where the rotation profile is very differential. No island modes are

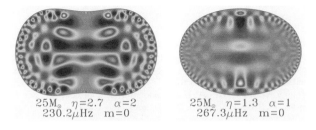

$25M_\odot$ $\eta=2.7$ $\alpha=2$ $25M_\odot$ $\eta=1.3$ $\alpha=1$
230.2μHz   m=0 267.3μHz   m=0

Figure 2: The left figure corresponds to a chaotic mode and the right one to what appears to be a whispering gallery mode in SCF models with a highly differential rotation profile. No island modes are shown as they seem to have disappeared in the models (for low m).

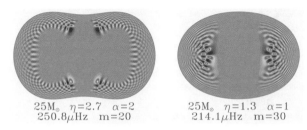

$25M_\odot$ $\eta=2.7$ $\alpha=2$ $25M_\odot$ $\eta=1.3$ $\alpha=1$
250.8μHz   m=20 214.1μHz   m=30

Figure 3: The two figures corresponds to pulsation modes with high azimuthal orders in SCF modes with a highly differential profile. The left figure corresponds to a whispering gallery mode and the right one to an island mode. As can be seen in these plots, pulsation modes become less chaotic with increasing azimuthal order.

shown as they seem to have disappeared. In the more distorted configurations, even whispering gallery modes become difficult to find. Instead, most of the modes show a very chaotic behaviour.

One way to counteract the effects of stellar distortion is to increase the azimuthal order m. As can be seen in Fig. 3, highly regular whispering gallery modes exist even in the most deformed configurations. Also, for models with less distortion, it is possible to find some island modes.

Conclusion

The numerical calculations presented here show that similar results apply to acoustic pulsation modes in SCF models as to those in uniformly rapidly rotating polytropic models, provided that the rotation profile is close to uniform. Pulsation modes fall in the same three categories (island, chaotic and whispering gallery) and the frequencies of the island modes follow the same asymptotic behaviour. When the rotation profile becomes highly differential, stellar deformation seems to destroy the island modes and perhaps even the whispering gallery modes for low azimuthal orders. Increasing the azimuthal order counteracts the effects of stellar distortion, thus leading to the reappearance of more regular modes.

Acknowledgments. Many of the numerical calculations were carried out on the Altix 3700 of CALMIP ("CALcul en MIdi-Pyrénées") and on Iceberg (University of Sheffield), both of which are gratefully acknowledged. DR gratefully acknowledges support from the UK Science and Technology Facilities Council through grant ST/F501796/1, and from the European Helio- and Asteroseismology Network (HELAS), a major international collaboration funded by the European Commission's Sixth Framework Programme. The National Center for Atmospheric Research is sponsored by the National Science Foundation.

References

Jackson, S., MacGregor, K. B., & Skumanich, A. 2004, ApJ, 606, 1196

Jackson, S., MacGregor, K. B., & Skumanich, A. 2005, ApJS, 156, 245

Dintrans, B., & Rieutord, M. 2000, A&A, 354, 86

Lee, U. 2006, MNRAS, 365, 677

Lignières, F., & Georgeot, B. 2008, Phys. Rev. E, 78, 016215

Lignières, F., Rieutord, M., & Reese, D. 2006, A&A, 455, 607

MacGregor, K. B., Jackson, S., Skumanich, A., & Metcalfe, T. S. 2007, ApJ, 663, 560

Meynet, G., & Maeder, A. 2005, ASPC 337, 15

Reese, D. 2008, JPhCS, 118, 012023

Reese, D., Lignières, F., & Rieutord, M. 2006, A&A, 455, 621

Reese, D., Lignières, F., & Rieutord, M. 2008a, A&A, 481, 449

Reese, D., MacGregor, K. B., Jackson, S., et al. 2008b, SF2A, 531

Reese, D., MacGregor, K. B., Jackson, S., et al. 2008c, in GONG 2008/SOHO XXI, submitted

Rieutord, M. 2006, A&A, 451, 1025

Townsend, R. H. D. 2005, MNRAS, 360, 465

Zahn, J.-P. 1992, A&A, 265, 115

DISCUSSION

Shibahashi: Could you tell us the counterparts of "whispering gallery modes", "chaotic modes", and "island modes" in non-rotating stars, please? Do they have the same origin or a different one? What is the correspondence between the usual quantum numbers (n,l,m) and the modified quantum numbers?

Reese: Geometrically, modes in non-rotating stars are whispering gallery modes. Based on how modes evolve from low to high rotation, we have the following correspondence:

-Island modes: *the rotating counterpart to low l modes*. The relationship between old and new quantum numbers is described in the HELAS II proceedings (see Reese, 2008)

-Chaotic modes: *the rotating counterpart to medium l modes*. There is no easy classification of modes so no quantum number.

-Whispering gallery modes: *the rotating counterpart to high l modes*. They are probably well classified but not studied yet.

Comm. in Asteroseismology
Vol. 158, 2009, 38th LIAC/HELAS-ESTA/BAG, 2008
A. Noels, C. Aerts, J. Montalbán, A. Miglio and M. Briquet., eds.

Theoretical amplitudes of solar-like oscillations in classical pulsators

K. Belkacem[1], M.J. Goupil[1], R. Samadi[1], and M.A. Dupret[1]

[1]Observatoire de Paris, LESIA, CNRS UMR 8109, 92195 Meudon, France

Abstract

Seismology based on oscillation mode amplitudes allows a different probing of turbulent convection zones than usual seismology based on frequencies as shown, for instance, by Belkacem et al. (2006) for the Sun. Going a step further, we now turn to investigations of stochastic excitation of solar-like oscillations in superficial convective layers as well as in convective cores of stars more massive than the Sun. Issues are the frequency domain where solar-like oscillations can be excited, the expected magnitude of these oscillation amplitudes, and whether these amplitudes are detectable with the CoRoT mission. This is an important task since the detection of solar-like oscillations will provide strong seismic constraints on the dynamical properties of the convective layers. The detection of solar-like oscillations in stars such as β Cephei or SPB stars will also help to determine their fundamental stellar parameters.

Introduction

Every turbulent convective region is theoretically able to excite modes through the Reynolds stresses. One issue therefore is to determine the amplitudes of damped modes excited by turbulent convection in main sequence classical pulsators. Hybrid pulsators are already known to exist, some stars present both γ Doradus and δ-Scuti type oscillations (e.g., Rowe et al. 2006) or β Cephei and SPB type oscillations (e.g., Jerzykiewicz et al. 2005). In particular, stars oscillating with both types of unstable and stable (stochastically excited) p modes would probe the uppermost region of the stars and would provide additional seismic contraints.

For instance, the knowledge of a large frequency separation would help identify unstable modes and would be a powerful tool to get more insight into the convective turbulent region through the study of the driving mechanisms.

With its high quality data (Michel et al. 2008b), CoRoT is the only available photometric mission able to detect stochastically excited modes that have very low amplitudes compared to unstable ones. Hence the objective of this paper is to provide some theoretical estimations of the amplitudes for damped stochastically excited modes for main sequence classical pulsators such as SPB and β Cephei stars.

Computation of amplitudes

The mean square amplitude ($\langle |A|^2 \rangle$) of a stochastically excited mode at the surface of a star is given by:

$$\langle |A|^2 \rangle = \frac{P}{2\,\eta\,I\omega_0^2} \tag{1}$$

where P is the excitation rate, η the damping rate, ω_0 the mode frequency, and the mode inertia (I) is defined by

$$I = \int_0^R \vec{\xi}^* \cdot \vec{\xi} \, r^2 \, \rho \, \mathrm{d}r \tag{2}$$

where $\vec{\xi}$ is the mode displacement, R the star radius and ρ the density.

From Eq. (1), the computation of mode amplitudes then needs the determination of both the excitation rates and damping rates. These are calculated using the non-radial nonadiabatic code developed by Dupret (2002), while the excitation rates are computed according to the formalism developed by Belkacem et al. (2008).

The typical convective length-scales are poorly known for main sequence massive stars. Hence, the classical mixing-length theory is used to get the injection length-scale (*i.e.* the scale at which the turbulent kinetic energy spectrum is maximum) and a parameter β is introduced (see also Balmforth 1992) such that the associated wavenumber is $k_0 = 2\pi/\beta\Lambda$, where Λ is the mixing-length. Note that in the Sun, Samadi et al. (2003) have shown that $\beta \approx 5$ using 3D numerical simulations. The sensitivity of the results to the parameter β is presented in the following sections. Apart from the length-scale, one also has to specify the way the turbulent eddies are time-correlated by defining an eddy-time correlation function. A Lorentzian function had successfully been used, in the solar case (Belkacem et al. 2006) as well as for the star α Cen A (Samadi et al. 2008) and reproduces the observational data. Consequently, such a modeling will be used in the following.

Slowly Pulsating B-stars

The problem of the theoretical determination of solar-like oscillations for in δ-Scuti stars has already been addressed by Samadi et al. (2002). Here, we then focus on more massive stars on the main sequence, *i.e.* Slowly Pulsating B-stars (SPB).

Driving regions

In such stars, one can distinguish three convective regions; the convective core and the superficial regions associated with He II and He I opacity bumps (at $T \approx 40\,000K$ and $T \approx 20\,000K$, respectively). The more favorable in terms of available turbulent kinetic energy flux is the convective core due to its high density. The less favorable is the He I convective zone located in the uppermost layers. Nevertheless, the matching between convective time-scales and mode periods plays an important role for determining the efficiency of the excitation. Convective turn-over time-scales, evaluated using the MLT, are found to be of several months in the core, several days in the He II region and several hours in the most superficial convective zone.

Hence, we find that acoustic modes are only weakly excited by turbulent convection, since in the region where there is a matching between the mode period and the convective time-scales, the He I region, the available turbulent kinetic energy is small due to low densities. The computation shows that the amplitudes of those modes are of the order of 0.1 ppm, that is lower than the CoRoT detection threshold (Lefèvre, 2008 same volume). Gravity modes of lower frequencies are found to be more efficiently excited by the convective core. The excitation rates increase with decreasing frequency, because of a better time-matching between mode periods and the convective core turn-over time-scale.

Amplitudes of stochastically excited gravity modes

Thus, we focus our attention to low-frequency gravity modes for which most of the supplied energy comes from the convective core. The damping rates are found to be of the order of $\eta \approx 10^{-5}\,\mu$Hz, the minimum is found near the instability strip and they increase with frequency. They are dominated by the radial part of the perturbation of the radiative flux in the radiative regions. Hence, following Eq. (1) the computed mode amplitudes are presented in Fig. 1.

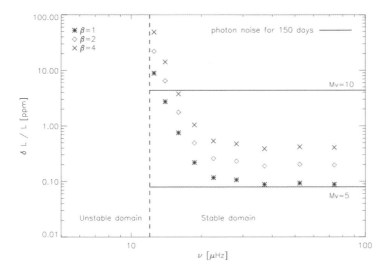

Figure 1: Relative amplitude in terms of luminosity as a function of frequency, for $\ell = 1$ g modes and three differents values of the parameter β. The dashed vertical line corresponds to the limit between the domain of unstable and stable modes while the black thick horizontal lines correspond to the photon noise for an observation of 150 days and for stars with a magnitude of $M_v = 10$ and $M_v = 5$. Those amplitudes are computed for a model of main sequence 5 $M_\odot$ star with a mixing-length parameter of $\alpha = 1$ using a Böhm-Vitense mixing-length formalism (see Samadi et al. 2006, for details) and a central hydrogen content of $X_c = 0.2$.

The maximum is found for high-order $i.e.$ low-frequency gravity modes, near the instability frequency domain. This is the result, as explained above, of the excitation rates that increase for modes with periods closer to the core convective turn-over time-scale. The amplitude of those modes are of several ppm depending on the values of the parameter β. Such low amplitudes, even with a high value of β, are still close to the CoRoT detection threshold (around 1 ppm, Lefèvre, same volume). It then makes their detection a challenge. Note that for modes of higher angular degrees, i.e. $\ell > 1$, the damping rates increase, leading to smaller amplitudes thus making the detection even more difficult.

However, near the instability domain, theoretical amplitudes are found to be high enough to be detectable. Nevertheless, the line-width are of the order of 3×10^4 yrs, much longer than the observing time duration. Hence, they will appear in the Fourier spectrum as unresolved modes which will be difficult to discriminate from unstable modes.

β Cephei Stars

The next step consists in considering more massive main sequence stars such as β Cephei type stars. Their masses range between ≈ 9 and $\approx 25\,M_\odot$. The overstable modes are found in the vicinity of the fundamental radial one. They are excited by the κ-mechanism due to the iron opacity bump that is even more developed than for *SPB* type stars, (see Pamyatnykh, 1999) and generates a convective region. The issue then is to determine whether or not such an unstably stratified region is efficient to excite stable modes to detectable levels.

Excitation by the iron convective region

Excitation of radial modes by the iron convective region is found to be efficient for two reasons; first, for a 10 $M_\odot$ model, the iron opacity bump is located deep into the star compared to the helium bump. Thus, the density is higher and the turbulent kinetic-energy flux which can be transferred to the modes is more important. Second, the efficiency of the excitation depends on the involved time-scales, *i.e.* the convective time-scale and the period of the mode. The period of radial modes is of several hours while, using mixing-length arguments, we find the convective time-scale to be also of several hours. Hence, excitation is nearly resonant. The damping rates are found to be dominated by the radial part of the radiative flux divergence, as for stable *SPB* modes.

The final result is presented in Fig. 2. The amplitudes, as shown in Fig. 2, are sensitive to the value adopted for β but are found to vary from several *ppm* to 20 *ppm*. These amplitudes lie well above the detection threshold of CoRoT for such stars, which is around one *ppm* (see Lefèvre same volume). The amplitude level indicates a quite efficient excitation process if one compares with the amplitudes for the Sun that are around 2.5 *ppm* (Michel et al. 2008a).

Fig. 2. also shows that the mode amplitudes oscillate with the frequency of the mode. This is mainly due to oscillations of the excitation rates. The reason is due to the dependence of P on the radial derivative of the radial component of the displacement eigenfunction. Hence depending on the location of the node (*i.e.* relatively to the iron convection zone) the derivative is more or less important explaining this oscillation. Such a behavior of the amplitude is of importance since it gives a possible diagnostic about the spatial extent of the iron convective region. In addition, the detection of such stable modes would give a signature of the driving region.

We find that along the main sequence in a Hersprung-Rüssell diagram, stars with masses around 10 $M_\odot$ are the most favorable candidates to present detectable solar-like oscillations. For lower masses, the convective region due to the iron opacity bump becomes too thin making the convection weak or inexistent. For higher masses, due to an increase of the effective temperature, the iron convective zone is located in layers with smaller densities. Then, the kinetic energy flux is smaller making the excitation less efficient, resulting in lower amplitudes.

Conclusion

To summarize, the purpose of this work was to investigate the possibility that stable (with respect to the κ-mechanism) modes can be stochastically excited by turbulence and in such a case to determine if those modes can be detected according to the CoRoT detection threshold. We find that detection of solar-like oscillations in *SPB* stars will be difficult since their theoretical amplitudes are expected below the *ppm*. In contrast, β Cephei stars are more promising. Theoretical computations for these stars yield amplitudes of radial *p* modes excited stochastically by the iron convective region well above the CoRoT detection threshold. The optimum mass is found to be around 10 $M_\odot$, as discussed above and near the *SPB* and β Cephei instability strip. Hence, the detection of a star simultaneously pulsating on *SPB*, β Cephei and solar-like oscillations would be the first *Chimeræ*[1] star.

---

[1] monstrous creature made of multiple parts of different animals. The denomination has been introduced by S. Talon.

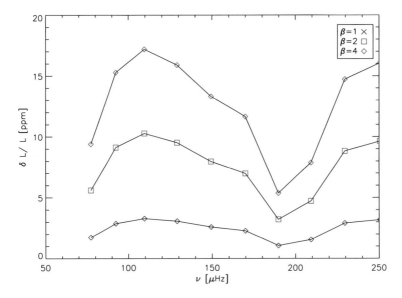

Figure 2: Relative amplitude in terms of luminosity as a function of frequency, for $\ell = 0$ p modes and three different values of the parameter β, for a 10 $M_\odot$ model on the main sequence.

References

Balmforth, N.J. 1992, MNRAS, 255, 603

Belkacem, K., Samadi, R., Goupil, M. J., et al. 2006, A&A, 460, 183

Dupret, M.A. 2002, A&A, 366, 166

Michel, E., Samadi, R., Baudin, F., et al. 2008a, submitted to A&A

Michel, E., et al. 2008b, Science, 322, 558

Rowe, J. F., Matthews, J. M., Cameron, C., et al. 2006, CoAst, 148, 34

Jerzykiewicz, M., Handler, G., Shobbrook, R. R., et al. 2005, MNRAS, 360, 619

Samadi, R., Kupka, F., Goupil, M. J., et al. 2006, A&A, 445, 233

Samadi, R., Belkacem, K., Goupil, M. J., et al. 2008, A&A, 489, 291

Samadi, R., Nordlund, A., Stein, R. F., et al. 2003, A&A, 404, 1129

Samadi, R., Goupil, M.J., & Houdek, G. 2002, A&A, 395, 563

Pamyatnykh, A. A. 1999, Acta Astronomica, 49, 119

Comm. in Asteroseismology
Vol. 158, 2009, 38th LIAC/HELAS-ESTA/BAG, 2008
A. Noels, C. Aerts, J. Montalbán, A. Miglio and M. Briquet., eds.

On the possible resonance of very high order pulsation modes in Cepheids

E. Antonello[1], L. Speroni[1], S. Kanbur[2], I. Richter[3] and R. Szabó[4]

[1] INAF-Osservatorio Astronomico di Brera, Milano-Merate, Italy
[2] Department of Physics, SUNY Oswego, USA
[3] Department of Phyiscs, University of Rochester
[4] Konkoly Observatory, Hungary

Abstract

Researches on resonances between pulsation modes in Cepheids in the past twenty years allowed to interpret the characteristics of Fourier parameters of the light and radial velocity curves of such variable stars. In these works it was generally assumed that a resonance could occur between the fundamental (or first overtone) radial mode and a pulsationally stable low-order radial overtone. Some observational and theoretical model results suggest however the presence of other possible effects due to resonances between the fundamental mode and high order overtones.

Observational Data

The amplitude ratios R_{i1} obtained from the decomposition of good light curves of some selected Cepheids in the Milky Way,

$$V = \Sigma A_i \cos(i\omega t + \phi_i), \quad R_{i1} = A_i/A_1, \quad \phi_{i1} = \phi_i - i\phi_1,$$

do not decrease regularly with increasing order i. When the value of R_{i1} is comparable with the respective formal error, the value of R_{i+1} can be significantly larger than such an error. This odd result is not related to a bad phase coverage of the data or other observational problems.

We have verified the presence of the same effect in some tens Cepheids of the Large Magellanic Cloud (LMC) using the OGLE dataset (I-band magnitudes, Udalski et al. 1999; Fourier decomposition of OGLE data was performed for example also by Ngeow & Kanbur, 2006, and by Abdel-Sabour, 2005). In Table 1 we show some examples; the columns include the name of the LMC star, the number of data points, the period P, the standard deviation of the fit, the mean error in the various R_{i1} values, and the R_{i1} values for i=2, ..., 10.

The effect is generally small, and there is no clear correlation with the period as in the well-known cases of fundamental mode Cepheids with P close to 10 days and first overtone mode Cepheids close to 3 days (for the latter, see however the remarks by Kienzle et al. 1999). Table 2 shows the number of stars with a negligible R_{i1} component for given i and for different P ranges; N is the total number of stars in each bin. This distribution suggests a sort of trend, that is a concentration of such stars in a P range between 4 and 8 days (log P between 0.6 and 0.9), and for increasing i with P.

Star/Field	N	P [d]	σ_{fit}	error(R_{i1}) R_{61}	R_{21} R_{71}	R_{31} R_{81}	R_{41} R_{91}	R_{51} $R_{10,1}$
184038 / 7	295	13.513	0.012	0.005	0.127	0.124	0.122	0.085
				0.047	0.037	**0.008**	0.017	0.034
47348 / 2	341	6.291	0.009	0.004	0.411	0.149	0.056	0.023
				0.001	0.022	0.013	0.017	0.010
47399 / 2	351	4.206	0.011	0.004	0.459	0.249	0.148	0.078
				0.047	0.017	0.018	**0.007**	0.016

Table 1: Some LMC Cepheids with a negligible intermediate Fourier component

		i-component										
log P range	N	3	4	5	6	7	8	9	10	11	12	13
0.2 - 0.3	11					1						
0.3 - 0.4	26				2	4		1				
0.4 - 0.5	160		1	1	3	7	5	4	1			
0.5 - 0.6	199			1	2	5	2	2		1		
0.6 - 0.7	148		1	9	6	3	1	1		1	1	
0.7 - 0.8	74		4	7	9	2	2		1			
0.8 - 0.9	56	3	4	6	6	10	7		1			
0.9 - 1.0	28		2	2	1	6	3					
1.0 - 1.1	17			1								
1.1 - 1.2	17					1	1					1

Table 2: Number of LMC Cepheids with a negligible intermediate Fourier component

Nonlinear models

The large set of nonlinear models of Aikawa & Antonello (2000) allowed us to verify the possible presence of similar effects in nonlinear model light curves. Those models were intended for a detailed comparison with Cepheids of the Milky Way and rather long P, above 10 days. The R_{i1} values of the models against P show a deep minimum, and the P corresponding to such a minimum increases with the order i, starting from $i = 7$ (see Fig. 1). The ϕ_{i1} values show a small jump when R_{i1} is close to the minimum value. On the one hand, the R_{i1} behavior of the theoretical models reminds of that suggested by the observations summarized in Table 2, but in the models it occurs at longer P. On the other hand, the R_{i1} and ϕ_{i1} features, though much smaller, remind of the characteristics observed in the fundamental mode Cepheids with P close to 10 days, that is they could be the effects of resonances between pulsation modes. If this were the case, a resonance between the fundamental mode and a very high radial overtone should be expected.

Unfortunately, the linear analysis of such models was performed only up to the 5*th* overtone. Whatever is the cause, such effects are very sensitive to the stellar parameters, since completely different results for sequences of models are obtained with systematically lower temperatures or different mass-luminosity relation.

Conclusion

Since the effect observed in LMC Cepheids is close to the error limit, it is important to confirm the result by means of more precise data; a preliminary check with the DIA OGLE data (difference image analysis; Zebrun et al. 2003) gave different results from those of the

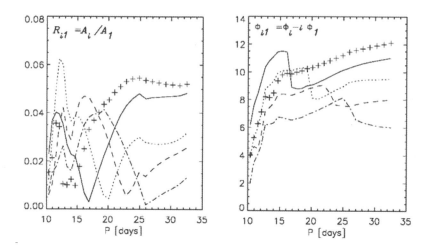

Figure 1: High order Fourier parameters of nonlinear models of galactic Cepheids. Plus signs: i=6; continuous line: i=7; dotted line: i=8; dashed line: i=9; dashed-dotted line: i=10.

standard image analysis.

Detailed linear and nonlinear models are required to verify the presence of the resonances and their dependence on the stellar physical parameters. Preliminary results of a large set of linear nonadiabatic models indicate that, in principle, resonances between the fundamental mode and high order overtones could indeed occur in LMC Cepheids at the expected period range.

Buchler et al. (1997) have discussed the possible existence of Cepheid strange modes of very high order, that are pulsationally unstable. If our interpretation in terms of resonances is correct, in the present case the very high order mode should be pulsationally stable.

Acknowledgments. IR and SK wish to acknowledge support from NSF grant OISE - 0755646.

References

Abdel-Sabour, M.A. 2005, Study of Cepheid variable stars using large photometric databases, Ph.D. thesis

Aikawa, T., & Antonello, E. 2000, A&A, 363, 593

Antonello, E. 1994, A&A, 282, 835

Buchler, J.R., Yecko, P.A., & Kolláth, Z. 1997, A&A, 326, 669

Kienzle, F., Moskalik, P., Bersier, D., & Pont, F. 1999, A&A, 341, 818

Ngeow, C., & Kanbur, S.M. 2006, MNRAS, 369, 723

Udalski, A., Soszynski, I., Szymanski, M., et al. 1999, Acta Astron., 49, 223

Zebrun, K., Soszynski, I., Wozniak, P.R., et al. 2003, Acta Astron., 51, 317

Comm. in Asteroseismology
Vol. 158, 2009, 38th LIAC/HELAS-ESTA/BAG, 2008
A. Noels, C. Aerts, J. Montalbán, A. Miglio and M. Briquet., eds.

Ledoux's convection criterion in evolution and asteroseismology of massive stars

Y. Lebreton[1,2], J. Montalbán[3], M. Godart[3], P. Morel[4], A. Noels[3], M.-A. Dupret[5]

[1] Observatoire de Paris, GEPI, CNRS UMR 8111, 92195 Meudon, France
[2] IPR, Université de Rennes 1, 35042 Rennes, France
[3] Institut d'Astrophysique et de Géophysique, University of Liège, Belgium
[4] Observatoire de la Côte d'Azur, CASSIOPEE, Nice, France
[5] Observatoire de Paris, LESIA, CNRS UMR 8109, 92195 Meudon, France

Abstract

Saio et al. (2006) have shown that the presence of an intermediate convective zone (ICZ) in post-main sequence models could prevent the propagation of g-modes in the radiative interior and hence avoid the corresponding radiative damping. The development of such a convective region highly depends on the structure of the star in the μ-gradient region surrounding the convective core during the main sequence phase. In particular,the development of this ICZ depends on physical processes such as mass loss, overshooting (Chiosi & Maeder 1986, Chiosi et al. 1992, see also Godart et al., 2009) and convective instability criterion (Schwarzschild's or Ledoux's criteria). In this paper we study the consequences of adopting the Ledoux's criterion on the evolution of the convective regions in massive stars (15 and 20 $M_\odot$), and on the pulsation spectrum of these new B-type variables (also called SPBsg).

Stellar models: evolution of convective regions during main sequence and post-main sequence phases

Stellar evolution models have been calculated with the CESAM code (Morel & Lebreton, 2008) for masses $M = 15$ and 20 $M_\odot$, a solar chemical composition ($Z/X = 0.0245$, $Y = 0.27$) without and with overshooting of convective zones ($\alpha_{ov} = 0.2H_p$). For all these models, we calculated the pre-main sequence evolution. In models including overshooting, we assumed that $\nabla = \nabla_{ad}$ in the overshoot region.

As the star evolves on the main sequence (MS), a gradient of chemical composition (∇_μ) develops at the outer border of the convective core. In the context of Schwarzschild's criterion (convective instability if $\nabla_{rad} \geq \nabla_{ad}$), the outwards increase of opacity leads to the formation of a region of semiconvective instability outside the convective core (CC) and therefore to the mixing of matter until the neutrality of gradients is reached ($\nabla_{rad} = \nabla_{ad}$). During the post-MS this region becomes an ICZ which develops as H starts burning in a shell in the μ-gradient region.

When the Ledoux's criterion for convection (convective instability if $\nabla_{rad} \geq \nabla_{ad} + \frac{\beta}{4-3\beta}\nabla_\mu$ where β is the ratio of the gas to the total pressure) is used instead of the Schwarzschild's criterion, the role of μ-gradients on the stability against convection is taken into account. Adopting the Ledoux's criterion does not change the size of the CC during the MS. Nevertheless, in models based on Ledoux's criterion, a convective region located outside the CC appears during the MS phase at the base of the homogeneous region at $m/M = 0.45$ in the

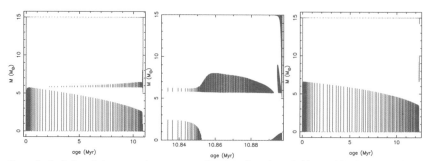

Figure 1: Evolution of the convective zones as a function of age in a 15 $M_\odot$ model calculated with the Ledoux's criterion for convection: (i) models without overshooting during the MS (left) and post-MS (center), and (ii) models with overshooting during the MS (right).

20 $M_\odot$ model and $m/M = 0.39$ in the 15 $M_\odot$ one (Fig.1, left panel). As can be seen in Fig.2 (central panel) this convective region results from the outwards increase of opacity in a region where $\nabla_\mu = 0$ and where the Schwarzschild and Ledoux criteria are hence equivalent. The location of the ICZ corresponds to that of the maximum extension of the CC during the MS and it remains the same during the post-MS phase until the onset of He burning. The location and the size of the ICZ during the post-MS in models adopting the Ledoux's criterion are different from those found in models based on the Schwarzschild's criterion (see also Godart et al, 2009). With Ledoux's criterion, the thickness of this ICZ can reach $15 - 20\%$ of the total star's mass (Fig.1). On the other hand, when overshooting is included, no ICZ appears during the MS for $M = 15M_\odot$ (right panels of Figs.1 and 2) while for $M = 20M_\odot$ the ICZ appears later than in models without overshooting.

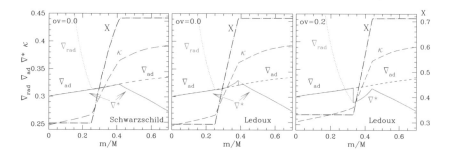

Figure 2: Temperature gradients (radiative one: dotted lines; adiabatic one: short-dashed lines; effective stellar gradient: solid lines), opacity κ (long-dashed lines), and hydrogen profile, X (dash-dotted lines), for a $M = 15M_\odot$ model near the middle of the MS calculated with the Schwarzschild's criterion (left) without overshooting, and with the Ledoux's criterion without (center) and with (right) overshooting.

Asteroseismology of post-MS B-type stars

We have investigated the seismic characteristics of our models according to the recipes described by Dupret et al. (2009). Excitation and damping of p and g modes highly depend on the location and thickness of the ICZ, hence on the change of luminosity as the star becomes cooler. In models computed with the Ledoux's criterion, the μ-gradient region located below the ICZ brings a large contribution to the Brunt-Väisälä frequency N_{BV} which leads to strong damping of the modes. For instance, we find that at log $T_{eff} = 4.27$ the kinetic energy of a $\ell = 1$ mode in the radiative core of post-MS models of 15 $M_\odot$ is much higher than in 20 $M_\odot$ models. As a consequence, the mode in the 15 $M_\odot$ star is more damped in the radiative centre and its amplitude in the driving region at log $T \sim 5.2$ is too low for the mode to be effectively excited.

The frequency range of excited modes and the T_{eff} domain of the instability strip are shown in Fig.3.

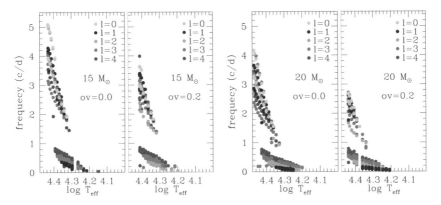

Figure 3: Excited g and p modes along the post-MS for 15 $M_\odot$ (left panel) and 20 $M_\odot$ (right panel) models based on Ledoux's criterion without (left) and with (right) overshooting.

Conclusions

In models computed with the Schwarzschild's criterion, the ICZ which is closely related to the H-burning shell is located within the μ-gradient region. In the ICZ, $N_{BV} = 0$ which corresponds to less radiative damping. On the other hand, in models computed with the Ledoux's criterion, the ICZ is thin and is located at higher values of m/M, at the base of the homogeneous region and therefore N_{BV} remains high in the μ-gradient region which leads to more radiative damping. As a consequence more modes are excited in models computed with the Schwarzschild's criterion than in models computed with the Ledoux's criterion.

Acknowledgments. YL is grateful to the European Helio- and Asteroseismology Network HELAS for financial support. JM acknowledges financial support from the Prodex-ESA Contract Prodex 8 COROT (C90199).

References

Chiosi, C., & Maeder, A. 1986, ARA&A, 24, 329

Chiosi, C., Bertelli, G., & Bressan, A. 1992, ARA&A, 30, 235

Dupret, M.-A., Godart, M., & Noels, A. 2009, CoAst, 158, 239

Godart, M., Dupret, M.-A., & Noels, A. 2009, CoAst, 158, 308

Morel, P., & Lebreton, Y. 2008, ApSS, 316, 61

Saio, H., Kuschnig, R., Gautschy, A., et al. 2006, ApJ, 650, 1111

Comm. in Asteroseismology
Vol. 158, 2009, 38th LIAC/HELAS-ESTA/BAG, 2008
A. Noels, C. Aerts, J. Montalbán, A. Miglio and M. Briquet., eds.

Testing the forward approach in modelling β Cephei pulsators: setting the stage

A. Miglio, J. Montalbán, and A. Thoul

Institut d'Astrophysique et de Géophysique
Université de Liège, Allé du 6 Août 17 - B 4000 Liège - Belgique

Abstract

The information on stellar parameters and on the stellar interior we can get by studying pulsating stars depends crucially on the available observational constraints: both seismic constraints (precision and number of detected modes, identification, nature of the modes) and "classical" observations (photospheric abundances, effective temperature, luminosity, surface gravity). We consider the case of β Cephei pulsators and, with the aim of estimating quantitatively how the available observational constraints determine the type and precision of our inferences, we set the stage for Hare&Hound exercises. In this contribution we present preliminary results for one simple case, where we assume as "observed" frequencies a subset of frequencies of a model and then evaluate a seismic merit function on a dense and extensive grid of models of B-type stars. We also compare the behaviour of χ^2 surfaces obtained with and without mode identification.

Tools

In order to set the stage for Hare&Hound exercises, the following three main components need to be defined:

- **Theoretical predictions:** The grid of models we use is BetaDat (Thirion & Thoul 2006). Stellar models and adiabatic frequencies are computed, respectively, with CLES (Scuflaire et al. 2008a) and LOSC (Scuflaire et al. 2008b). The masses considered in the grid span the domain between 7.8 and 18.5 $M_\odot$, a metal mass fraction Z between 0.01 and 0.025, an initial hydrogen mass fraction $X = 0.70$, and four values of the overshooting parameter (α_{ov}) in the range 0-0.25. Frequencies of low-order oscillation modes of degree up to $\ell = 2$ are computed for main sequence models.

- **Observational constraints:** we consider only seismic constraints, i.e. a subset of theoretical eigenfrequencies of a model M0 in the grid. The effects of rotation on the oscillation modes are not considered in this first step, thus all modes are assumed to be axisymmetric ($m = 0$). Concerning the degree of the observed modes, we consider the case where ℓ is unknown as well as the one where ℓ is available as a constraint.

- **Merit function:** In order to compute a merit function at each point of the grid, we use a double optimisation procedure similar to the one extensively adopted in sdB astero-seismology (see e.g. Brassard et al. 2001 and Charpinet et al. 2005). For each model in the grid we find the best global match between "observed" and theoretical frequencies by using a standard χ^2 merit function. Then we study the properties of good-fit

Table 1: Theoretical oscillation frequencies of model M0 considered as observational constraints, the uncertainty on the frequencies is assumed to be 0.1 μHz.

ℓ	ν (μHz)
0	57.78
1	58.93
1	80.29
2	39.46

models looking at minima in the χ^2 as a function of the stellar parameters/properties (e.g. location in an HR diagram, mass, central hydrogen mass fraction (X_c), mean density, ...).

First test

We consider 4 oscillation modes of the model M0 (see Table 1) as seismic constraints, with frequencies in the typical domain of the pulsation modes observed in β Cephei stars. The main parameters and properties defining M0 are: $M = 10\,M_\odot$, $X_c = 0.2$, $\alpha_{OV} = 0$, $Z = 0.02$, $\log T_{eff} = 4.34$ and $\log L/L_\odot = 4.02$.

We then compute the merit function on a sub-grid of models of BetaDat, where α_{OV} and Z are the same as in M0. The behaviour of χ^2 for main-sequence models of different mass and evolutionary status is shown in Fig. 1. In the case where ℓ is given as an observational constraint for all the modes (right panels), the properties and parameters of the input model M0 are easily recovered due to the appearance of an isolated global minimum in the χ^2 function. The increase of the number of χ^2 minima in the left panels (ℓ is unknown) compared to that in the right panels allows us to estimate the loss of information when mode identification is not available. Nevertheless, even if no mode identification is available, the frequencies of 4 modes allow to constrain the parameter space in regions close to χ^2 minima.

The results of this simple test show that the approach presented here is viable tool to determine the number and precision of observational data required to constrain the properties of β Cephei pulsators. However, an extensive and thorough study is needed to investigate the effect of considering both additional uncertainties (e.g. on parameters such as overshooting and initial chemical composition, on rotational splittings and identification of m) as well as other constraints (e.g. luminosity, T_{eff}, $\log g$, photospheric abundances, nonadiabatic constraints).

Acknowledgments. A.M. and J.M. acknowledge financial support from the Prodex-ESA Contract Prodex 8 COROT (C90199). A.M. is a *Chargé de Recherches* and A.T. is a *Chercheur qualifié* of the FRS-FNRS.

References

Brassard, P., Fontaine, G., Billères, M., et al. 2001, ApJ, 563, 1013

Charpinet, S., Fontaine, G., Brassard, P., et al. 2005, A&A, 437, 575

Scuflaire, R., Théado, S., Montalbán, J., et al. 2008a, ApSS, 316, 83

Scuflaire, R., Montalbán, J., Théado, S., et al. 2008b, ApSS, 316, 149

Thirion, A., & Thoul, A. 2006, ESA SP-1306, 383

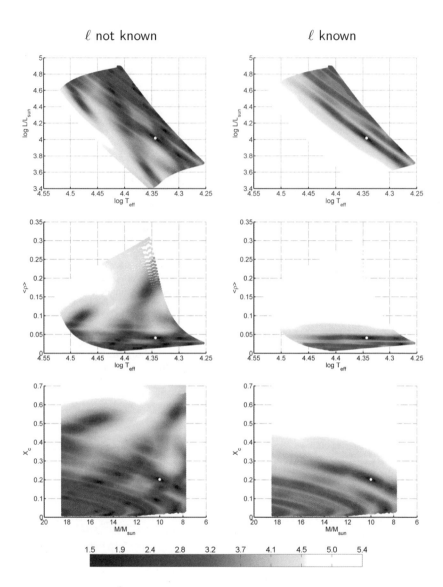

Figure 1: Colour-coded χ^2 (logarithmic scale) in three different planes: $\log T_{\mathrm{eff}}$-$\log L/L_\odot$ (*upper panels*), $\log T_{\mathrm{eff}}$-mean density (*middle panels*), and mass-X_c (*bottom panels*). White dots represent the position of the input model M0 in each parameter space. Right panels show χ^2 computed including the identification of ℓ as a constraint, whereas in left panels ℓ is considered unknown ($\ell= 0,1$ or 2).

Comm. in Asteroseismology
Vol. 158, 2009, 38th LIAC/HELAS-ESTA/BAG, 2008
A. Noels, C. Aerts, J. Montalbán, A. Miglio and M. Briquet., eds.

Domain of validity of a 1D second order perturbative approach for the effects of rotation on stellar oscillations

R-M. Ouazzani[1], M-J. Goupil[1], M-A. Dupret[1], and D. Reese[2]

[1]Observatoire de Paris, LESIA, CNRS UMR 8109, 92195 Meudon, France
[2]Department of Applied Mathematics, University Of Sheffield, Sheffield S3 7RH, United Kingdom

Abstract

At rotational velocities such as that of upper Main Sequence stars – between 50 to 200 km/s – the effects of rotation on oscillation frequencies must be included. Considering the accuracy reached by available ground-based and space observations, the aim of this study is to determine the limits – in terms of rotational velocity – of a perturbative approach to model the effects of rotation on oscillation frequencies. We thus compare the oscillation frequencies computed by 1D second order perturbative methods to the ones obtained in Reese et al. (2006) – direct integration of a 2D eigenvalue system. To do so, we use polytropic models (N=3) in uniform rotation, and we discuss the results for a β Cephei star (8.2 M$_\odot$, 5.04 R$_\odot$).

Equilibrium state of a 1D uniformly rotating polytrope

The 1D polytropic model is computed by solving the Lame-Emden equation in which the spherical part of the centrifugal acceleration is included in an effective gravity:

$$\frac{1}{x^2}\frac{d}{dx}(x^2\frac{dy}{dx}) + y^N = \Omega'^2 \tag{1}$$

Where y is defined by $\tilde{\rho} = \rho_c y^N$ and $\tilde{p} = p_c y^{N+1}$, x is the radius normalised to the critical radius of the polytrope, and $\Omega'^2 = \Omega^2/2\pi G\rho_c$, - Ω the rotational angular velocity. The resolution of the Lame-Emden equation yields the spherically symmetric equilibrium quantities – $\tilde{\rho}(r)$, $\tilde{p}(r)$, and $\tilde{\Phi}(r)$. Their second order non-spherically symmetric parts are computed as perturbations of the spherical case to the second order, such that for a quantity f: $f(r, \theta) = \tilde{f}(r) + f_{22}(r)P_2(cos\theta)$, P_2 being the second order Legender polynomial.
In the nonperturbative case, structure variables are expanded on around 50 spherical harmonics using a code developed by M. Rieutord.

Computation of oscillation frequencies

In Reese et al. (2006) the nonperturbative oscillation frequencies are calculated by direct integration of a 2D eigenvalue system. In the perturbative approach, we apply a perturbation about an axisymmetric steady configuration. After ignoring the resonant interaction due to near degeneracy, we derive the perturbed oscillation frequency by obtaining an additional correction of the order of $O(\Omega^2)$ to the frequency without rotation.

Implicit assumptions made in the perturbative method are no longer valid when two or three modes are close to each other in terms of frequencies. Those modes are then coupled according to selection rules, and the frequencies are then modified (Dziembowski & Goode 1992, Soufi et al. 1998, Suarez et al. 2006, and references therein).

Results for a β Cephei star $(8.2\ M_\odot, 5.04\ R_\odot)$

We treated the 2D nonperturbative frequencies like observation results, and calculated the discrepancies due to the perturbative approach. We present in Figure 1 the discrepancies on the splittings between the two models, where the error is defined as:

$$\delta = \frac{1}{2\ell}[(\omega_{n,\ell,-m} - \omega_{n,\ell,m})_{Npert} - (\omega_{n,\ell,-m} - \omega_{n,\ell,m})_{Pert}] \tag{2}$$

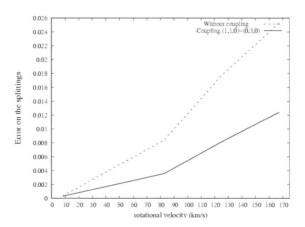

Figure 1: Comparison of the splitting computed by the perturbative and nonperturbative approach for the pressure mode P_1, $\ell = 1$, $m = \pm 1$.

Coupling due to near degeneracy clearly improves the accuracy of the perturbative method: if one accepts an observational relative error of 1.310^{-3}, which corresponds to 20 days of observation, the discrepancies are above this limit around 15 km s^{-1}. When coupling is included, this limit is pushed up to about 20 km s^{-1}. Considering the splittings, the discrepancy is much lower, and is under the error bars until around 40 km/s. Moreover, we studied the impact of these discrepancies on, for example, the measurement of a rotational angular velocity. We found that if we accept an accuracy of 1%, on the splittings, the error made by the perturbative approach on the rotational velocity is around 0.5%.

Acknowledgments. I would especially like to thank Juan-Carlos Suarez for discussions.

References

Monaghan, J., & Roxburgh, I.W., 1964, MNRAS, 131,13

Dziembowski, W.A., & Goode, P. 1992, ApJ, 394, 670

Reese, D., Lignières, F., & Rieutord, M. 2006, A&A, 455, 621

Soufi, F., Goupil, M.J., & Dziembowski, W.A. 1998, A&A, 334, 911

Suarez, J.C., Goupil, M.J., & Morel, P. 2006, A&A, 449,673

Comm. in Asteroseismology
Vol. 158, 2009, 38th LIAC/HELAS-ESTA/BAG, 2008
A. Noels, C. Aerts, J. Montalbán, A. Miglio and M. Briquet., eds.

Relative weights of the observational constraints on the determination of stellar parameters

N. Ozel, M-A. Dupret, A. Baglin

Observatoire de Paris, LESIA, CNRS UMR 8109, 92195 Meudon,France

Abstract

We study the effect of using different observed quantities (oscillation frequencies, binarity, interferometric radius) and the impact of their accuracy on constraining the uncertainities of global free stellar parameters (i.e. the mass, the age etc.). We use the Singular Value Decomposition (SVD) formalism to analyse the behavior of the χ^2 fitting function around its minimum. This method relates the errors in observed quantities to the precision in the model parameters. We apply this tool to the α Cen A for which, seismic, binarity and interferometric properties are known with high accuracy. We determine how changes of the accuracy of the observable constraints affect the precision obtained on the global stellar parameters.

Individual Objects: α Cen A

The Method

Given a set of n measurements $y_{obs,i}$ (e.g. T_{eff}, L, $\Delta\nu$, etc.) with associated error bars and a set of m free parameters x_j (e.g. τ (age), α, M, etc.), we first determine the reference model (RM) which minimizes the χ^2 fitting function. The linear transformation of the model ($y_{the,i}$) around the minimum of the reference set of parameters (x_{j0}) produces a derivative matrix which is the so-called design matrix $\mathbf{D}$. This matrix relates small changes in the parameters to corresponding changes in the observables. This minimisation problem is most conveniently solved by using the SVD method as $\mathbf{D}$ may be decomposed as $D_{n\times m} = U_{n\times m}W_{m\times m}V_{m\times m}$. $\mathbf{U}$ and $\mathbf{V}$ are orthonormal and $\mathbf{W}$ is a diagonal matrix. By studying the behaviour of χ^2 around its minimum using the SVD method, we analyse the sensitivity of the observables to the parameters. The behaviour of χ^2 around its minimum is expressed by an m dimensional ellipsoidal equation $\Delta\chi^2$.

$$\Delta\chi^2 = \frac{\|V^{(1)}\delta x\|^2}{W_1^{-2}} + ... + \frac{\|V^{(m)}\delta x\|^2}{W_m^{-2}}, \quad Cov(\delta x_j, \delta x_k) = \sum_i^N \frac{V_{ji}V_{ki}}{W_i^2} \quad (1)$$

where $\delta\mathbf{x} = \mathbf{x_j} - \mathbf{x_{j0}}$. A major advantage of the SVD method is that the columns of $\mathbf{V}$ are precisely the principal axes of the error ellipsoid, while the corresponding values of $\mathbf{W}^{-1}$ are the lengths of these axes. The estimation of the variances-covariances matrix $Cov(\delta x_j, \delta x_k)$ of the free parameters due to the measurement errors on the n observables is expressed above. The solution of this minimisation problem by the SVD method was proposed in the case of solar-like oscillation by Brown et al. (1994) and applied also by Creevey et al. (2007). To study the relation between different observables and free stellar parameters (the mixing length parameter α, the mass M, and the metalicity Z/X), we analyse the error ellipsoids in the parameter space by the SVD method.

Application to Different Astrophysical Situation Based on the α Cen A

The α Cen A is used here as a template. Its nature, its proximity, and the detection of solar-like oscillation constraints provide a unique opportunity to test the SVD method.

The stellar models of the α Cen A are computed with the stellar evolution code CESAM (Morel 1997) starting from the ZAMS. The adopted physical description is the standard MLT for convection calculation (Böhm- Vitense 1958); the OPAL opacities (Iglesias & Roger 1996) completed at low temperatures with the opacities of Alexander & Ferguson (1994); the OPAL equation of state; and an Eddington atmosphere as the surface boundary condition. The adiabatic oscillation frequencies are calculated for ℓ = 0-3 and n = 15-25 using the adiabatic oscillation code Losc (Scuflaire et al. 2008). To construct the derivative matrix D, we vary each of the parameters x_j. Each derivative is computed from differences centered on the reference parameter values given in Table 1. The interval δx has to be sufficiently small such that the linear approximation is good, yet still large enough to avoid numerical problems. The increments used are 20 Myr for (τ) age, 0.05 for α, 0.005 for M(M$\odot$), 0.003 for initial helium abundance Y_0 and 0.0005 for initial metallicity Z/X_0. The constraints of α CenA and all characteristics of our RM are summarized in Table 1. The technique used in this study to estimate all charracteristics of our RM follows that of Miglio & Montalbán (2005).

Table 1: Observations of α Cen A and the properties of the RM

	T_{eff} [1]	$L/L_\odot$ [1]	Z/X [2]	$R/R_\odot$ [3]	$\Delta\nu$ [4]	$\delta\nu$ [4]	$M/M_\odot$ [5]
α CenA	5810	1.522	0.039	1.224	105.5	5.6	1.105
σ	50K	0.030	0.06	0.003	$0.1\mu Hz$	$0.7\mu Hz$	0.007
RM	5782	1.516	0.039	1.229	105.5	5.7	1.099
x_j of RM		$\tau(Gyr)$	α	M	Y_0	Z/X_0	
		5.65	1.6747	1.099	0.280	0.039	

Apart from the real α Cen system, we study several different cases to cover a large range of realistic situations. Case 1 describes α Cen itself (d = 1.3pc), while Case 2 corresponds to a system located ten times further away (d = 13pc). For distant objects where the binary and interferometric data are unavailable, the seismic data are the major source of information (Case 3). We discuss also the influence of the seismic data precision.

Table 2 shows the results from these different cases. In all these cases, the mass is the best constrained parameter with these observables. We therefore focus here on its uncertainity. In Case 1 for α Cen itself, we see that the mean large separation $\Delta\nu$ and the radius $R/R_\odot$ both give more or less the same precision on the mass parameter (ϵ = 2.09%, and 2.20% ,respectively). If these two constraints are considered together, the relative precision on the mass is much higher (ϵ = 0.75%). This comes from the fact that $M \propto \Delta\nu^2 R^3$. Using the mean $\Delta\nu$ or the individual $\Delta\nu_i$ large separation give about the same precision on the mass, as we are close to the asymptotic regime. Comparing Case 2 with Case 1 allows us to estimate the effect of increasing distance on determination of the precision of mass parameter for the same combination of observables. The observables depending on the distance (i.e, mass, radius, luminosity) become less effective to constrain the mass parameter. For example the uncertainity on the mass parameter increases from ϵ = 0.75% to ϵ = 2.13% when the radius and the seismic data are considered together. In Case 3, the seismic information alone gives ϵ = 2.19%. As the precision on the seismic data does not depend on distance, the ϵ does not change (ϵ = 2.09% for d=1.3pc, ϵ = 2.19% for d=13pc). Contrary to Case 1, in Case 3, even if the mass or radius information were available, they would not reduce this value significantly (ϵ = 2.13% if the radius is available, ϵ = 2.06% if the mass). We also reduce the precision of seismic data taking $\sigma_{\Delta\nu}$ = 2 μHz as the worst case scenario. For this precision, the uncertainty on the mass parameter increases only slightly because of the flatness of the error ellipsoid as we increase the error on the oscillation frequencies by a factor of 20.

Table 2: The rms error on the mass ($\epsilon(M)$), taking into account different sets of observables for α Cen A and for the case if it were located at 13pc. $Q_i = (T_{eff}, L/L_\odot, Z/X_0)$ represent the set of classical observables and are included in all cases. The symbol ($\sqrt{}$) indicates that the observable is included, and the symbol ($\times$) indicates that the observable is not included in the SVD analyse.

| Observables | | | | | d=1.3pc | d=13pc |
| Classic | | Seismic (μHz) | | | | |
$R/R_\odot$	$M/M_\odot$	$(\Delta\nu, \delta\nu)$	$(\Delta\nu_i, \delta\nu_i)$	$\sigma_{\Delta\nu}$	$\epsilon(M)$ (%)	$\epsilon(M)$(%)
$\times$	$\times$	$\times$	$\times$	–	2.21	2.26
$\times$	$\times$	$\sqrt{}$	$\times$	0.1	2.09	2.19
$\times$	$\times$	$\times$	$\sqrt{}$	–	1.73	1.75
$\times$	$\times$	$\sqrt{}$	$\times$	2	2.12	2.20
$\sqrt{}$	$\times$	$\times$	$\times$	–	2.20	2.24
$\sqrt{}$	$\times$	$\sqrt{}$	$\times$	0.1	0.75	2.13
$\times$	$\sqrt{}$	$\times$	$\times$	–	0.61	2.12
$\times$	$\sqrt{}$	$\sqrt{}$	$\times$	0.1	0.61	2.06

Conclusion

We approach the problem of the accuracy on the determination of the stellar parameters by computing the error ellipsoids in the parameter space. The impact of each constraint is obtained by performing the SVD analysis for different cases taking into account or not each observable. We apply this method to the real α Cen system and for different cases in changing the distance and the precision on the constraints.

We have shown that, except for α Cen itself for which all the seismic and classical constraints are available to high precision, the seismic constraints play a dominant role in determining stellar free parameters for the distant objects.

References

Alexander, D. R., & Ferguson, J. W. 1994, ApJ, 437, 879

Böhm-Vitense, E. 1958, Zeitschrift Astrophys., 46, 108

Brown, T.M., Christensen-Dalsgaard, J., Weibel-Mihalas, B., & Gilliland, R.L. 1994, ApJ, 427, 1013

Bouchy, F., & Carrier, F. 2002, A&A, 390, 205 ([4] in Table 1)

Creevey, O.L., Monteiro, M.J.P.F.G., Metcalfe, T.S., et al. 2007, A&A, 659, 616

Eggenberger, P., Charbonnel, C., Talon, S., et al. 2004, A&A, 417, 235 ([1] in Table 1)

Iglesias, C. A., & Rogers, F. J. 1996, ApJ, 464, 943

Kervella, P., Thévenin, F., Ségransan, D., et al. 2003, A&A, 404, 108 ([3] in Table 1)

Miglio, A., & Montalbán, J. 2005, A&A, 615, 629

Morel, P., 1997, A&AS, 124, 597

Neuforge-Verheecke, C., & Magain, P. 1997, A&A, 328, 261 ([2] in Table 1)

Pourbaix, D., Nidever, D., McCarthy, C., et al. 2002, A&A, 386, 280 ([5] in Table 1)

Scuflaire, R., Montalbán, J., Théado, S., et al. 2008, Ap&SS, 316, 149

Söderhjelm, S. 1999, A&A, 341, 121

Comm. in Asteroseismology
Vol. 158, 2009, 38th LIAC/HELAS-ESTA/BAG, 2008
A. Noels, C. Aerts, J. Montalbán, A. Miglio and M. Briquet., eds.

Uncertainties in the chemical composition of B-type stars: effects on the opacity and on the excitation of pulsation modes

J. Montalbán, A. Miglio, T. Morel

Institut d'Astrophysique et de Géophysique
Université de Liège, Allée du 6 Août 17 - B 4000 Liège - Belgique

Abstract

Recent determinations of element abundances in early-B stars have shown a significant discrepancy with those derived for the Sun. β Cep pulsators are early-B stars whose pulsations are driven by the opacity mechanism operating at the Fe-group opacity bump (e.g. Dziembowski & Pamyatnykh 1993). Since each element contributes differently to the stellar opacity in the Z-bump region, differences in the metal mixture may affect the pulsation properties of these early-B stars. Here we study the role of different elements in the stellar opacity of B-type stars, and the consequences of the uncertainties in the composition of the metal mixture for the instability strip of β Cep stars.

Introduction

In the new solar mixture derived by Asplund et al. (2005, AGS05), the abundances of the elements that contribute the most to the metal mass fraction (Z), that is C, N, O, and Ne, decreased by 38% with respect to the Grevesse & Noels (1993) mixture (GN93), while that of Fe (generally used in deriving stellar metallicities) decreased by 15%. The solar metallicity (Z/X) has changed from 0.0245 down to 0.0167. For a fixed Z, the Fe mass fraction in the AGS05 mixture has increased by $\sim$25% with respect to the one in the standard GN93. While the lower O abundance in AGS05 ruins the good agreement between the solar model and the seismic one, a larger Fe mass fraction favourably affects the excitation of β Cep pulsation modes in early-type stars. But, is the solar mixture adequate to describe the chemical composition of B-type stars?

The examination of a sample of 150-200 early-B stars carefully selected in the literature (Morel 2009) has revealed that the element abundances in early-B stars in the solar environment are systematically and significantly lower even than those derived by AGS05 for the Sun. Is this discrepancy real or, on the contrary, are the B-star abundances underestimated due to theoretical uncertainties in atomic models or due to problems in the derivation of atmospheric parameters as claimed by Przybilla et al.(2008)? In addition to questions concerning the chemical evolution of the Galaxy, the pattern of element abundances derived by Morel (2009) (hereafter B-dwarfs Mix.) has important implications for the pulsation studies of β Cep stars.

Results

We have computed the contribution of each metal to the total opacity in a typical β Cep star. The results for the most relevant elements are shown in Fig. 1.

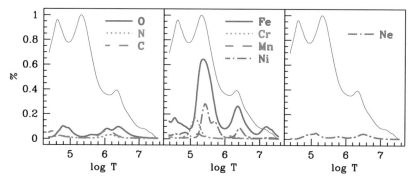

Figure 1: Opacity profile inside a 10 $M_\odot$ model (solid thin line) in the middle of the main sequence, and relative contribution to opacity of different elements: CNO (left panel), Fe-group elements (center) and Ne (right).

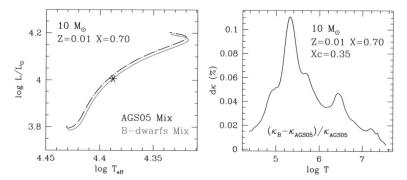

Figure 2: Left: main-sequence evolutionary tracks of 10 $M_\odot$ models computed with OPAL opacity tables for two different metal mixtures: AGS05 (dash-dotted line) and the one derived for B-dwarfs in the solar vicinity (within 1 kpc, in solid grey line). The asterisk marks the location of the model used in the right panel plot. Right: relative difference of opacity between AGS05 and B-dwarf mixtures for a fixed stellar structure of a 10 $M_\odot$, $X = 0.70$, $Z = 0.01$ model at the middle of the main sequence.

For a given metal fraction Z, the B-dwarfs Mix contains less C (17%), N (15%), O (2%), Al (41%), Si (55%) and Mg (2%) than the AGS05 mixture, and more Ne (26%), S (34%) and Fe (17%). For the other elements contributing to Z, the abundances of AGS05 are used, implying an increase of 30% for the remaining Fe-group elements (Cr, Mn, and Ni).

The effect of the difference of opacity on the evolutionary track is shown in the left panel of Fig. 2 where two main-sequence evolutionary tracks of a 10 $M_\odot$ star are plotted. The tracks were computed for fixed mass fractions of hydrogen (X) and metals (Z). The metal composition is different in the two cases and corresponds to AGS05 and to the B-dwarfs mixture. The higher opacity of the latter leads to a cooler track. Since the main contribution to κ comes from Fe-group elements, the difference shown in the right panel of Fig. 2 reflects the different relative contribution of the Fe-group elements to Z.

A higher and larger κ-peak at $\log T \sim 5.3$ implies a larger instability strip as well as a larger frequency domain of excited modes. In Fig. 3 we show the results of the nonadiabatic computation of oscillation frequencies for the models in the two evolutionary tracks of Fig. 2.

The abundance of Ne has recently been a matter of debate (see e.g. Morel & Butler 2008

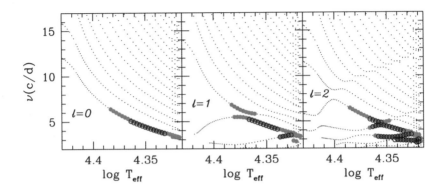

Figure 3: Frequencies of $\ell = 0, 1$ and 2 pulsation modes as a function of log T_{eff} for MS models of 10 $M_\odot$ in Fig. 2. Excited modes are symbolized by large-open circles (AGS05-mixture models) and large-solid circles (B-dwarfs).

and references therein). The contribution of Ne to κ at the bottom of the solar convective region is of the order of 20%, therefore, an increase of Ne could help to recover the agreement with helioseismic constraints as suggested by Bahcall et al. (2005). For early-B stars, however, the density at log $T_{eff} \sim 6.3$ is almost two orders of magnitude smaller than in the Sun and the role of Ne in the opacity is almost negligible (Fig. 1). The effect of Ne on pulsation properties of β Cep stars is only a collateral one. In fact, varying Ne for a given Z will change the Fe-group relative abundance. The difference in Ne abundance between Cunha et al. (2006) and AGS05 has only a slight effect on the instability strip of B-type stars.

Conclusions

Since each element contributes differently to the stellar opacity in the Z-bump region, differences in the metal mixture for a given Z may significantly affect the excitation of β Cep pulsation modes. The determination of the detailed element abundances is essential for the stability analysis of β Cep stars.

Acknowledgments. JM and AM acknowledge financial support from the Prodex-ESA Contract Prodex 8 COROT (C90199).

References

Asplund, M., Grevesse, N., & Sauval, J.A. 2005, ASCP, 336

Bahcall, J.N., Basu, S., & Serenelli, A.M. 2005, ApJ, 618, 1281

Cunha, K., Hubeny, I., & Lanz, T. 2006, ApJ, 647, 143

Dziembowski, W, & Pamyatnykh, A.A. 1993, MNRAS, 262, 204

Grevesse, N., & Noels, A. 1993, AVCP

Morel, T., & Butler, K. 2008, A&A, 487, 307

Morel, T. 2009, CoAst, 158, 122

Przybilla, N., Nieva, M-F., & Butler, K. 2008, ApJ, 688, 103

Session 6
What about real stars ?

Comm. in Asteroseismology
Vol. 158, 2009, 38th LIAC/HELAS-ESTA/BAG, 2008
A. Noels, C. Aerts, J. Montalbán, A. Miglio and M. Briquet., eds.

Ground-based observations of the β Cephei CoRoT main target HD 180642

M. Briquet[1], K. Uytterhoeven[2,3], C. Aerts[1,4], T. Morel[1,6], P. De Cat[5], A. Miglio[6], P. Mathias[7], K. Lefever[1], E. Poretti[2], S. Martín-Ruiz[8], M. Paparó[9], M. Rainer[2], F. Carrier[1], J. Gutiérrez-Soto[10], J.C. Valtier[7], J.M. Benkő[9], Zs. Bognár[9], P.J. Amado[8], J.C. Suarez[8], C. Rodriguez-Lopez[8] and R. Garrido[8]

[1]Instituut voor Sterrenkunde,
Katholieke Universiteit Leuven, Celestijnenlaan 200 D, B-3001 Leuven, Belgium
[2]INAF-Osservatorio Astronomico di Brera, Via E. Bianchi 46, I-23807 Merate, Italy
[3]Instituto de Astrofísica de Canarias,
Calle Via Láctea s/n, E-38205 La Laguna, TF, Spain
[4]Department of Astrophysics,
University of Nijmegen, IMAPP, PO Box 9010, 6500 GL Nijmegen, the Netherlands
[5]Koninklijke Sterrenwacht van België, Ringlaan 3, B-1180 Brussel, Belgium
[6]Institut dAstrophysique et de Géophysique de l'Université de Liège,
Allée du 6 Août 17, B-4000 Liège, Belgium
[7]UMR 6525 H. Fizeau, UNS, CNRS, OCA, Campus Valrose,
F-06108 Nice Cedex 2, France
[8]Instituto de Astrofisica de Andalucia, CSIC, Apdo. 3004, 18080 Granada, Spain
[9]Konkoly Observatory, P.O. Box 67, 1525 Budapest, Hungary
[10]GEPI, Observatoire de Paris, CNRS, Université Paris Diderot,
place Jules Janssen 92195 Meudon Cedex, France

Abstract

We present the preliminary results of a detailed study of ground-based photometric and spectroscopic observations dedicated to the β Cephei CoRoT main target HD 180642. Besides the non-linear dominant radial mode several low-amplitude modes are detected in both kinds of datasets. Our aim is to derive the wavenumbers (ℓ, m) of these modes, as additional constraints to the CoRoT pulsation frequencies, for forthcoming asteroseismic modelling of the star.

Individual Objects: HD 180642

Introduction

The B 1.5 II-III star HD 180642 (V1449 Aql, HIP 94793, $m_v = 8.27$) is the only known β Cephei star in CoRoT's core programme. It has been observed by the satellite during a long run (some 150 days) from May to October 2007. The star presents a dominant non-linear mode with an amplitude of ~40 mmag together with several modes of much lower amplitudes (below 4 mmag). Fourier periodograms of its CoRoT lightcurve are displayed in Michel et al. (2008) and are not repeated here. Numerous frequencies are found in the CoRoT lightcurve, among which harmonics and combination frequencies. Its full analysis will be published elsewhere (Degroote et al., in preparation).

To complement the space white light data, both ground-based photometric and spectro-scopic data were collected, to which we added several archival observations. In total, we

Table 1: Logbook of the spectroscopic observations. For each instrument we give the number of spectra N, the time-span ΔT in days, the average S/N-ratio and the resolution of the spectrograph.

Instrument	N	ΔT	$<S/N>$	resolution
FEROS	223	967.5	186	48 000
SOPHIE	35	10.1	106	40 000
Aurélie	22	8.1	93	25 000

made use of 507 high-quality multi-colour photometric data, 172 white light data, and 280 high-resolution high S/N spectra. In this paper, we describe the first results of the analysis of these datasets.

Ground photometry

Several telescopes with multi-colour photometric instruments are involved in this project. Strömgren *ubvy* photometry was collected with the 90cm telescope at Sierra Nevada Observatory and with the 1.5m telescope at San Pedro Mártir Observatory; Geneva 7-colour photometry ($UB_1 BB_2 V1 VG$) with the 1.2m Mercator telescope at the Observatorio Roque de los Muchachos and with the 70cm Swiss telescope at La Silla Observatory; and Johnson V photometry with the 50cm telescope at Konkoly Observatory. Hipparcos measurements are also available. A logbook of the observations is shown in Table 2 of Uytterhoeven et al. (2008).

We combined the datasets for the filters that allow to do so, as follows. First, we merged the Strömgren data of both sites. Second, we constructed ultraviolet, blue, and visual ground-based lightcurves (combined U, B, V data hereafter) by merging the Geneva U and Strömgren u, B and v, and V and y data, respectively. Finally, we also constructed a more extensive visual band lightcurve by adding also the Johnson V data and the Hipparcos H_p data to the Geneva V and Strömgren y data.

As an illustration of our frequency analysis of the merged datasets, Scargle periodograms calculated from the combined U data are shown in Fig. 1. The variations due to the main mode occur with $f_1 = 5.48694$ d^{-1}, $2f_1$ and $3f_1$. The corresponding amplitudes in the combined U filter are 82.8, 8.3 and 4.3 mmag. Note that HD 180642 has one of the highest amplitudes among the known β Cephei stars (see catalog of Stankov & Handler 2005 for comparison). The dominant mode of HD 180642 was already identified by Aerts (2000) as a radial mode by means of Geneva photometry.

In the combined Strömgren and ultraviolet datasets, we also detect two additional significant frequencies, i.e. their corresponding amplitudes reach at least 4 times the noise level in the periodogram, following the criterion derived empirically by Breger et al. (1993). The results from the CoRoT data (Michel et al. 2008) allow us to pick out the right peak. The second frequency present in our photometric data is $f_2^p = 0.30818(5)$ d^{-1} instead of its one-day-alias whose amplitude is higher in the periodogram (see Fig. 1). Afterwards, we find the frequency $f_3^p = 7.36673(7)$ d^{-1}.

Spectroscopy

In the framework of the CoRoT ground-based Large Programme (e.g. Uytterhoeven & Poretti 2007; Uytterhoeven et al. 2008), aimed at the follow-up of selected CoRoT targets, HD 180642 was observed with the FEROS spectrograph at the 2.2m telescope at ESO La Silla. The target was also monitored with the SOPHIE (1.93m telescope) and Aurélie (1.52m telescope) spectrographs, both situated at the Haute-Provence observatory. The logbook of the observations is shown in Table 1.

A phase diagram of the first moment $< v^1 >$ (radial velocity placed at average zero) is displayed in Fig. 2 and clearly shows the non-linear behaviour of the dominant pulsation f_1 in HD 180642. The peak-to-peak amplitude is ~ 90 km s^{-1} and harmonics up to $5f_1$ are present in $< v^1 >$. This makes HD 180642 the β Cephei star with the third largest radial-velocity amplitude, BW Vul having a peak-to-peak amplitude of ~ 200 km s^{-1} (Aerts et al. 1995) and σ Sco of ~ 110 km s^{-1} (Mathias et al. 1991).

In the radial velocity measurements, we also clearly detect $f_2^s = 8.408(1)$ d^{-1} as well as f_2^p, with an amplitude of 1.5 km s^{-1} and 1.1 km s^{-1}, respectively. In a 2D frequency search across the line profile, we additionally find $f_3^s = 6.325(1)$ d^{-1}. The amplitude and phase distributions for the Si III 4553 Å line for f_2^s and f_3^s, also detected in the CoRoT lightcurve, are shown in Fig. 3.

We note that the difference between f_2^p, f_2^s and f_3^s is ~ 1.041 d^{-1}. The frequency precision of our observations as well as the uninterrupted CoRoT photometry exclude the possibility of dealing with one-day-aliases.

Conclusions

Besides the dominant radial mode, low-amplitude non-radial modes are detected in our ground-based observations of HD 180642. Our aim is to perform a careful mode identification for them. Because of the highly non-linear behaviour of the radial mode, available mode identification techniques cannot be applied straightforwardly, as they are implemented for a linear pulsation. In particular for spectroscopy, the moment method (Briquet & Aerts 2003) and the FPF method (Zima 2006) need to be adapted. This work is ongoing and the final results will be described in a forthcoming paper.

Acknowledgments. MB and FC are Postdoctoral researchers of the Fund for Scientific Research, Flanders. KU acknowledges financial support from a *European Community Marie Curie Intra-European Fellowship*, contract number MEIF-CT-2006-024476. The FEROS data are being obtained as part of the ESO Large Programme: LP178.D-0361 (PI: E. Poretti). This work was supported by the Italian ESS project, contract ASI/INAF I/015/07/0, WP 03170, by the Hungarian ESA PECS project No 98022, and by the European Helio- and Asteroseismology Network (HELAS), a major international collaboration funded by the European Commission's Sixth Framework Programme.

References

Aerts, C., Mathias, P., Van Hoolst, T., et al. 1995, A&A, 301, 781

Aerts, C. 2000, A&A, 361, 245

Breger, M., Stich, J., Garrido, R., et al. 1993, A&A, 271, 482

Briquet, M., & Aerts, C. 2003, A&A, 398, 687

Mathias, P., Gillet, D., & Crowe, R. 1991, A&A, 252, 245

Michel, E., Baglin, A., Weiss, W., et al. 2008, CoAst, 156, 73

Stankov, A., & Handler, G. 2005, ApJS, 158, Issue 2, 193

Uytterhoeven, K., Poretti, E., & the CoRoT SGBOWG 2007, CoAst, 150, 371

Uytterhoeven, K., Poretti, E., Rainer, M., et al. 2008, in 'HelasII international conference: Helioseismology, Asteroseismology and MHD Connections', Journal of Physics: Conference Series, IOP Publishing, in press (astro-ph 0710.4068)

Zima, W. 2006, A&A, 455, 227

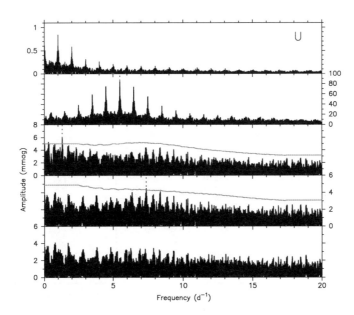

Figure 1: Scargle periodograms calculated from the combined Geneva U and Strömgren u data. The uppermost panel shows the spectral window of the data. All subsequent panels show the periodograms at different stages of prewhitening. From top to bottom: periodogram of the observed light variations, of the data prewhitened with f_1 and its two harmonics $2f_1$ and $3f_1$, of the data subsequently prewhitened with f_2^p, and of the data subsequently prewhitened with f_3^p. The three frequencies found are indicated by dashed gray lines. The light gray lines indicate the 4 S/N level.

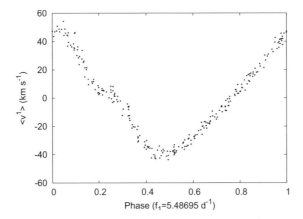

Figure 2: Phase diagram of the first moment computed from the Si III 4553 Å line, for $f_1 = 5.48695$ d^{-1}.

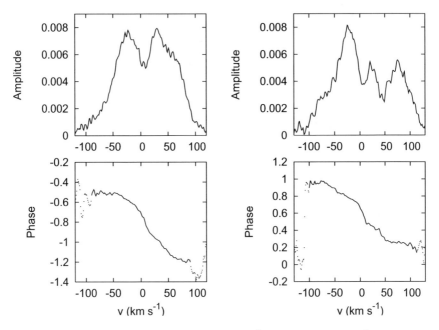

Figure 3: Amplitude and phase distributions for $f_2^s = 8.408$ d^{-1} (left) and $f_3^s = 6.325$ d^{-1} (right) for the Si III 4553 Å line. The amplitudes are expressed in units of continuum and the phases in π radians.

Comm. in Asteroseismology
Vol. 158, 2009, 38th LIAC/HELAS-ESTA/BAG, 2008
A. Noels, C. Aerts, J. Montalbán, A. Miglio and M. Briquet., eds.

Observations of the Wolf Rayet star WR123 with MOST

L. Lefèvre[1], A.N. Chené[2]

[1] Observatoire de Paris
LESIA, 5 place Jules Janssen, 92190, Meudon, France
[2] NRC - CNRC, Victoria, BC, Canada

Abstract

We present an analysis of the intensive visual-broadband photometric monitoring of the highly variable WN8 Wolf-Rayet star WR123, obtained by the MOST satellite[1]. The Canadian space telescope observed WR123 for 38 days non-stop during June/July 2004. To complement previous investigations (Lefèvre et al. 2005), we show spectroscopic data taken a year before the MOST observations and a supplementary analysis of the dataset. This work shows that it is possible to fit quite reliably this supposedly unstable signal with a few sinusoidal components and may lead us to new insights in the variability of Wolf-Rayet stars.

Individual Objects: WR123

Observations & Analysis

WR123 is a presumably single, relatively bright WN8 star ($V_{GSC2}=11.39 \pm 0.10$) which was observed both in photometry and spectroscopy.

MOST (Walker et al., 2003) is a low-cost microsatellite designed to detect low-amplitude stellar oscillations with a precision of a few micro-magnitudes. The telescope focuses light on photometric and guiding CCDs in two modes of observation: Fabry Imaging (for targets with $V \lesssim 6$ mag) or Direct Imaging (for fainter objects).

Spectroscopy

Spectroscopy was carried out in 2003 (a year before the MOST observations) on WR123 (Chené et al 2008, submitted). A temporal variance spectrum analysis (TVS: Fullerton, Gies & Bolton, 1996) was used to determine the deviations in spectral features in a statistically rigorous way. The TVS spectrum clearly showed that *all the lines* were significantly variable at the 99% level($HeI\lambda4471$, $HeII\lambda5412$, $HeI\lambda5876$, $HeII\lambda6560$).

Figure 1 shows the variations in the absorption and emission parts of the $HeII\lambda5412$ line phased with the ~10h period previously discovered in photometry. Note that the equivalent width variability curve has first been "cleaned" of large-scale variability to isolate variations of a few hours. We also remark that the amplitude of the variations are the same for all the studied lines and are of about 10% for emission features while they reach 25% for the absorption features. These variations indeed appear to be related to those seen in ground-based and space-based photometry and lend support to this ~10h period.

---

[1] Microvariability and Oscillations of STars : This study is based on data from the MOST satellite, a Canadian Space Agency mission, jointly operated by Dynacon Inc., the University of Toronto Institute for Aerospace Studies and the University of British Columbia, with the assistance of the University of Vienna.

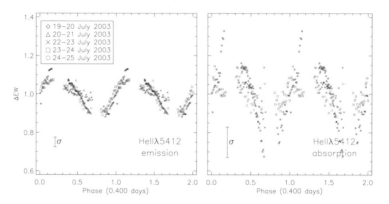

Figure 1: Measure of the equivalent width of the HeII λ5412 line phased with the P∼10h period. Different nights of observation have been plotted with different symbols (Chené A.N., Private Communication).

Photometry

WR123 was also observed by MOST in the Direct Imaging mode in ∼ 25s exposures every 30s for 38 days as well as with the 1.3 m Robotically Controlled Telescope (RCT) at KPNO[1] (Gelderman et al., 2004). Comparing the RCT and MOST data, we found a good match, with both telescopes showing the same trends in the overlapping time intervals.

The entire data set was reduced with a custom software developed by L. Lefèvre (LL) using fixed aperture photometry. The high levels of stray light have however shortened the duty-cycle to 52%. The final rms error is roughly 5 mmag per point for WR123.

The 10h period

Lefèvre et al (2005) analysed the dataset with the CLEAN algorithm (Roberts et al., 1987) to avoid misinterpretations due to relatively damaging window and background levels. Figure 2 represents the simple Fourier spectrum of WR123 where the dramatic effects of the window can easily be seen. The dataset was sliced into smaller subsets to check for the presence of persistent frequencies in the Fourier domain. Eventually, a Time-Frequency analysis confirmed the presence, throughout the whole set, of the ∼ 10h period.

Further analysis

To check the stability of other components, we used a sinusoids fitting algorithm (Lefèvre et al., in preparation) in the time domain. This technique has the advantage of enabling close frequencies to be resolved in a way that would be impossible using a conventional DFT or FFT. It is also scalable as it can be performed relatively quickly if only a small number of frequencies is being analysed. In addition, because it employs a best-fit algorithm, it can be used to determine the amplitude and phase of frequency components that do not fully complete a cycle within the sample window.

The sinusoids fitting method used here was originally developed for the analysis of the CoRoT data and is based on a Levenberg-Marquardt algorithm and a minimization in the least-squares sense. It uses the frequency domain as a *first guess* for the fitting of sinusoids

[1]Kitt Peak National Observatory is operated by AURA, Inc., under contract with the National Science Foundation.

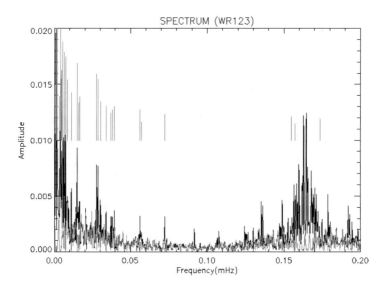

Figure 2: Fourier spectrum of the complete MOST dataset for WR123. The y-axis represents the amplitude of the variations relative to the mean flux. In black is the original Fourier spectrum of the dataset, overplotted in red are (above) the peaks that the algorithm fitted and (below) the residual spectrum after "cleaning". Note that the effect of the window has been completely removed by the "cleaning" of low frequency peaks.

in the time domain. It fits the *highest amplitude sinusoid*, creates a parallel signal from which the sinusoid is subtracted, and can then access the next highest sinusoidal signal, while keeping the main signal complete. We also developed a *sliding-medallion* method which fits only a medallion at a time and prewithens the remaining parameters only after fitting them a few times within the medallion. This method enables us to analyse classical pulsators such as δ Scuti stars with thousands of frequencies (cf. LL 2009). It can overcome the RAM and CPU shortcomings that can be encountered when processing 1000 sinusoids $\times$ 3 parameters $\times$ 405000 points (the typical number of points for 150d of observations with the CoRoT satellite for ex.). It is also faster and can reach a good precision ($< 1\%$) in a reasonable amount of time.

Every existing signal can be reproduced by sinusoids, provided the amount of sinusoids is large enough. A fit of less than 30 sinusoids (figure 3 shows the 29 peaks signal overplotted on the original WR123 data) was necessary to significantly reduce the power in the fourier spectrum (see the residual spectrum in figure 2), i.e. the signal does not appear as unstable as initially assumed. The low frequency part of the signal ($f < 1c/d$) could be explained by the presence of *large-scale structures* in the wind of Wolf-Rayet stars (de Jong et al., 2001). Moreover, there is a peak, located at $\sim$ 1.3 c/d (1.2976$\pm$ 0.0006 c/d $\sim$ 18.5h) which is easily represented by the fit of a single, constant sinusoid. Note that this peak, within the usual frequency resolution of a Fourier transform, is harder to distinguish from twice the 10h period. However, the fitting algorithm enables a much higher resolution and seems to suggest that both signals are not related. The initial time frequency (TF) analysis, optimized for the detection of a possible period around 10h, was not able to single it out. A new TF analysis should help assessing the stability of this component.

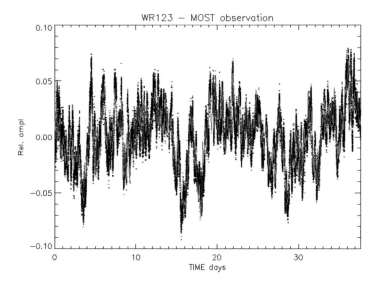

Figure 3: MOST WR123 data (in black) with a fit of 29 sinusoids overplotted in red (peaks indicated in figure 2). The x-axis is the number of days since the beginning of the run and the y-axis represents the amplitude of variations relative to the mean flux of the star.

Discussion

Table 1 summarizes the known parameters for WR123. If we suppose a mass of 26 $M_\odot$ for WR123 with a radius of 15 $R_\odot$ (mean values of lines 6 and 7, table 1) , nothing with a period lower than 1.27 days can be attributed to an orbiting companion, unless the latter is orbiting inside the star. On the other hand, periods lower than approximately 1.32 days cannot be attributed to rotation as the star would be well over its break-up velocity. Thus we can say that neither the $\sim$ 10h nor the $\sim$ 18.5h periods can be attributed to either phenomenon. They clearly seem to be related to pulsational instabilities.

However, there are remaining questions. First, the complex relations between what we observe in photometry (i.e. integrated flux) and in spectroscopy (different lines correspond to different depths in the stellar wind) and also their respective relations with the pulsations that are supposed to occur at the wind base. Continuous observations with a good temporal coverage (such as a COROT long run) would definitely help to disentangle those effects and thus find the impact of the wind on the pulsations. A longer timespan would also enable the study of time variations of the observed frequencies. Then, a consistent model that would reproduce most of WR123's observables would greatly help in our understanding of Wolf-Rayet stars. It certainly seems conceivable, as W. Glatzel mentioned a viable solution in his talk at this very conference.

Table 1: WR123 parameters in the literature

Parameters	T2006 WNL	T2006 WNE	D2006	H2006	C1995
$M_{initial}$ $(M_\odot)$	100	100			
Age (Myr)	3.57	3.74			
X	0.117	0	0.35	0.0	$< 0.5\%$
Z	0.02	0.02	0.02		
$log(L_\star/L_\odot)$	5.77	5.62	5.45	6.05	5.45
$M_\star/M_\odot$	20.6	16.2	25	35	
$R_\star/R_\odot$	10.3	2.1		17.7	15.3
$R/R_\odot$			15.38		25.5
$T_\star$ (K)				44700	33900
T_{eff} (K)			33900		26300
$log(\dot{M})$					-4.04
V_∞ (km/s)				970	970

T2006 : Townsend & McDonald 2006 - D2006 : Dorfi et al. 2006
H2006 : Hamman et al. 2006 - C1995 : Crowther et al. 1995

References

Crowther, P.A., Hillier, D.J., & Smith, L.J. 1995, A&A, 293, 403

de Jong, J. A., et al. 2001, A&A, 368, 601

Dorfi, E.A., Gautschy, A., & Saio, H. 2006, A&A, 453, L35

Gelderman, R., et al. 2004, Astronomische Nachrichten, 325, 559

Fullerton, A. W., Gies, D. R., & Bolton, C. T. 1996, AJS, 103, 475

Hamann, W.-R., Gräfener, G., & Liermann, A. 2006, A&A, 457, 1015

Lefèvre, L., et al. 2005, ApJL, 634, L109

Lefèvre L., Michel, E., Aerts, C., et al. 2009, CoAst, 158, 189 (LL 2009)

Roberts, D. H., Lehar, J., & Dreher, J. W. 1987, AJ, 93, 968

Townsend, R.H.F., & MacDonald, J. 2006, MNRAS, 368, L57

Walker, G., et al. 2003, PASP, 115, 1023

DISCUSSION

Rauw: The MOST signal consists of white light where a mixture of continuum and lines influence the signal. WN8 stars are among those that have the largest optical depth among WN stars and so I wonder how you can get a consistent signal out of the wind especially since different parts of your signal arise in different parts of the wind.

Lefèvre: First, the total influence of the emission lines does not exceed about 10 %. Second, the variations of the continuum observed with MOST are of about 3 % while the lines show 20 % variations, i.e. the line variations do not come from the continuum (3 % for the 10h period). The lines do show a similar period as the total flux, but there is an unknown delay between the two, so we don't know if the variations do interfere. In conclusion, we need really simultaneous observations in spectroscopy and in photometry.

Cox: Have the WR stars been to the red or have they just evolved a bit from the main sequence?

Lefèvre: It depends on the evolution scenario we adopt. Did WR123 go through on LBV phase? We don't know yet.

Noels: I do not understand why models computed by Townsend and MacDonald (2006) should be rejected since they found g-modes excited through the κ-mechanism acting in the deep opacity bump (DOB) in models with hydrogen at the surface and also in models without hydrogen at the surface. Of course, if the requirement is to have a large radius together with a very low hydrogen abundance at the surface, the solution proposed by Saio, which I favor, of an instability coming from strange modes is much better and it has the advantage of nicely proposing a possible origin of mass loss in WR stars.

Comm. in Asteroseismology
Vol. 158, 2009, 38th LIAC/HELAS-ESTA/BAG, 2008
A. Noels, C. Aerts, J. Montalbán, A. Miglio and M. Briquet., eds.

Simultaneous MOST photometry and high-resolution spectroscopy of Spica, a binary system with a massive β Cephei star component.

M. Desmet[1], G. Walker[2,3], S. Yang[4], D. Bohlender[3], M. Briquet[1]*,
R. H. Østensen[1], C. Cameron[5,6†], J. Matthews[5], and C. Aerts[1,7]

[1] Institute of Astronomy - KULeuven
Celestijnenlaan 200D, B 3001 Leuven, Belgium
[2] 1234 Hewlett Place, Victoria, BC V8S 4P7, Canada
[3] NRC Herzberg Institute of Astrophysics
5071 West Saanich Road, Victoria, BC V9E 2E7, Canada
[4] Department of Physics and Astronomy, University of Victoria
Victoria, BC V8W 3P6, Canada
[5] Department of Physics & Astronomy, University of British Columbia
6224 Agricultural Road, Vancouver, V6T 1Z1, Canada
[6] Department of Astronomy and Physics, Saint Mary's University
923 Robie Street, Halifax, NS B3H 3C3, Canada
[7] Department of Astrophysics, University of Nijmegen
PO Box 9010, 6500 GL Nijmegen, The Netherlands

Abstract

We present the data from a unique observational study of Spica, a binary system with a primary component of β Cephei type. We exploit simultaneous high-resolution spectroscopic observations and high-precision photometry obtained by the MOST satellite. By disentangling the spectra of this binary, we get an accurate determination of the orbit. Future work includes a full seismic analysis of the system.

Individual Objects: Spica (α Virginis)

Introduction

The bright star Spica (α Virginis, $m_v = 1.04$) has long been known to be a spectroscopic binary (Vogel 1890) with an orbital period of almost exactly 4 days and a relatively high eccentricity of $e = 0.18$ (Batten, et al. 1989). The eccentricity causes the line of apsides to rotate with a period of 139 days (Aufdenberg et al. 2007). Shobbrook et al. (1969, photometry) found the primary component to be a β Cephei variable with a pulsation frequency of $5.75\,d^{-1}$. Several studies of the β Cephei variation were conducted from 1967 until 1974. Lomb (1978, photometry and radial velocities) found that the pulsation amplitude had been decreasing and, by 1972, had become undetectable. Balona (1985, photometry and radial velocities) suggested that the amplitude decrease is a geometric effect and that the amplitude varies over the precession cycle of the system (~ 200 years). Smith (1985a,b, spectroscopy) observed Spica spectroscopically and suggested a number of short-period modes among which toroidal modes. These are not visible in photometry due to their high degree. Recently, Dukes et al. (2005) obtained a number of photometric observations from 1996 until

---

*MB is Postdoctoral Fellow of the Fund for Scientific Research, Flanders.
†CC is partially supported by a CITA National Fellowship.

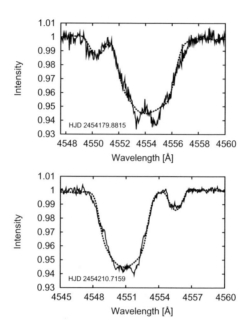

Figure 1: Spectra (full lines) taken with the Euler (top) and DAO (bottom) telescope. In both cases the phase of the system is such that the two stellar spectra are well separated. The dashed line shows the combined KOREL fit of both components for that particular phase.

2004 and concluded that there is no sign of the original $5.75\,\mathrm{d}^{-1}$ term even though Balona's model (Balona 1985) predicts that the amplitude during that time of observations should be detectable. In order to get an accurate determination of the orbit and to give conclusive arguments concerning the variability of the system, we have set up a unique ground-based spectroscopic and space-based photometric campaign.

Data

Ground-based data

The ground-based data originate from a high-resolution spectroscopic bisite campaign. We used both the CORALIE spectrograph with a resolution of 50000 attached to the 1.2-m Euler telescope at La Silla (Chile) and the coudée spectrograph of the DAO 1.2-m telescope at Victoria (Canada) with a resolution of 45000. In total 1856 observations were gathered using 2 different telescopes. The average S/N ratio near 4500 Å ranged between 300 and 800. Table 1 summarizes the logbook of our spectroscopic data. All data were subjected to the normal reduction process, which consists of de-biasing, background subtraction, flat-fielding and wavelength calibration. Finally, the heliocentric corrections were computed. The common technique of normalizing the spectra to the continuum by fitting a cubic spline function gave considerable scatter in the equivalent width of the lines. Therefore we used a different strategy (see e.g. Telting & Schrijvers 1998). We normalized the spectra night per night. We used the mean spectrum as a template and divided each individual spectrum by this nightly average. We normalized these quotient spectra by fitting low-degree polynomials, which turned out

Table 1: Log of our spectroscopic bisite campaign. The Julian Dates are given in days (-2454000), ΔT denotes the number of nights, N is the number of spectra and S/N denotes the average signal-to-noise ratio for each observatory measured at the continuum between 4500 and 4551 Å.

Long.	Lat.	Telescope	Julian Date		Amount and quality		
			Begin	End	ΔT	N	S/N
Dominion Astrophysical Observatory (DAO)							
$-123°25'$	$+48°31'$	1.2-m	191	210	7	105	800
Euler telescope in La Silla (Chile)							
$-70°44'$	$-29°15'$	1.2-m	178	191	11	788	300
			208	220	12	801	
			304	312	8	162	
Total					31	1856	

to be much easier than fitting the raw spectra. Subsequently, we constructed a final set of normalized spectra using the fits to the quotient spectra and a normal continuum fit of the template spectrum. For our study, we considered the Si III triplet around 4567 Å because these spectral lines are sufficiently strong and are dominated by temperature broadening, such that the intrinsic profile can simply be modelled with a gaussian. Moreover, they are not too sensitive to temperature variations (see De Ridder et al. 2002; Aerts & De Cat 2003), so that neglecting these variations remains justified. Fig. 1 shows 2 spectra of Spica taken in 2 opposite orbital phases of the system.

Space-based photometry

The space-based data originate from the MOST satellite (Walker et al. 2003) which has a 15 cm Rumak-Maksutov optical telescope feeding a CCD photometer. Because of its brightness, Spica was an excellent target for the MOST satellite and was observed in the principal mode of MOST, which is Fabry imaging (see Walker et al. 2003). The light curve of Spica was sampled using ~ 30 sec. exposures over a baseline of ~ 22.92 days resulting in a duty cycle of $\sim 44\%$. The MOST data reveal the orbital frequency and many of its harmonics, as well as the $5.75\,\mathrm{d}^{-1}$ frequency, already reported in the literature. This frequency has a very low amplitude (0.6 mmag), which explains why it was missed in some earlier studies. Moreover, clear line-profile variability is detected in our spectroscopic data (see Fig. 1), pointing out the multiperiodicity of the star, in agreement with the findings by Smith (1985a,b). The reported amplitude change of the photometric mode is therefore likely a result of unresolved beating with very low amplitude modes only visible in spectroscopy. A full analysis of the MOST light curve will be published elsewhere.

Orbital determination

In order to derive the orbital parameters of Spica and to analyse the possible variability of both components, we used the technique of spectral disentangling. This technique determines the contributions of both components to the composite spectra and the orbital parameters in a self-consistent way. We used the code KOREL (version 2.12.04), made publicly available by Hadrava (1995, and references therein). KOREL cross-correlates the input spectra with the disentangled component spectra to provide relative radial velocities. It is important to mention that the technique of spectral disentangling is found to be robust and successful also in the presence of complex variations of line shapes due to non-radial oscillations (Frémat et al. 2005, Harmanec et al. 2004, Uytterhoeven et al. 2004, De Cat et al. 2004). The

Table 2: Final solution of the orbital parameters of Spica from KOREL. P is the orbital period, T_0 is the time of nodal passage, e is the eccentricity, ω is the longitude of periastron, $K_{1,2}$ are the amplitudes in radial velocity of both components (1 denotes the primary, 2 the secondary component), q is the mass ratio and $d\omega/dt$ is the advance of periastron (U is the corresponding apsidal period).

Parameter	KOREL
P	4.0145 days
T_0	HJD 2440690.05
e	0.13
ω	128.9 deg
K_1	115.9 km s^{-1}
K_2	182.5 km s^{-1}
$q(m_2/m_1)$	0.63
$d\omega/dt$	0.0071 deg/day (U=139 years)

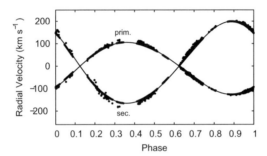

Figure 2: Radial velocities of the spectra for both components after disentangling (black dots) as a function of orbital phase together with the theoretical model (black lines) with orbital parameters listed in Table 2.

orbital parameters were derived applying KOREL to all 1856 spectra in the Si III triplet spectral region. During the fitting procedure, eight free parameters were allowed to describe the orbit of the close binary system. Table 2 lists the final solution for these unknowns. Fig. 1 shows the combined KOREL fit together with the observations for 2 different phases of Spica and Fig. 2 shows the radial velocities of both components after disentangling, together with the theoretical model with orbital parameters of Table 2.

Future work

We are presently trying to improve the orbital parameters of Spica by fitting the radial velocities of the disentangled spectra simultaneously with the complete light curve from MOST through a Wilson-Devinney algorithm. The results from two independent public codes, FO-TEL(Hadrava 2004) and PHOEBE(Prša & Zwitter 2005), will be compared with each other. Afterwards we will investigate the variability of the system in two ways. First of all, we will subtract the binary model from the light curve and look for variability in the residuals. Secondly, we will investigate the disentangled spectra which can give us information on the variability of both components separately. Subsequently, this will allow us to apply spectroscopic mode identification techniques on the found modes with state-of-the-art codes, such as FAMIAS(Zima 2008).

References

Aerts, C., & De Cat, P. 2003, SSRv, 105, 453

Aufdenberg, J. P., Ireland, M. J., Mérand, A., et al. 2007, IAUS, 240, 271

Balona, L. A. 1985, MNRAS, 217, 17

Batten, A. H., Fletcher, J. M., & MacCarthy, D. G. 1989, PDAO, 17, 1

De Cat, P., de Ridder, J., Hensberge, H., & Ilijic, S. 2004, ASPC, 318, 338

De Ridder, J., Dupret, M.-A., Neuforge, C., & Aerts, C. 2002, A&A, 385, 572

Dukes, R. J., Jr., Sonnett, S. M., & Mills, L. R. 2005, AAS, 37, 1364

Frémat, Y., Lampens, P., & Hensberge, H. 2005, MNRAS, 356, 545

Hadrava, P. 1995, A&AS, 114, 393

Hadrava, P. 2004, PAICz, 92, 1

Harmanec, P., Uytterhoeven, K., & Aerts, C. 2004, A&A, 422, 1013

Lomb, N. R. 1978, MNRAS, 185, 325

Prša, A., & Zwitter, T. 2005, ApJ, 628, 426

Shobbrook, R. R., Herbison-Evans, D., Johnston, I. D., & Lomb, N. R. 1969, MNRAS, 145, 131

Smith, M. A. 1985a, ApJ, 297, 206

Smith, M. A. 1985b, ApJ, 297, 224

Telting, J. H., Schrijvers, C. 1998, A&A, 339, 150

Uytterhoeven, K., Telting, J. H., Aerts, C., & Willems, B. 2004, A&A, 427, 593

Vogel, H. C. 1890, AN, 125, 305

Walker, G., Matthews, J., Kuschnig, R, Johnson, R., et al. 2003, PASP, 115, 1023

Zima, W. 2008, CoAst, 155, 17

Comm. in Asteroseismology
Vol. 158, 2009, 38th LIAC/HELAS-ESTA/BAG, 2008
A. Noels, C. Aerts, J. Montalbán, A. Miglio and M. Briquet., eds.

Unveiling the internal structure of massive supergiants: HD 163899

M. Godart[1], M.A. Dupret[2] and A. Noels[1]

[1] Institut d'Astrophysique et de Géophysique
Université de Liège, Allé du 6 Août 17 - B 4000 Liège - Belgique
[2] Observatoire de Paris, LESIA, CNRS UMR 8109, 5 place J. Janssen, 92195 Meudon, France

Abstract

Supergiant massive stars are post-main sequence stars. During the H shell burning phase of evolution, they present a radiative core in which a strong damping prevents the pulsation modes from being excited. However Saio et al. (2006) have recently highlighted p and g pulsation modes in a post-main sequence star (HD 163899) observed by MOST. They suggest that the presence of an intermediate convective region (ICZ) at the top of the radiative core allows a partial or total reflexion of the mode. Through some numerical results achieved with CLES (Scuflaire et al. 2008) and MAD (Dupret et al. 2003) codes, we show that this scenario depends on the evolution stage of the star and on the considered mass loss rate and overshooting parameter.
Individual Objects: HD 163899

Introduction

p and g mode pulsations have been detected in a B supergiant star observed by MOST: HD 163899 (B2 Ib/II Schmidt & Carruthers 1996). Lefever et al. 2007 have also suggested the presence of non-radial pulsations in a sample of B supergiants. At first sight this is quite unexpected. Indeed supergiant stars present a high density radiative helium core surrounded by a hydrogen rich envelope. A strong damping occurs if the modes propagate into the radiative core. In that case g-modes are not observed. However Saio et al. (2006) have shown that an intermediate convective zone (ICZ) at the top of the radiative core could prevent the propagation of g-modes into the core. The κ-mechanism due to the iron opacity bump at $\log T_{\rm eff} = 5.2$ is then sufficient to excite some g-modes. We show that mass loss and overshooting can prevent the formation of an ICZ and therefore the excitation of g-modes.

Effect of mass loss

The presence of an ICZ during the post-main sequence (post-MS) phase is a result of the MS evolution (Dupret et al. 2009): a region where the neutrality of the temperature gradient is reached during the MS is needed for the formation of the ICZ. However, when taking mass loss into account, the central temperature and the radiation pressure increase less rapidly during central H-burning. As a result, the adiabatic temperature gradient decreases less and less with time as the mass loss rate increases. If a large enough mass loss rate is assumed during the MS, no region where $\nabla_{\rm rad} \cong \nabla_{\rm ad}$ is formed and no ICZ appears during the supergiant phase (Chiosi & Maeder 1986).

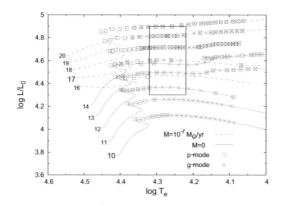

Figure 1: Evolutionary tracks computed without mass loss (10 to 14M$_\odot$) and with $\dot{M} = 10^{-7}$M$_\odot$/yr (16 to 20M$_\odot$) in the HR diagram. All sequences computed without mass loss present excited g-modes on the supergiant phase, but the sequences computed with mass loss present excited g-modes only for a mass higher or equal to 17M$_\odot$. The error box of the MOST star is also shown (Saio et al. 2006).

We adopted a mass loss rate of 10^{-7} M$_\odot$/yr to emphasize the effect of mass loss on the excitation of g-modes. According to Vinck et al. (2001) the mass loss rate should be between 10^{-9} and 10^{-7} M$_\odot$/yr, depending on the location of the star with respect to the bistability jump and on the mass of the star. We have computed evolutionary tracks with and without mass loss. Their location in the HR diagram is shown on fig. 1: tracks from 10 to 14M$_\odot$ (respectively tracks from 16 to 20M$_\odot$) are computed without (respectively with) mass loss. We performed nonadiabatic computations to determine whether unstable modes were present or not. On the one hand, for the sequences computed without mass loss, there are indeed excited g-modes during the supergiant phase for models within the error box for the MOST star (Saio et al. 2006). Even at 10M$_\odot$ the supergiant phase is characterized by excited g-modes. This is true for all higher masses. On the other hand, in the sequences computed with mass loss, we find excited g-modes only for stars more massive than about 17M$_\odot$. With an even higher mass loss rate, this value becomes larger and larger. The sequences without any excited g-modes are characterized by the absence of ICZ on the supergiant phase, due to the mass loss effect during the MS phase. Hence, all the modes enter the radiative core and suffer the strong radiative damping. The frequency distribution of the theoretical excited modes (fig. 2) is in good agreement with the observed frequencies.

Effect of overshooting

An enlargement of the mixed core makes the star more luminous. It also increases the core H burning lifetime while, in the HR diagram, the MS track reaches lower effective temperature values. When assuming a large enough overshooting during MS, the zone where $\nabla_{rad} \cong \nabla_{ad}$ does not exist especially if the radiative gradient is assumed in the overshooting region and this can prevent the formation of an ICZ during the post-MS phase. No g-modes are thus found to be excited. We have checked the presence of an ICZ by progressively increasing the overshooting parameter (fig. 3) from one evolutionary sequence to the next. For a model of 12M$_\odot$ the ICZ is well-developed for a moderate amount of overshooting $\alpha_{ov} = 0.2$, it is much smaller for 0.3 and it completely disappears for $\alpha_{ov} = 0.4$. From preliminary computations the extent of the ICZ is not very different if the adiabatic temperature gradient is chosen in

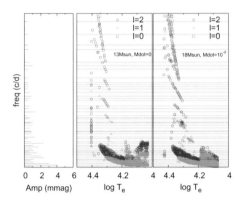

Figure 2: Frequency distribution of excited p and g-modes for supergiant models of $13M_\odot$ computed without mass loss (middle panel) and $18M_\odot$, computed with $\dot{M} = 10^{-7}M_\odot/yr$ (right panel). The observed frequency of MOST are also shown (left panel) from Saio et al. 2006.

the overshooting region.

However, another possibility to solve the problem of the presence of excited g-modes would be to bring MS evolutionary tracks into the error box of HD 163899. This can be achieved by including larger overshooting in MS models. We have computed evolutionary tracks with different overshooting parameters, ranging from $\alpha_{ov} = 0.2$ to 0.5 (fig. 4). MS evolutionary tracks cross the error box for an overshooting parameter equal to or larger than 0.3. Since MS massive stars present a convective core surrounded by a radiative envelope, the Brunt-Väisälä frequency is therefore zero in the core and no damping can occur. The κ-mechanism in the superficial layers excites p and g-modes. We performed nonadiabatic computations which revealed excited g-modes in all the sequences. The spectrum of the theoretical excited modes is shown on fig. 5 for $\alpha_{ov} = 0.4$ and 0.5 during the MS (decreasing effective temperatures) and near the turn off (increasing effective temperatures). The agreement in the mode spectrum is however not as good as it was for the 'true' supergiant model.

Conclusions

The presence of excited g-modes in B supergiant stars depends on physical processes during the MS: we have shown that large amounts of either overshooting or mass loss affect the formation of an ICZ and therefore can prevent the excitation of g-modes. The supergiant star could also be helium burning, but in that case again, as long as the ICZ is present, there should be excited g-modes. Another phenomenon which affects the formation of the ICZ is the convective criterion which can be either the Schwarzschild or the Ledoux's criterion (Lebreton et al. 2009). In a future work we shall extend this preliminary analysis in order to define an instability strip depending on those physical aspects and compare it to the observations. In this confrontation process, it is clear that asteroseismology of massive supergiant stars can give us a better understanding on the physical processes not only during the supergiant phase but also during the MS phase.

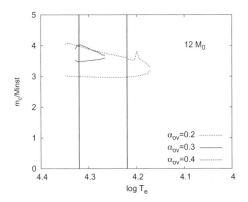

Figure 3: Evolution of the mass extension of the ICZ ($m_c/Minst$: ratio of the mass of the convective zone to the instantaneous mass of the star) during the supergiant phase (effective temperature decreasing on x-axis) for a $12M_\odot$ star. For a moderate amount of overshooting 0.2, the ICZ is well-developed, it is much smaller for 0.3 and it disappears for 0.4.

References

Chiosi, C., & Maeder, A. 1986, ARA&A, 24, 329

Dupret, M.-A., De Ridder, J., De Cat, P., et al. 2003, A&A, 398, 677

Dupret, M.-A., Godart, M., & Noels, A. 2009, CoAst, 158, 239

Lebreton, Y., Montalbán, J., Godart, M., et al. 2009, CoAst, 158, 277

Lefever, K., Puls, J., & Aerts, C. 2007, A&A, 463, 1093

Saio, H., Kuschnig, R., Gautschy, A., et al. 2006, ApJ, 650, 1111

Schmidt, E., & Carruthers, G. 1996, ApJ, 104, 101

Scuflaire, R., Théado, S., Montalbán, J., et al. 2008, Ap&SS, 316, 83

Vinck, J. S., de Koter, A., & Lamers, H. J. G. L. M. 2001, A&A, 369, 574

DISCUSSION

Shibahashi: Is the presence and/or extent of the intermediate convective zone dependent on the treatment of semi-convection?

Godart: Yes. We have no numerical treatment of the semiconvection in our evolutionary code, but with the Schwarzschild's criterion, small ICZs appear and disappear during the main sequence. So a partial mixing occurs at the top of the convective core and finally we do have a region where the temperature gradients (∇_{rad} and ∇_{ad}) are really close to each other.

Noels: I would like to comment on the mass loss rate adopted in this analysis. Mélanie chose indeed a too high value for $\dot{M}$ to show that this can make the ICZ disappear. But that's good news since excited g-modes are observed in that part of the HRD! That means that the ICZ is indeed present. This is a nice asteroseismic constraint on the mass loss rate.

Meynet: Could HD 163899 be on a blue loop after having evolved into a red supergiant stage? In that case, the star would have a helium convective core.

Godart: It certainly could be, either coming back from the red or having started burning helium in the blue. We are currently investigating this aspect.

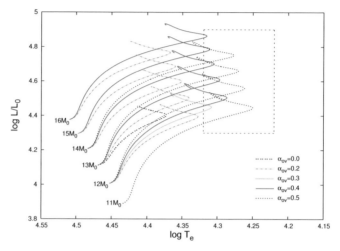

Figure 4: Main sequence evolutionary tracks computed with different overshooting parameters ranging from $\alpha_{ov} = 0.2$ to 0.5. The black box is the error box of HD 163899. Main sequence evolutionary tracks with at least $\alpha_{ov} = 0.3$ cross this error box.

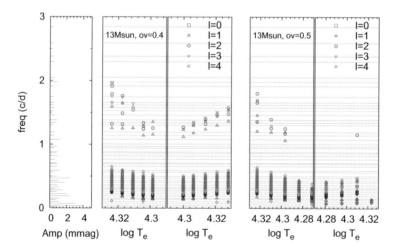

Figure 5: Frequency distribution of the excited p and g-modes during the MS (decreasing T_{eff}) and near the turn off (increasing T_{eff}) for $13M_{\odot}$ models computed with $\alpha_{ov} = 0.4$ (middle panels) and 0.5 (right panels). The agreement between the observed frequencies (left panel) and the theoretical mode spectrum is not as good as it was for the supergiant models.

Comm. in Asteroseismology
Vol. 158, 2009, 38ᵗʰ LIAC/HELAS-ESTA/BAG, 2008
A. Noels, C. Aerts, J. Montalbán, A. Miglio and M. Briquet., eds.

The discrepant mean density of μ Eridani

M. Jerzykiewicz

Instytut Astronomiczny, Uniwersytet Wrocławski,
Kopernika 11, 51622 Wrocław, Poland

Abstract

The star μ Eri = HR 1520 = HD 30211 (B5 IV, $V = 4.00$ mag) is an SB1 and EA system with an SPB primary. Using the available radial-velocity and photometric time-series observations, I revise the spectroscopic orbital elements and solve the eclipsing light curve. From the orbital solution, I derive the primary's mean density. This mean density is smaller than the value obtained from the star's position in the HR diagram. The discrepancy amounts to about 3σ. I show that the discrepancy can be removed by adding a "third light" to the system. However, the "third light" would have to be so bright that it is difficult to understand why it has not been detected long ago.
Individual Objects: μ Eri, ν Eri, 16 (EN) Lac

Introduction

The star was discovered to be a single-lined spectroscopic binary by Frost & Lee (1910) at the Yerkes Observatory. An orbit was derived by Blaauw & van Albada (1963) from radial velocities measured on 19 spectrograms taken in November and December 1956 with the 82-inch McDonald telescope. The orbit was revised by Hill (1969) who supplemented the McDonald observations with the Yerkes discovery radial-velocities (published by Frost et al. 1926), the Lick Observatory data of Campbell & Moore (1928) and the Dominion Astrophysical Observatory (DAO) observations of Petrie (1958). Since the data used by Hill (1969) span more than half a century, he has derived a value of the orbital period accurate to six digits, viz., 7.35886 d.

The star was discovered to be variable in light by Handler et al. (2004) while used as a comparison star in the 2002-2003 multi-site photometric campaign devoted to ν Eri. A frequency analysis of the campaign data revealed a dominant frequency of 0.616 d^{-1} and a very strong $1/f$ component, indicating a complex variation. Handler et al. (2004) concluded that μ Eri should be classified as an SPB variable. The frequency of 0.616 d^{-1} was confirmed by Jerzykiewicz et al. (2005) from observations obtained in 2003-2004. In addition, these authors found five further frequencies, ranging from 0.568 to 1.206 d^{-1}. The multiperiodicity, the values of the frequencies and the decrease of the uvy amplitudes with increasing wavelength implied high-radial-order g modes, strengthening the conclusion of Handler et al. (2004) that μ Eri is an SPB variable.

In addition to confirming the SPB classification, Jerzykiewicz et al. (2005) showed μ Eri to be an eclipsing variable. As can be seen in figs. 3 and 5, the eclipse is a transit, probably total, the secondary is fainter than the primary by several magnitudes, and the system is widely detached. The eclipse ephemeris, derived from the combined 2002-2003 and 2003-2004 data,

Table 1: Orbital elements of μ Eri from the McDonald radial-velocities of Blaauw & van Albada (1963).

P_{orb} (assumed)	7.3806 d
T	HJD 2435790.01 $\pm$ 0.20 d
ω	$127^{\circ} \pm 12^{\circ}$
e	0.418 ± 0.059
γ	23.2 ± 1.1 km/s
K_1	22.8 ± 2.0 km/s
$a_1 \sin i$	$(2.10 \pm 0.19) \times 10^6$ km
$f(M)$	0.0068 ± 0.0019 M$_\odot$

can be expressed as follows:

$$\text{Min. light} = \text{HJD } 2452574.04\,(4) + 7.3806\,(10)\,E. \tag{1}$$

Here, E is the number of minima after the epoch given (which is that of the middle of the first eclipse observed in 2002), and the numbers in parentheses are estimated standard deviations with the leading zeroes suppressed.

Spectroscopic orbital elements

The photometric orbital period in Eq. (1) differs by more than 20 standard deviations from Hill's (1969) period, mentioned in the Introduction. Hill's (1969) value is certainly less secure than the photometric one because it is based on a small number of observations, distributed unevenly over an interval of 53.2 years. In the periodogram of these observations that I computed, the three highest peaks in the period range of interest, all of nearly the same height, occur at 7.35889, 7.38116, and 7.38962 d. The first of these numbers corresponds to Hill's (1969) value, while the second is equal to the photometric period to within half the standard deviation of the latter. Since both these numbers represent the radial-velocity data equally well, I conclude that the photometric period is the correct one.

Using the photometric value of the orbital period and the McDonald data of Blaauw & van Albada (1963), I re-computed the spectroscopic orbit of μ Eri by the method of differential corrections of Schlesinger (1908). The elements of the orbit are given in Table 1. Fig. 1 shows the McDonald data and the radial-velocity curve computed from these elements.

The eclipse light curve solution

In order to derive the eclipse light curve, I pre-whitened the 2002-2003 and 2003-2004 differential magnitudes of μ Eri with the six SPB frequencies mentioned in the Introduction. I then phased the residuals using Eq. (1) and computed normal points in adjacent intervals of 0.0025 in phase. The y-filter normal points are plotted in Fig. 2 as a function of the orbital phase. In this figure, the whole range of orbital phase is shown in the upper panel, while the lower panel covers the phase interval from -0.08 to 0.08.

For computing the parameters of μ Eri from the eclipse light curve I used Etzel's (1981) computer program EBOP. The program, based on the Nelson-Davis-Etzel model (Nelson & Davis 1972, Popper & Etzel 1981) is well-suited for dealing with detached systems such as the present one. The spectroscopic parameters ω and e, needed to run the program, I took from Table 1. The primary's limb-darkening coefficient I interpolated from Table II of Wade & Rucinski (1985) for $T_{eff} = 15\,670$ K and $\log g = 3.5$ of μ Eri (Handler et al. 2004). For the secondary, I assumed no limb-darkening. Finally, I had to derive the central

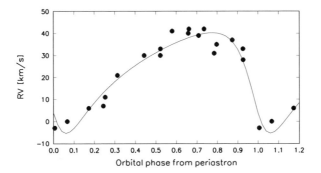

Figure 1: The Blaauw & van Albada (1963) McDonald data (circles) and the radial-velocity curve of μ Eri (line) computed from the elements listed in Table 1

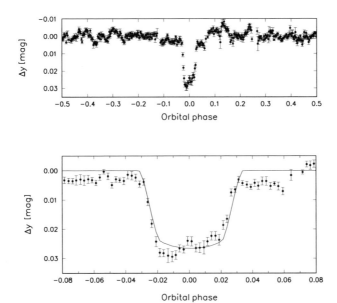

Figure 2: Top: y-filter normal points (circles with error bars), computed from differential magnitudes 'μ Eri $-$ ξ Eri' pre-whitened with the SPB frequencies. Bottom: The normal points and EBOP synthetic light curve computed with the following parameters: $i = 74.^{\circ}3 \pm 1.^{\circ}8$, $k = R_s/R_p = 0.1520 \pm 0.0023$, and $r_p = R_p/a = 0.287 \pm 0.010$ (line). Although all normal points were used in the solution, a limited range of phase is displayed to avoid crowding. Beyond this range, the synthetic light curve continues at zero ordinate.

surface brightness of the secondary in units of that of the primary, J, which EBOP uses as a fundamental parameter. Assuming that the effective temperature scales as the square root of the radius, $T_{\mathrm{eff},s} = \sqrt{k}\,T_{\mathrm{eff},p}$, I found J to be equal to 0.024. This exercise resulted in the following parameters: $i = 74.^{\circ}3 \pm 1.^{\circ}8$, $k = R_s/R_p = 0.1520 \pm 0.0023$, and $r_p = R_p/a =$

0.287 ± 0.010, where R_p and R_s are the radii of the primary and the secondary component, respectively, and a is the semi-axis major of the relative orbit. The synthetic light curve, computed with these parameters, is shown in the lower panel of Fig. 2 as the solid line. At the phases not shown in the figure, i.e., from -0.50 to -0.08 and from 0.08 to 0.50, the line continues at zero ordinate. The standard deviation of the fit to all points is equal to 0.0027 mag; the standard deviation of the fit to observations falling between the first and the fourth contact is equal to 0.0030 mag.

The discrepancy

As noted by Pigulski & Jerzykiewicz (1988), in case of a binary with the primary's mass much larger than that of the secondary, the spectroscopic and photometric elements can be combined to derive a mass-radius relation

$$M_p \sim R_p^{3+x}, \tag{2}$$

where x is a small positive number, slowly varying with M_p, the assumed mass of the primary. A consequence of this relation is that the mean density of the primary, $\langle \rho \rangle_p$, is only weakly dependent on M_p. This fact has been used by Dziembowski & Jerzykiewicz (1996) to derive the mean density of the β Cephei primary of 16 (EN) Lac, an SB1 and EA system rather similar to μ Eri. Note that the mean density derived in this way is independent of spectroscopic, photometric or any other calibrations.

In the present case, the spectroscopic parameters $a_1 \sin i$ and $f(M)$ from Table 1 and the photometric parameters r_p and i (see the preceding section) give $\langle \rho \rangle_p = 0.0130 \pm 0.0049$ and 0.0133 ± 0.0051 g cm^{-3} for $M_p = 4$ and 8 M$_\odot$, respectively. For $M_p = 6$ M$_\odot$, a mass read off the evolutionary track crossing the star's position in the theoretical HR diagram (see Handler et al. 2004, fig. 5), $\langle \rho \rangle_p = 0.0132 \pm 0.0050$ g cm^{-3}. The problem is that the mean density of the star consistent with its position in the same HR diagram, $\langle \rho \rangle_{HR}$, is equal to 0.038 ± 0.008 g cm^{-3}. Thus, $\langle \rho \rangle_{HR} - \langle \rho \rangle_p = 0.025 \pm 0.009$ g cm^{-3}, a 3σ discrepancy.

The discrepancy would disappear if the standard deviation of $\langle \rho \rangle_{HR} - \langle \rho \rangle_p$ were increased by a factor of three. But this standard deviation is mainly determined by the standard deviation of the radius read off the HR diagram in fig. 5 of Handler et al. (2004) which, in turn, is defined by the standard deviation of the star's luminosity. The latter was estimated from the standard deviation of Hipparcos parallax and the uncertainty of the bolometric correction. I do not believe there is room for increasing the standard deviation of $\langle \rho \rangle_{HR} - \langle \rho \rangle_p$ by a factor greater than 1.5.

In addition to random errors, $\langle \rho \rangle_p$ may be affected by systematic errors arising from inadequate photometric precision or time sampling during eclipse. However, such systematic errors would *decrease* the mean density (see Seager & Mallén-Ornelas 2003), so that the discrepancy would be increased.

A cure worse than the disease?

The discrepancy encountered in the preceding section can be traced back to the fact that the primary's radius obtained for $M_p = 6$ M$_\odot$ from the spectroscopic and photometric parameters of the system is equal to $R_p = 8.6 \pm 1.2$ R$_\odot$, while the star's position in the HR diagram implies 6.3 ± 1.3 R$_\odot$.

In an attempt to tackle the problem, let us try to decrease R_p. This can be accomplished by increasing i: the minimum value of R_p, equal to 6.6 ± 0.9 R$_\odot$, will be obtained for $i = 90^\circ$. This minimum value of R_p agrees with the HR-diagram value to within 0.2σ. The corresponding value of $\langle \rho \rangle_p$ is 0.030 ± 0.012 g cm^{-3}, so that the mean density discrepancy would be reduced to below 1σ. However, $i = 90^\circ$ requires the depth of the eclipse to be equal

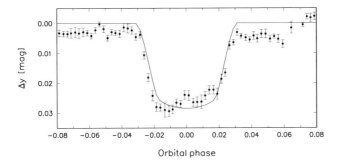

Figure 3: The normal points and EBOP synthetic light curve (solid line) computed with the postulated "third light," responsible for 0.67 of the system's flux.

to 0.080 mag, 0.054 mag deeper than observed (see Fig. 2). The required increase of the eclipse depth is much larger than maximum residuals outside eclipse (see Fig. 2, upper panel), and therefore is much too large to be accounted for by errors of observation or inadequate pre-whitening. The only remaining possibility of increasing the depth of the eclipse I am able to propose is light dilution by a third component of μ Eri, physical or not, which has not been detected previously.

Fig. 3 shows an EBOP solution with $i = 90^o$ and an assumed "third light," responsible for 0.67 of the system's flux. The fit is quite good. However, such a "third light" implies that the postulated third component of μ Eri would be about 0.8 mag brighter than the primary. It is therefore difficult to understand why its line spectrum escaped detection. One possibility is that the lines are shallow because of fast rotation, so that they are masked by the lines of the primary. However, rotation so fast that it would make the lines difficult to see would probably cause emission at $H\beta$. As far as I am aware, this has not been observed. On the contrary, the Strömgren-Crawford indices of μ Eri, including β (Hauck & Mermilliod 1998), are what one would expect for a slightly reddened B5 IV star (see Crawford 1978). In addition, if the lines of the primary were blended with lines having constant radial-velocity, the measured Doppler shifts would be affected. If the resulting radial-velocity semi-amplitude, K_1, were smaller than the true one (i.e., one that would be derived from unblended lines) by a factor of x, the true $a_1 \sin i$ and $f(M)$ would be equal to the observed ones divided by x and x^3, respectively. If $x > 1$, which seems to be the only possibility, the mean-density discrepancy would become even more severe.

Acknowledgments. I am indebted to Dr. Paul B. Etzel for providing me with the source code of his computer program EBOP and explanations. The use of NASA's Astrophysics Data System Abstract Service is acknowledged. This work was supported by MNiSW grant N203 014 31/2650.

References

Blaauw, A., & van Albada, T.S. 1963, ApJ, 137, 791

Campbell, W.W., & Moore, J.H. 1928, Publ. Lick Obs., 16, 62

Crawford, D.L. 1978, AJ, 83, 48

Dziembowski, W.A., & Jerzykiewicz, M. 1996, A&A, 306, 436

Etzel, P.B. 1981, in Carling E.B., Kopal Z., eds, Photometric and Spectroscopic Binary Systems. Reidel, Dordrecht, p. 111

Frost, E.B., & Lee, O.J. 1910, Science, 32, 876

Frost, E.B., Barrett, S.B., & Struve, O. 1926, ApJ, 64, 1

Handler, G., Shobrook, R. R., Jerzykiewicz, M., et al. 2004, MNRAS, 347, 454

Hill, G. 1969, Publ. DAO Victoria, 13, 323

Hauck, B., & Mermilliod, M. 1998, A&AS, 129, 431

Jerzykiewicz, M., Handler, G., Shobrook, R.R., et al. 2005, MNRAS, 360, 619

Nelson, B., & Davis, W.D. 1972, ApJ, 174, 617

Petrie, R.M. 1958, MNRAS, 118, 80

Pigulski, A., & Jerzykiewicz, M. 1988, AcA, 38, 401

Popper, D.M., & Etzel, P.B. 1981, AJ, 86, 102

Schlesinger, F. 1908, Publ. Allegheny Obs., 1, 33

Seager, S., & Mallén-Ornelas, G. 2003, ApJ, 585, 1038

Wade, R.A., & Rucinski, S.M. 1985, A&AS, 60, 471

Note added in proof: Using WIRE observations, Bruntt & Southworth, J. Phys. Conf. Ser. **118**(2008)012012, confirm the orbital period of 7.381 d. Moreover, they suggest that the depth of the eclipse may be variable. However, from their figure 4 it can be seen that the range of the variation is of the order of 0.010 mag, which is much too small to explain the discrepancy discussed in the present paper.

DISCUSSION

Aerts: Could it be that the discrepancy between the radius of μ Eri from the binarity and HR diagram comes from the rapid rotation of the primary?

Jerzykiewicz: No, I do not think so. The evolutionary tracks from which the mass was estimated were computed assuming $V_{rot} \sim 200$ km/s on the ZAMS.

Comm. in Asteroseismology
Vol. 158, 2009, 38th LIAC/HELAS-ESTA/BAG, 2008
A. Noels, C. Aerts, J. Montalbán, A. Miglio and M. Briquet., eds.

Preliminary results on the pulsations of Be stars with CoRoT

C. Neiner[1], J. Gutiérrez-Soto[1,2], M. Floquet[1], A.-L. Huat[1], A.-M. Hubert[1], B. de Batz[1], B. Leroy[2], Y. Frémat[3], L. Andrade[4], P. D. Diago[5], M. Emilio[6], J. Fabregat[5], E. Janot-Pacheco[4], C. Martayan[1,3], T. Semaan[1] and J. Suso[5]

[1] GEPI, Observatoire de Paris, CNRS, Université Paris Diderot; 5 place Jules Janssen, 92190 Meudon, France
[2] LESIA, Observatoire de Paris, CNRS, Université Paris Diderot; 5 place Jules Janssen, 92190 Meudon, France
[3] Royal Observatory of Belgium, 3 avenue circulaire, B-1180 Brussels, Belgium
[4] University of Sao Paulo, Brazil
[5] Observatori Astronòmic de la Universitat de València, ED. Instituts d'Investigatió, Poligon La Coma, 46980 Paterna, València, Spain
[6] Universidade Estadual de Ponta Grossa, Brazil

Abstract

We present preliminary results of the study of pulsations of Be stars from CoRoT observations. CoRoT detected many pulsation frequencies in all observed Be stars, usually interpreted in terms of g-modes but sometimes also corresponding to p-modes. Even in late Be stars, for which the detection of pulsation frequencies is very difficult from the ground, because of their very low amplitude, the quality and precision of the CoRoT data allows us to detect many frequencies. Seismic models have not yet been calculated but the calculations are ongoing.

Individual Objects: HD 181231, HD 175869, CoRoT-ID 102719279, HD 49330

Observing Be stars with CoRoT

Be stars are non-supergiant B stars which have shown at least once emission in their Balmer lines (see Neiner & Hubert 2009). Be stars display short-term variations associated with rotation and pulsations of β Cephei and SPB types. They are thus interesting targets for asteroseismic missions such as CoRoT.

CoRoT observes on average 1 bright Be star in each of its seismology field and a few tens of fainter Be stars in each exoplanet field. The length of CoRoT runs vary from about 3 weeks (short runs) to 5 months (long runs). Up to now 5 bright and a few tens of faint Be stars have already been observed by CoRoT.

We present herewith the preliminary results of a seismologic analysis for 3 bright Be stars and 1 faint Be star observed by CoRoT.

HD 175869

HD 175869 is a bright B8IIIe star observed in the SRC1 short run of CoRoT. The observations lasted 27.2 days and were uninterrupted. Although this star was classified as non-variable in the Hipparcos catalog, the CoRoT light curve displays variations with an amplitude of 0.2 mmag (see Fig. 1). A main frequency at 0.64 c.d^{-1} and its 5 harmonics are detected. The main frequency is compatible with the rotation frequency expected from a determination of

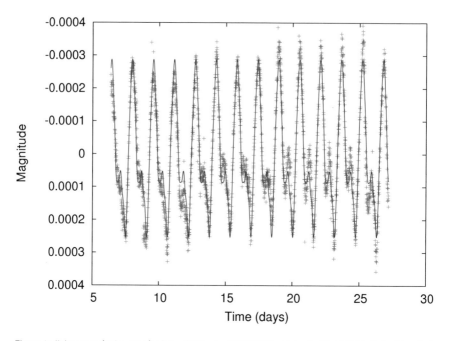

Figure 1: light curve (red crosses) of the B8IIIe star HD 175969 obtained with CoRoT. A fit (blue line) with the main period and its 5 harmonics is overplotted.

fundamental parameters through spectral fitting (Frémat et al. 2006). Additional frequencies are detected with lower amplitudes of a few ppm. They can be interpreted in terms of pulsations. See more details about that star in Gutierrez-Soto et al. (2009).

HD 181231

HD 181231 is a bright B5IVe star observed in the LRC1 long run of CoRoT. The observations lasted 156.6 days and were uninterrrupted. HD 181231 was known to be variable from ground-based observations with a frequency at 0.67 c.d^{-1} (Gutierrez-Soto et al. 2007). The CoRoT light curve shows clear variations and a clear beating of several frequencies. About 30 significant frequencies are detected with conservative search criteria and various Fourier-based methods. Two main groups of frequencies are found around 0.62 c.d^{-1} and 1.24 c.d^{-1}. Following models of Saio et al. performed for Be stars observed by the MOST satellite (e.g. Saio et al. 2008) these 2 groups could be attributed to prograde g modes with $|m| = 1$ and $|m| = 2$ respectively, with a rotation frequency just below 0.62 c.d^{-1}. However such a rotation frequency for HD 181231 can only be reproduced if extra mixing is assumed, either core overshooting or rotational mixing (H. Saio, private communication). Note that the second group of frequencies probably also includes harmonics of the first group of frequencies, since the second group is located around twice the frequencies of the first group.

Another clear single frequency peak is detected at 0.69 c.d^{-1}. This frequency is also detected in ground-based spectroscopic data obtained in the frame of the CoRoT project with FEROS (PI Poretti) and NARVAL (PI Neiner). From the spectroscopic data, following

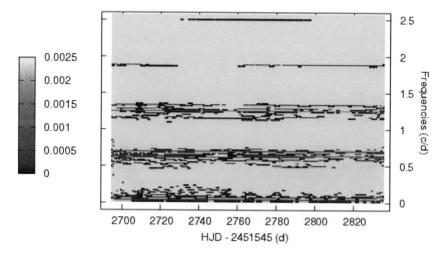

Figure 2: Periodogram of the B5IVe star HD 181231 from CoRoT data. The frequencies have been determined using a sliding window. The colors indicate the amplitude of the frequency peaks.

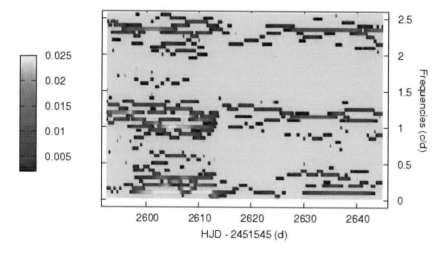

Figure 3: Same as Fig 2 but for the B5e star CoRoT-ID 102719279.

Telting & Schrijvers (1998) we can identify this frequency with a mode represented mainly by $l = 3$ or 4. Other lower amplitude frequencies are also detected in the CoRoT data, with frequencies up to 2.5 c.d^{-1}.

The link between pulsations and outbursts in Be stars

CoRoT-ID 102719279

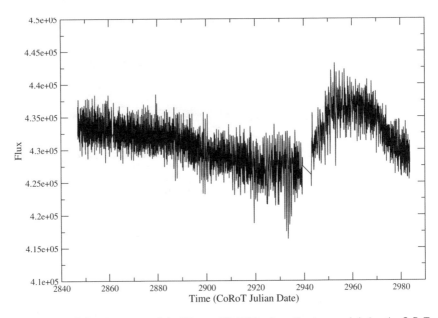

Figure 4: The CoRoT light curve of the B0e star HD 49330. An outburst occured during the CoRoT observations.

CoRoT-ID 102719279 is a faint B5e star observed in the exoplanet field of the IR1 run. The observations lasted 57.7 days. The light curve of this star shows a beating of frequencies with increasing amplitude until an outburst (seen as a fading in the light curve because of the inclination angle of the star compared to the observer) occurs. Then the variations almost disappear and the amplitude starts to grow again until the end of the light curve (see Fig.2 in Gutiérrez-Soto et al. 2008). The outburst produces a decrease in average magnitude of about 0.1 mag.

Two groups of frequencies are detected around 1.16 c.d^{-1} and 2.32 c.d^{-1}. Other frequencies are also observed around 0.98 c.d^{-1}. The amplitude of the frequency peaks around 1.16 and 0.98 c.d^{-1} are very strong before the outburst and about four times fainter after (see Fig.3). The amplitude of the frequencies around 2.32 c.d^{-1}, however, remains comparable before and after the outburst. The CoRoT data of this faint Be star suggest that there is a direct link between the observed pulsation amplitudes and the Be outburst.

HD 49330

A fainter (0.03 mag) but clear outburst has been observed in the bright B0e star HD 49330 observed by CoRoT in its LRA1 field during 136.9 days. There is a short data interruption of 3.5 days in the light curve (Fig.4). Preliminary results for this star show groups of frequencies around 1.5 and 3 c.d^{-1}, as well as individual peaks at 3.66 and 5.03 c.d^{-1}, associated with pulsation g modes visible mainly during the outburst, while a few strong other peaks at higher frequencies ($\sim$11.9 and 16.9 c.d^{-1}) associated with p modes are mainly visible before and after the outburst. These higher frequencies have also been detected in ground-based spectroscopic observations obtained with FEROS and NARVAL in the frame of the CoRoT project. They are identified with modes corresponding mainly to $l = 4$ and 6.

Careful seismic modelling of the two states of the star should provide useful insight about the role of pulsations as a trigger for the Be phenomenon and/or the role of the Be phenomenon on the excitation and visibility of pulsation modes. The calculations of such seismic models are ongoing.

Conclusions

The high level of photometric accuracy, signal-to-noise, time sampling and duty cycle of CoRoT data provides unprecedented photometric datasets of early and late Be stars. The CoRoT light curves of Be stars are very complex and often show beatings. The data allow us to detect pulsation modes in all Be stars studied with CoRoT so far, even late-type ones for which the pulsation frequency peaks have a very low amplitude. However, mode identification requires supplementary spectroscopic data as regular patterns cannot be recognized in the power spectra because of rapid rotation. The spectroscopic data also allow us to determine the fundamental parameters of the studied Be stars. The results we obtained with CoRoT and from spectroscopy will then allow us to perform seismic modelling of Be stars.

Moreover, the occurrence of outbursts in some of the Be stars during the CoRoT observations allows us to directly study the link between the Be phenomenon and pulsations. In particular, we will test the role of pulsations as a trigger of matter ejections from Be stars to form their circumstellar disks.

Acknowledgments.   We thank the CoRoT team for their work on the data acquisition and reduction, E. Poretti for his contribution via the CoRoT ground-based spectroscopic programme, as well as H. Saio for useful discussions.

References

Frémat, Y., Neiner, C., Hubert, A.-M., et al. 2006, A&A, 451, 1053

Gutierrez-Soto, J., Fabregat, J., Suso, J., et al. 2007, A&A, 476, 927

Gutierrez-Soto, J., Floquet, M., Neiner, C., et al. 2009, CoAst, 158, 208

Neiner, C., & Hubert, A.-M. 2009, CoAst, 158

Saio, H., Cameron, C., Kuschnig, R., et al. 2007, ApJ, 654, 544

Telting, J., & Schrijvers, C. 1997, A&A, 317, 723

Comm. in Asteroseismology
Vol. 158, 2009, 38th LIAC/HELAS-ESTA/BAG, 2008
A. Noels, C. Aerts, J. Montalbán, A. Miglio and M. Briquet., eds.

The driving mechanism of roAp stars : effects of local metallicity enhancement

S., Théado[1], M.-A., Dupret[2], and A., Noels[2]

[1] LATT, Université de Toulouse, CNRS, 14 avenue Edouard Belin, 31400 Toulouse, France
[2] Observatoire de Paris, LESIA, CNRS UMR 8109, 5 place J. Janssen, 92195 Meudon, France
[3] Institut d'Astrophysique et de Géophysique de Liège, , Allé du 6 Août 17, 4000 Liège, Belgium

Abstract

We have investigated the influence of a local metallicity enhancement on the excitation mechanism of roAp star pulsations. Our computations show that such accumulations poorly affect the position of the theoretical roAp star instability strip although the opacity in the driving region of roAp modes is affected by metal accumulation.

Context

In the past, several studies investigated the excitation of pulsations in roAp stars (e.g. Dolez & Gough 1982, Dziembowski & Goode 1996, Gautchy & Saio 1998, Balmforth et al. 2001, Cunha 2002). Chemically homogeneous models and models with stratified helium and hydrogen compositions have been proposed to account for the roAp star properties but they failed to account for the extent of the roAp star instability strip.

Up to now the effects of a metal stratification (probably induced by microscopic diffusion) have not been considered in roAp star models. However abundance determinations suggest a relation between the excitation mechanism of pulsations in roAp stars and their heavy element distribution (Gelbmann 1998, Kochukhov 2003, Ryabchikova et al. 2004) and numerous cool magnetic Ap stars show vertical stratification and evidences for accumulations of some heavy elements in the deeper atmosphere (Bagnulo et al. 2001, Ryabchikova et al. 2002, 2005, Kochukhov 2003, Kochukhov et al. 2004).

Computations and results

With this in mind, we have computed with the code Clés (Scuflaire et al. 2008) grids of stellar evolutionary models adequate for A stars, including a local metal accumulation profile in their external layers, to test if such an accumulation could explain the location and the extent of the roAp instability strip. Following Balmforth et al. (2001), we assumed that the convective motions which take place in the envelope of A stars, are suppressed by the magnetic field: our models have then a fully radiative envelope. They include the solar metal mixture (Asplund, Grevesse and Sauval 2005, thereafter AGS05) with X=0.71. The opacity tables (computed with the AGS05 mixture) are those of OPAL96 (Iglesias & Rogers 1996) completed at low temperature with tables based upon calculations from Ferguson et al. (2005). As outer boundary conditions, Kurucz atmospheres (Kurucz 1998) are joined to the interior at an optical depth equal to 1. Each grid of models (labelled pZ1 to pZ5) includes a Z-accumulation profile centered on a given temperature (as shown in fig. 1a). The stability of these models has been computed using the nonadiabatic pulsation code MAD (Dupret 2002).

Fig. 1b shows that the position of the roAp theoretical instability strip is poorly sensitive to metal accumulations whatever their location in the stellar envelope, although these accumulations strongly affect the opacity profile. Our models account for the blue edge of the instability strip but they lead to too hot red edges. These results, which are discussed in details in Théado et al. (2009), fail to explain the correlation observed between the excitation mechanism and the metal distribution in roAp stars and cannot account for the roAp modes excitation in cool roAp stars.

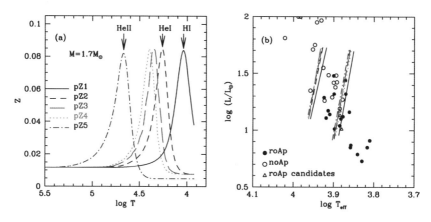

Figure 1: a) Z-accumulation profiles introduced in our models, each grid of models includes a different profile labelled pZ1 to pZ5. The vertical arrows show the position of the H and He ionization regions. b) Theoretical instability strips for the five grids of models, the line convention are the same for the 2 figures. Observational points are from Kochukhov & Bagnulo (2006) and North et al. (1997)

References

Asplund, M., Grevesse, N., & Sauval, A. J. 2005, ASPC, 336, 25a

Bagnulo, S., Wade, G. A., Donati, J.-F., et al. 2001, A&A, 369, 889

Balmforth, N. J., Cunha, M. S., Dolez, et al. 2001, MNRAS, 323, 362

Cunha, M. S. 2002, MNRAS, 333, 47

Dolez, N., & Gough, D.O. 1982, in Pulsations in Classical and Cataclysmic Variable Stars, p. 248, Eds Cox J. P., Hansen C. J.

Dupret, M.-A. 2002, PhD thesis, Bull. Soc. Roy. Sc. Liège, 5-6, 249-445

Dziembowski, W. A., & Goode, P. R. 1996, ApJ, 458, 33

Ferguson, J. W., Alexander, D. R., Allard, F., et al. 2005, ApJ, 623, 585F

Gautschy, A., Saio, H., & Harzenmoser, H. 1998, MNRAS, 301, 31

Gelbmann, M. J. 1998, CoSka, 27, 280

Iglesias, C. A., & Rogers, F. J. 1996, ApJ, 464, 943

Kochukhov, O. 2003, A&A, 404, 669

Kochukhov, O., Drake, N. A., Piskunov, N., & de la Reza, R. 2004, A&A, 424, 935

Kochukhov, O., & Bagnulo, S. 2006, A&A, 450, 763

Kurucz, R. L. : Grids of model atmospheres (1998), http://kurucz.harvard.edu/grids.html

North, P., Jaschek, C., Hauck, B., et al. 1997, ESA SP-402, p. 239-244

Ryabchikova, T. A., Piskunov, N., Kochukhov, O., et al. 2002, A&A, 384, 545

Ryabchikova, T., Nesvacil, N., Weiss, W. W., et al. 2004, A&A, 423, 705

Ryabchikova, T., Leone, F., & Kochukhov, O. 2005, A&A, 438, 973

Scuflaire, R., Théado, S., Montalbán, et al. 2008, Ap&SS, 316, 83

Théado, S., Dupret, M.-A., Noels, A., & Ferguson, J. W. 2009, A&A, 493, 159

Special Session
Future asteroseismic missions

Comm. in Asteroseismology
Vol. 158, 2009, 38th LIAC/HELAS-ESTA/BAG, 2008
A. Noels, C. Aerts, J. Montalbán, A. Miglio and M. Briquet., eds.

The Kepler mission

J. Christensen-Dalsgaard[1,2], T. Arentoft[1,2], T. M. Brown[3], R. L. Gilliland[4], H. Kjeldsen[1,2],
W. J. Borucki[5] & D. Koch[5]

[1] Department of Physics and Astronomy, University of Aarhus, DK-8000 Aarhus C, Denmark
[2] Danish AsteroSeismology Centre
[3] Las Cumbres Observatory, 6740B Cortona Dr, Goleta, CA 93117, USA
[4] Space Telescope Science Institute, Baltimore, MD 21218, USA
[5] NASA Ames Research Center, Moffett Field, CA 94035, USA

Abstract

The Kepler mission will provide a vast improvement in the characterization of extrasolar planetary systems, and in addition give a dramatic increase in the data available for asteroseismology. The present paper gives a brief overview of the mission, emphasizing the asteroseismic aspects, and with references to more detailed presentations.

Introduction

Kepler is a NASA Discovery mission, scheduled for launch in March 2009, whose principal purpose is to characterize extra-solar planetary systems, through the detection of planetary transits across their parent star. An important goal is to obtain statistics on the number of planets with conditions that may support the development of life, in the habitable zones around their central stars (Borucki et al. 2007). The required photometric precision, better than 20 parts per million averaged over 6.5 hours of observations of a 12th magnitude star, also makes the mission very well suited for asteroseismology. This is included as a part of the mission, an important purpose being the characterization of the central stars in planetary systems. In addition, our general understanding of stellar structure and evolution will be greatly enhanced through detailed asteroseismic investigations of a broad range of stars, covering much of the Hertzsprung-Russell diagram and including a substantial number of solar-like oscillators.

Kepler will observe a field spanning around 105 square degrees, in the region of Cygnus and Lyra, nearly continuously for the duration of the mission. The main group of targets observed to search for planetary transits, around 170,000 stars, will be observed at a 30-minute cadence. However, 512 stars will be selected for observation with a one-minute cadence; the selection of these short-cadence targets can be changed during the mission. They will be used in part for detailed characterization of planetary transits in already discovered systems, but a substantial fraction is reserved for asteroseismology of short-period pulsating stars, including solar-like stars near the main sequence. The mission was described in the recent paper by Borucki et al. (2008); further details are available at the Kepler web site: http://kepler.nasa.gov.

The Kepler Asteroseismic Investigation

To organize the asteroseismic use of the Kepler data a Letter of Direction has been set up from the Science Principal investigator to the asteroseismic community; this defines the

Kepler Asteroseismic Investigation (KAI). The high photometric precision will allow the study of solar-like oscillations down to magnitude 12 in stars near the main sequence, and exquisite data will be obtained also for other types of variables stars, including a large number of stars that can be observed at the long cadence. The asteroseismic analysis of central stars in planetary systems will constrain the stellar radii, an essential parameter to determine planetary radii from the depths of transits. However, for the stars for which extensive data are obtained we expect to move beyond the overall properties of the stars towards investigations of stellar interiors, stellar rotation and perhaps effects of stellar magnetic activity. The asteroseismic aspects of the project were further discussed by Christensen-Dalsgaard et al. (2007, 2008).

The KAI will be based on asteroseismic data made available at the Kepler Asteroseismic Science Operations Centre (KASOC). As discussed in more detail by Christensen-Dalsgaard et al. (2008) the data will be pre-filtered to remove evidence for planetary transits, in a manner having minimal impact on the asteroseismic investigations, before being released; this is in accordance with the Letter of Direction. In return, it is possible to operate with an open data policy within the KAI. The activities are organized in the Kepler Asteroseismic Science Consortium (KASC), a large international group established to make sure that full use is made of the Kepler data. Members of KASC will have full access to all the data provided through KASOC; the final analysis and presentation of the results will be organized through working groups established within the KASC.

The initial target selection has taken place on the basis of proposals made by the KASC members. These were based on the expected properties of the mission and the oscillation properties of the proposed targets, including extensive surveys of the available information on the stars in the Kepler field. An important aspect was to use existing information about the possible variability of the stars. The selection of Kepler targets for both investigations of planetary systems and asteroseismology also relied heavily on the Kepler Input Catalogue, the result of a huge undertaking involving both new observations and the use of existing data. It should be pointed out in the context of the present meeting, however, that the Kepler field contains a relatively small number of massive stars, owing to the fact that it has been chosen to lie somewhat above the plane of the Galaxy to avoid excessive crowding. Further details on the target selection, and references, were also provided by Christensen-Dalsgaard et al. (2008).

The Kepler mission will provide unique possibilities for high-precision near-continuous photometry over a very extended time. Thus there is no doubt that it will uncover entirely new phenomena, in addition to revolutionizing our knowledge about extra-solar planetary systems and stellar properties.

Acknowledgments. The establishment of the Kepler Asteroseismic Science Operations Centre is supported by the Danish Natural Science Research Council. We also gratefully acknowledge support from the European Helio- and Asteroseismology Network (HELAS), a major international collaboration funded by the European Commission's Sixth Framework Programme.

References

Borucki, W. J., Koch, D. G., Lissauer, J., et al. 2007, in Transiting Planets Workshop, eds C. Alfonso, D. Weldrake & T. Henning, ASP Conf. Ser. Vol. 366, p. 309

Borucki, W., Koch, D., Basri, G., et al. 2008, in Proc. IAU Symp. 249, Exoplanets: Detection, Formation and Dynamics, eds Y.-S. Sun, S. Ferraz-Mello & J.-L. Zhou, IAU and Cambridge University Press, p. 17

Christensen-Dalsgaard, J., Arentoft, T., Brown, T. M., et al. 2007, CoAst, 150, 350

Christensen-Dalsgaard, J., Arentoft, T., Brown, T. M., et al. 2008, CoAst, 157, 266

Comm. in Asteroseismology
Vol. 158, 2009, 38th LIAC/HELAS-ESTA/BAG, 2008
A. Noels, C. Aerts, J. Montalbán, A. Miglio and M. Briquet., eds.

PLATO: PLAnetary Transits and Oscillations of stars

C. Catala[1], on behalf of the PLATO Study Science Team[2]

[1] LESIA, Observatoire de Paris, CNRS, Université Pierre et Marie Curie, Université Paris Diderot ; 5 Place Jules Janssen 92190 Meudon, France

Abstract

The *PLAnetary Transits and Oscillations of stars* Mission (PLATO) is one of the four M-class mission proposals selected for an assessment study phase in the framework of the ESA "Cosmic Vision" programme. Its objective is to detect and characterize exoplanets by means of their transit signature in front of a very large sample of bright stars, and measure the seismic oscillations of the parent stars orbited by these planets in order to understand the properties of the exoplanetary systems.

PLATO is the next-generation planet finder, building on the accomplishments of CoRoT and *Kepler*: its targets will be three magnitudes brighter (hence the precision of the measurements will be correspondingly higher as will be those of post-detection investigations, *e.g.* spectroscopy, asteroseismology, and eventually imaging).PLATO will be capable of observing significantly smaller exoplanets, as well as planets orbiting hotter stars.

These goals will be achieved by a long-term (3 years), high-precision, high-time-resolution, high-duty-cycle monitoring in visible photometry of a sample of more than 100,000 relatively bright ($m_V \leq 11$) stars and another 400,000 down to $m_V = 13$. The space-based observations will be complemented by ground- and space-based follow-up observations.

In its current design, the PLATO payload includes an ensemble of 28 identical small, very wide-field telescopes, assembled on a single platform and all looking at the same 26° diameter field. Each one of these 100mm pupil telescopes has its own CCD focal plane, comprised of 4 CCDs with 3854 × 3854 pixels of 18 μm.

Introduction

The *PLAnetary Transits and Oscillations of stars* mission (PLATO) addresses the general question of the existence of life beyond Earth. In this context, the search for and study of planetary systems around other stars, in particular in the habitable zone, is a prerequisite to our understanding of planet formation and evolution, and of the distribution of life in the Universe.

We have recently entered a new era in exoplanet science, with the advent of CoRoT (Baglin et al. 2006), and with the prospect of the *Kepler* launch in 2009 (Borucki et al. 2003). These missions will study the distribution of planet sizes and orbits down to Earth-sized bodies, but they are targetting faint and distant stars which are difficult to characterize. Yet, in addition to the detection of a large number of exoplanets, we need to determine the characteristics of their host stars, such as radius, mass, age, and element abundances. The radii and masses of the host stars must be measured accurately in order to provide a precise measurement of sizes and masses of the detected planets, while measuring the ages of the central stars will

---

[2] see http://lesia.obspm.fr/cosmicvision/plato/pages/science_team.html

provide us with an estimate of the ages of their planetary systems. Asteroseismology is the best tool to provide this characterization.

Asteroseismology of a large sample of stars all across the HR diagram can also bring us a full and deep understanding of stellar evolution, which is central to astrophysics. In this area, a first important step will be taken by CoRoT, while *Kepler* will also include an asteroseismology programme, but they will both provide seismic data for only a modest number of targets, and a next generation mission is clearly needed.

Understanding the processes of star and planet evolution, by the study of stellar interiors and the distribution of planetary systems will constitute a major step for future progress in most areas of astrophysics and in the scientific and philosophical approaches towards the origin of life in the Universe. It is the cornerstone that will bring us from the first, still fairly unsystematic planet diskoveries towards a systematic survey, providing a global understanding of the richness and diversity of the "Other Worlds" that populate the Universe.

Science objectives of PLATO

In our conception of planetary systems, the formation and evolution of both components, stars and planets, are closely related. Stars and planets are born together from the same parental material, and therefore share a common initial history. In particular the initial protoplanetary disk and its central stellar core have the same chemical composition and their respective angular momentum reflects the angular momentum distribution in the protostellar/protoplanetary nebula. Also, in the early phases of evolution, young stars exchange angular momentum with their protoplanetary accretion disks, which eventually evolve into planets and transfer their angular momentum to planet orbits.

Even later in the evolution, various processes occur that result in mutual interactions between the stars and their planets. Stellar radiation impacts the planet atmospheres, while particle bombardment by stellar winds can also affect the chemical and biological evolution of the planets. Planets can also influence their parent stars, *e.g.* by colliding with them, and enriching them in various chemical elements. Giant planets in close-in orbits can also influence their star's rotation via tidal effects.

The study of planetary system evolution thus cannot be separated from that of stellar evolution. We cannot understand how planets are formed and how they evolve without a proper knowledge of stellar formation and evolution. Also, we cannot characterize planetary systems without characterizing precisely their host stars. The basic philosophy behind the PLATO mission is precisely to study complete planetary systems composed of planets and their host stars, these two components being observed together with the same technique.

The most important goals of the PLATO mission are therefore twofold:
- to open a new way in exoplanetary science, by providing a full statistical analysis of exoplanetary systems around stars that are bright and nearby enough to allow for simultaneous and later detailed studies of their host stars.
- to perform seismic analysis of stars all across the HR diagram, including, but not restricted to, those with detected exoplanetary systems.

A more detailed description of PLATO's science objectives is available in Catala et al. (2008)

Characterisation of planet host stars

The characterisation of exoplanets requires a characterisation of their host stars, including their radius, mass and age.

Stellar radii can be obtained if one can measure absolute luminosities, hence distances, as well as effective temperatures. Stellar parallaxes, such as soon provided by Gaia, as well as stellar effective temperatures obtained by detailed spectroscopic analysis can be used to determine stellar radii to within a couple of percent.

Classical methods of determination of masses and ages of stars rely on a precise loca- tion of the star in the HR diagram, and a comparison with theoretical evolutionary tracks and isochrones. Unfortunatey, this method suffers severe uncertainties. These are related to uncertainties in the effective temperatures and absolute luminosities, and most impor- tantly to uncertainties in the calculation of theoretical tracks, in particular due to our poor knowledge of internal chemical composition of stars. Even if the surface composition can be constrained by high resolution spectroscopy and detailed abundance analysis, uncertainties on physical processes in stellar interiors (microscopic diffusion, rotational mixing, etc), which make metallicities uncertain by up to a factor 2, are sufficiently severe to seriously affect mass estimates, as shown in Fig. 1. As a consequence, stellar masses cannot usually be determined to better than 15 to 20%.

In addition, stellar ages on the main sequence are often undetermined, even with a precise location of the stars in the HR diagram, as also shown in Fig. 1.

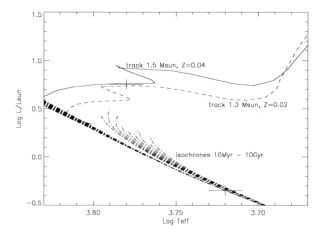

Figure 1: Examples of stellar evolutionary tracks and isochrones. The crosses represent typical observational error bars for stellar fundamental parameters. Note that stellar masses can be uncertain by 20%, while stellar ages on the main sequence are essentially undetermined.

The characterization of planet host stars, and in particular the measurement of their masses and ages, must therefore be obtained with another, more accurate and more reliable method. Seismic analysis is this much needed method.

The oscillation frequencies of a star are determined solely by the run of the sound speed $c^2(r) = \Gamma_1 P(r)/\rho(r)$ across its interior, where Γ_1 is the adiabatic exponent, P the pressure and ρ the density. The pressure can be obtained from the density using the condition of hydrostatic equilibrium. Since Γ_1 is a very slowly varying function of r, we can in principle derive $\rho(x)$ from the frequencies alone, where $x = r/R$, R being the stellar radius. Details of the physics of the stellar interior, such as chemical composition, energy flux, nuclear reactions, equation of state, etc, do not have to be known. Finally, a measurement of the radius R, e.g. obtained from Gaia as mentioned above, gives the mass M, typically to within a couple of percent, in a model independent way.

A precise and reliable estimate of stellar ages via asteroseismology is also possible. The sound speed in the stellar core depends on the mean molecular weight, which changes in the course of stellar evolution. Frequencies of low degree oscillation modes, which probe the sound speed down to the central regions of the stars, can be used to estimate the central hydrogen and helium content, and therefore constrain stellar ages. This asteroseismic age determination is somewhat dependent on the details of the physics in the stellar core, but oscillation frequencies are much more sensitive to evolutionary stage than classical observables, in particular on the main sequence, and can provide a reliable age estimate to within about 10%, which represents a major improvement.

Needed observations

The science objectives of PLATO require three types of observations:
- detection and characterisation of planetary transits, used to measure the planet-to-star radius ratio, the planet orbital period, as well as the ratio of the orbit semi-major axis to the stellar radius; a very high photometric precision is needed to detect transits by earth-sized planets in front of solar-type stars.
- asteroseismology of exoplanet host stars, which will provide measurements of the stellar masses, radii and ages, as well as detailed probes of their internal structure and rotation.
- complementary ground-based observations, in particular high precision radial velocity monitoring, which will be used to confirm the detected planets and to measure the planet-to-star mass ratios.

Further ground- and space-based follow-up observations will also provide more detailed information on the planets and their host stars, including information on the planet atmospheres in some favourable cases.

In order to reach these goals, PLATO will perform long duration (3 year), short cadence (30 sec) uninterrupted photometric monitoring of $\approx$100,000 bright stars of all spectral types. These primary targets of PLATO are bright stars ($m_V \leq 11$), observed with a photometric noise level better than 2.7×10^{-5} per hour, i.e. 1 ppm in 30 days, allowing the detection of transits by small planets, as well as solar-type oscillations. In addition PLATO will also perform an extensive survey of planetary transits in front of $\approx$400,000 stars down to $m_V = 13$, with a noise level better than 8×10^{-5} per hour, which is still sufficient for the detection of earth-size planets. On the bright end, PLATO will be capable of observing stars as bright as $m_V = 4$.

Two long runs of three years on primary target fields will be performed, followed by a set of shorter ones (of a few months each) on different target fields, either to revisit and confirm planetary transit candidates of the long runs or to study stars in open clusters and population II stars, with the goal to improve stellar evolution models.

Instrumental concept

The science objectives of PLATO require both a very wide field of view and a large collecting area. The very wide field of view requires a short focal length, yielding a large plate factor, which implies the use of a small pupil. In order to comply with the collecting area requirement, PLATO will use a large number of identical small size optics, each one coupled to its own focal plane, all of them observing the same stellar field.

A model payload was studied by ESA's "Concurrent Design Facility" (CDF) with the help of the PLATO Study Science Team. It includes a set of 28 identical off-axis telescopes of 10cm pupil, each one equipped with a focal plane including four CCDs of 3854^2 pixels of 18μm. These telescopes include 2 mirrors and a two-lens corrector. Their field-of-view is 557 deg^2.

Twenty-six out of these 28 telescopes have their focal planes working in full frame mode, and acquire images in white light at a cadence of 30 sec. The dynamical range of these observations is 8 − 14 mag. The remaining two telescopes include broadband filters (one red, one blue), and use their CCDs in frame-transfer mode at a cadence of 1 sec. Their dynamical range of 4 − 8 mag complements that of the 26 "normal" telescopes.

The PLATO data will be processed onboard the spacecraft. Each telescope is coupled to its own Digital Processing Unit, whose task is to produce an individual light curve for each star, using a weighted mask photometry algorithm. All individual light curves are then transmitted to a central onboard computer, where they are averaged before transmission to the ground.

The PLATO payload will be mounted on a dedicated platform and injected into a large Lissajous orbit around the L2 point. The satellite will be rotated by 90° every three months, with no change in the surveyed field.

Figure 2 presents a sketch of the PLATO payload.

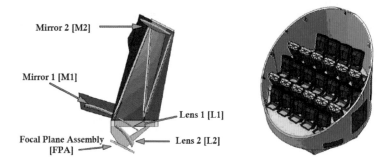

Figure 2: Sketch of the PLATO payload. The left panel shows one of the 28 telescopes, while the one on the right displays the whole set of telescopes mounted on their common optical bench, and protected by the sunshade.

Expected performances

The performance of the design presented above can be estimated in terms of number of accessible targets at a particular noise level. The ESA CdF study has included such estimates, based on a template field centered at ecliptic coordinates $(210°, -60°)$. The USNO-B1.0 catalogue was used as entry, both for the target stars and for the polluting nearby sources. Conservative estimates of the fraction of stars lost to confusion were used in the calculation. Table 1 summarizes these performances.

The performances for planet detection are summarized in Fig. 3, which shows the expected transit depth as a function of planet radius for various values of the stellar radius. Horizontal lines labelled with noise levels indicate the depth of transits detectable in 10 hrs at 3σ.

Finally, the expected noise performances of PLATO expressed in Fourier space are shown in Fig. 4, showing that detection of solar-like oscillations will be possible at least down to $m_V = 10.5$.

These expected performances are very close to the target performances of the mission,

noise level in 1 hr	# stars	# main sequence	m_V
2.7×10^{-5}	78,000	47,500	10.5
3.6×10^{-5}	98,000	60,000	11.0
8.0×10^{-5}	619,000	378,000	12.2

Table 1: Number of stars observable by PLATO with various levels of photometric noise. A constant 61% fraction of main sequence stars was assumed.

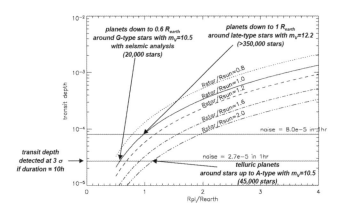

Figure 3: Expected performance of PLATO mission for exoplanet detection.

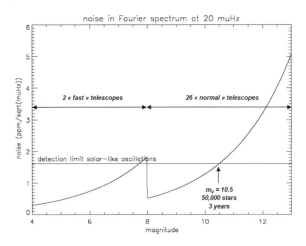

Figure 4: Expected noise level in Fourier space.

giving us strong confidence that a very efficient PLATO mission can indeed be designed.

Acknowledgments. We warmly thank the staff of ESA CdF, the ESA PLATO project team, as well as the PLATO Study Science Team, for their efficient work in this initial study of the PLATO mission.

References

Catala, A. 2009, Experimental Astronomy, 23, 329

Baglin, A., & Fridlund, M. 2006, *The CoRoT Mission, Pre-Launch Status*, ESA SP-1306, Eds. M. Fridlund, A. Baglin, J. Lochard, L. Conroy, p. 11-538

Borucki, W.J., Koch, D. G., Basri, G. B., et al. 2003, The Kepler Mission: Finding the Sizes, Orbits and Frequencies of Earth-size and Larger Extrasolar Planets, *ASP CS*-294, p. 427

DISCUSSION

Defise: What is the ideal orbit for an asteroseismic space mission?
Catala: The best orbit is around L2. Let us notice that Kepler will quickly be too far from the earth to allow high data rate.

Comm. in Asteroseismology
Vol. 158, 2009, 38th LIAC/HELAS-ESTA/BAG, 2008
A. Noels, C. Aerts, J. Montalbán, A. Miglio and M. Briquet., eds.

SIAMOIS: asteroseismology in Antarctica

B. Mosser[1], T. Buey[1], C. Catala[1], T. Appourchaux[2], S. Charpinet[3], Ph. Mathias[4],
JP. Maillard[5], and the SIAMOIS team[2,3,4]

[1]LESIA, CNRS, Université Pierre et Marie Curie, Université Denis Diderot, Observatoire de Paris,
F-92195 Meudon cedex, France - benoit.mosser@obspm.fr
[2] Institut d'Astrophysique Spatiale, UMR8617, Université Paris XI, Bâtiment 121, F-91405 Orsay cedex,
France,
[3] Laboratoire d'Astrophysique de Toulouse-Tarbes, CNRS, Université de Toulouse, 14 av. Ed. Belin,
F-31400 Toulouse, France
[4] Laboratoire FIZEAU, Université de Nice Sophia-Antipolis, CNRS-Observatoire de la Côte d'Azur,
F-06108 Nice cedex 2
[5] Institut d'Astrophysique de Paris, 98b bld Arago, F-75014 Paris

Abstract

SIAMOIS is a ground-based asteroseismology project, to pursue velocity measurements from
the Dome C Concordia station in Antarctica. The scientific program of SIAMOIS is based on
the very precise asteroseismic observation of nearby bright targets, focussing on the observations of solar-like oscillations in solar-like stars. Spectrometric observations with SIAMOIS
will be able to detect l=3 oscillation modes that cannot be analyzed with space-borne photometric observations. The Doppler data, less affected by the stellar activity noise, will yield a
more precise mode structure inversion, thus a high-precision determination of the stellar interior structure. The benefit of precise Doppler observations of nearby targets, with addition of
interferometric and high-resolution spectrometric measurements, will allow us to investigate
in detail the physical laws governing the stellar interior structure and evolution.
    Dome C appears to be the ideal place for ground-based asteroseismic observations as
it is capable of delivering a duty cycle as high as 90% during the three-month long polar
night. This duty cycle, a crucial point for asteroseismology, is comparable to space-borne
observations. The SIAMOIS concept is based on Fourier Transform interferometry, which
leads to a small instrument designed and developed for the harsh conditions in Antarctica.
The instrument will be fully automatic, with no moving parts, and it will require only a very
simple initial set up in Antarctica.

Introduction

Stellar evolution theory is central to most areas of modern astrophysics. It plays for example
a crucial role in the determination of distances and ages in the Universe. Stellar structure and
evolution of stars was recently identified as a major domain of research for the next 20 years
by the Panel "Evolution of Stars and Planets" of Astronet during the future prospectives
workshop held in January 2007. The theory of stellar evolution, not sufficiently tested by
observations, will benefit from the analysis of stellar oscillation modes: they propagate inside
stars down to the core, and give a powerful tool to probe their internal structure. Thus, they
provide the necessary observational basis for the theory.

Successful monosite asteroseismic observations have been performed recently with the échelle spectrometers developed for the search of exoplanets. Mostly sub-giants were observed. Individual eigenfrequencies were identified for the brightest targets; in other cases, only global oscillations parameters, such as the large separation, were measured. The window effect due to daily interruptions leads to large uncertainties and ambiguities in the frequency determination, and the frequency resolution remains limited by short observations. Two-site observations were performed on 4 bright targets (α Cen A and B, β Hyi, η Boo, see Bedding & Kjeldsen 2008), for about a week. The single network campaign for observing solar-like oscillation occurred in January 2007, on Procyon (F5, $m_V = 0.4$): its central part, achieving a duty cycle of about 90%, was limited to 10 days (Arentoft et al. 2008).

SIAMOIS (in French for Sismomètre Interférentiel À Mesurer les Oscillations des Intérieurs Stellaires), to be installed at Dome C in Antarctica, will constitute a major step in the development of asteroseismology from the ground, using spectroscopy. Indeed, it will be the first asteroseismometer able to acquire the several months of continuous observations on the same objects that are required to obtain the frequency resolution necessary to carry out inversions of stellar internal structure. Duration and stability of the measurements will have the highest priority. The core of the scientific case will be completed with the observation of bright targets.

In this paper, we examine the reasons motivating ground-based Doppler asteroseismic observation, following space-borne photometric missions (CoRoT, Baglin et al. 2002; Kepler, Basri et al. 2005). We present the scientific program, with the great advantage of observing bright targets, for convergence of very-precise measurements obtained in interferometry, high-resolution spectrometry and asteroseismometry. The advantages of observing with a single site in Antarctica are then reported: the Concordia station at Dome C provides unique condition for a high duty cycle during 3 months. A few words on the SIAMOIS project are finally given.

Asteroseismic signal: Doppler versus intensity

Any project for the purpose of asteroseismology must answer the question: "What can this project achieve that has not been done by the CoRoT or Kepler missions?". The asteroseismic core program of CoRoT (respectively Kepler) consists in the observation of 6 main targets (about a few hundreds), for 5 months each (from 3 months up to 4 years for a few of them). CoRoT and Kepler are designed for faint targets, respectively about $m_V \geq 6$ or 9, and will not observe the brightest stars.

Photometry and spectrometry observations are not sensitive to the same signal in the stellar photosphere. Photometry is sounding deeper regions, so that the photometric signal is corrupted by the different scales of granulation, whereas spectroscopic lines formed in higher regions are less affected by granulation. As a result, the low-frequency domain (≤ 2 mHz) of the Fourier spectrum is much less contaminated by the activity and granulation signals at low frequency when observed in velocity. This was clearly demonstrated in helioseismology, and is also observed on stars (Fig. 1).

Therefore, spectrometric observations with SIAMOIS will be able to detect oscillation modes in the low-frequency domain, with much longer lifetimes. Longer mode lifetimes yield more accurate frequencies, then a more accurate structure inversion. In the solar case, inversion is 4 times more precise with Doppler data (Gabriel et al 1997). This is due to the fact that these data give access not only to the small frequency separation between $\ell = 0$ and 2, but also between $\ell = 1$ and 3. These small separations are crucial for asteroseismic inversion (Provost et al 2005). As a consequence, Doppler observations give a better asteroseismic signal, especially for low mass stars. However, only slow rotators ($v \sin i \leq 20 \, \mathrm{km \, s^{-1}}$) present sufficiently thin lines for the benefit of spectrometric observations.

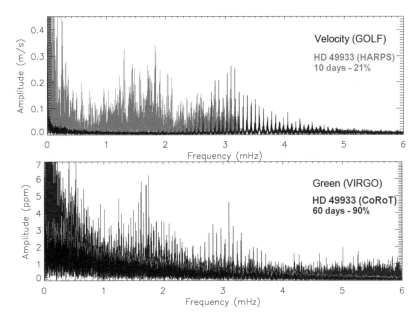

Figure 1: Comparison of spectrometric and photometric observations of the Sun (SOHO observations) and of HD 49933 (HARPS and CoRoT observations, Mosser et al 2005, Appourchaux et al 2008). For both targets, Doppler observations at 0.5 mHz remain photon noise limited, whereas the photometric signal include a significant contribution of stellar activity noise. The bushy aspect of the HD 49933 Doppler spectrum is due to a very low duty cycle (about 21 % of the single site observation of this CoRoT target)

Scientific case

The scientific program of SIAMOIS is based on the very precise asteroseismic observation of nearby bright targets, focussing on the observations of solar-like oscillations in solar-like stars. It includes also observations of classical pulsators (PMS, δ- Scuti), of solar-like oscillations in red giants, of gravity modes in γ Doradus targets. Possible circumpolar targets are represented in Fig. 2. Contrary to photometric observations, the list of targets includes a higher fraction of low-mass targets. Due to the instrumental requirement of observing thin lines, it concerns only slow rotators (Table 1).

This program is deliberately oriented towards bright targets. Contrary to the Kepler mission, which will focus on the statistical properties of a few hundreds of stars whose primary parameters are basically unknown, SIAMOIS will scrutinize in detail nearby stars. Thanks to astrometry, interferometry and high resolution spectroscopy, their distance, radius, T_{eff}, $\log g$ can be very accurately measured. Creevey et al. (2007) have demonstrated the advantage to constrain the stellar modeling with asteroseismic data with an independent interferometric measure of the radius. Such bright targets can therefore benefit from convergent measurements, for a scientific program aiming at the determination of a sample of standards. As a result, the program will permit to address in detail not only the internal structure of the stars, but the physical laws governing their interior structure and evolution (see Table 2; a more detailed version of the scientific program of SIAMOIS is given at http://siamois.obspm.fr/).

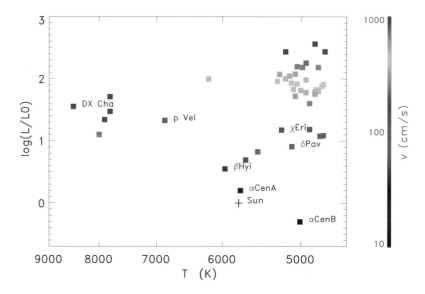

Figure 2: Identification of possible targets for SIAMOIS. All selected stars can be observed with a signal-to-noise ratio (in amplitude) greater to 6 after 5 days.

	space	ground
observation	photometry	spectrometry
max. degree	2	3
magnitude	dim	bright
$v \sin i$	–	$< 20\,\mathrm{km\,s^{-1}}$
inversion		4 times more precise

Table 1: Doppler versus intensity signal.

Observations at Dome C, Antarctica

Dome C is the only ground-based site allowing continuous observations for many weeks. A simulation of the observation at Dome C is shown in Fig. 3. On-site measurements during the 2005 and 2006 winters show a duty cycle better than 90%, with long periods of time with full time coverage (Mosser & Aristidi 2007). On bright stars at high negative declination (α Cen, β Hyi), observations will be possible even at dawn and twilight. The seeing quality appears to be limited by the lowest atmospheric layers, yielding a moderate seeing at ice level, but excellent 30 m above the ground (Trinquet et al 2008). However, due to the large area (5") collected on the sky by its fiber, SIAMOIS will be rather insensitive to seeing, taking full advantage of the high duty cycle, independently on the photometric conditions. The high altitude of Dome C (3200 m, equivalent to 3700 m barometric altitude) will ensure high-quality performance. Dome C will also give access to continuous observations for as long as 3 months (duty cycle greater than 90% for more than 90 days, centered on the winter southern solstice), which is otherwise feasible only from space. April and August provide each a month-long period with a duty cycle about 60%, for specific programs withstanding a reduced duty cycle.

    Network observations with non-dedicated sites present a duty cycle limited to less than

- High-precision determination of the interior structure of nearby stars: primary parameters, age determination, composition
- Convection; diagnostic of convective cores; depth of convection and of second helium ionization zone; damping, excitation mechanism
- Non-linear physics, saturation effects, mode coupling; stochastically excited modes
- Comparative study: photometry / Doppler

Table 2: The scientific case of SIAMOIS

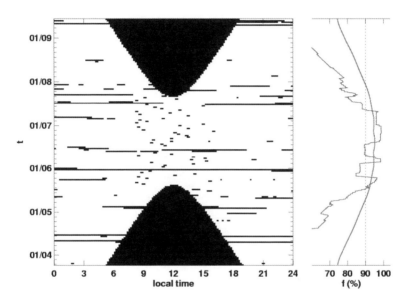

Figure 3: Left panel: simulation of the observation at Dome C, taking into account the Sun elevation, the bad meteorological conditions, and a daily 20-min interruption for operations such as telescope derotation. The Sun altitude limit has been fixed to $-4°$, for the observation of bright targets. Right panel: 10-day boxcar averaged duty cycle, and duty cycle integrated over the whole winter night.

70% (e.g. WET network, Dolez et al 2006). With at least 6 selected observatories in excellent sites, the performance of a network remains inferior to the Dome-C capability (Mosser & Aristidi 2007). It was stressed during the Astronet prospective workshop that both kinds of instrumentation are required for a better sky coverage. The Panel "Evolution of Stars and Planets" recommended a supporting facility instrument capable of performing long observations of high-precision stellar radial velocities for applying the tool of asteroseismology. Network observations as proposed by the SONG project (Grundahl et al 2007) are complementary to observations at Dome C. Like SIAMOIS, SONG proposes to use small collectors, and therefore its program is also focussed on bright stars, but with declinations near the equatorial plane. Due to the scientific specifications, both projects need an automatic and robust instrumentation, with the necessary incidence on both budgets. However, the SONG network intends to include 8 observing sites (8 domes and infrastructures, 8 telescopes, 8 instruments), when 1 site at Dome C is sufficient. Operations at Dome C will be of course demanding, but with better expected results (in SNR and duty cycle).

The ARENA network (http://arena.unice.fr/) has identified that the best niche for

Target	Limit magnitude m_V	
	40-cm tel	2-m tel
solar-like oscillations		
- solar-like stars	5	8.5
- red giants	7	10.5
classical pulsators	7	10.5

Table 3: Limit magnitude for different types of targets, according to the diameter of the telescope.

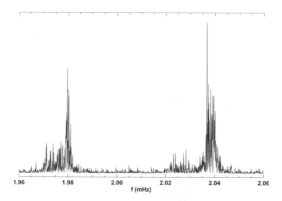

Figure 4: Photon-noise limited performance: 90-day observation, for a $mV = 4$ G0V target, and a 40-cm telescope (Mosser et al. 2003). The limited lifetime of the modes is responsible for the noisy aspect of the multiplets (from left to right: $\ell =2, 0, 3, 1$).

time series observation at Dome C is spectroscopy. In fact, even if photometric stability at Dome C is exceptional compared to other ground-based sites, duty cycle in spectroscopy will be better since spectroscopic measurements are much less sensitive to most of atmospheric noise, and can afford the presence of light clouds. Furthermore, the quality of ground-based photometric observations are surpassed by space-borne ones, whereas a Doppler instrument as SIAMOIS has no space competitor.

SIAMOIS

Échelle spectrometers such as HARPS at ESO used for single-site radial velocity measurements are complex and expensive instruments that are not suitable for Dome C, where only very limited technological support is currently available. Conversely, a Fourier Tachometer like SIAMOIS, with no moving parts, whose interferometer is designed to be installed without any fine tuning, and whose conception and design were directly dictated by the environment at Dome C is precisely what is required for observations in Antarctica.

The principle of a Fourier tachometer dedicated to asteroseismology is presented in Maillard (1996). The specifications of the project are derived from the analysis conducted in Mosser et al. (2003). The necessary stability of the instrument is provided by a monolithic interferometer with no moving part, which means that the optical setup at Dome C will be limited to basic image formation, without strong requirements. In the same way, the design is directly suited for quasi-automatic operations, allowing simple and reduced operations. The fiber-fed instrument is operated at room temperature. A low-resolution ($\mathcal{R} \simeq 1000$) post-

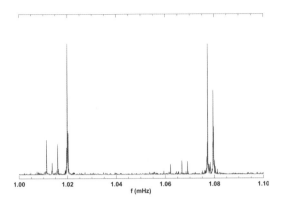

f (mHz)

Figure 5: Same as Fig. 4, but with a zoom on a lower frequency range. Due to longer lifetimes, the modes multiplets appear very clearly.

dispersion provides an efficiency greater by a factor of about 7 compared to a single bandpass. Photon noise limited performance insures a SNR better than 6, after 5 days observation, for all selected targets (and up to 30 for the brightest targets).

As continuous observations are mandatory for asteroseismology, the project includes a dedicated telescope. As the scientific goals are achieved with bright targets, a 40-cm class telescope is enough to provide "big science" at Dome C. First observations will run on bright targets such as α Cen A and B, β Hyi... with a small dedicated telescope (Table 3). Other specific programs may require the use of a 2-m class telescope (such as PILOT, Storey et al 2007), as for example observations of solar-like oscillations in an exoplanet-host star (Vauclair et al 2008). In fact, with a 2-m class telescope, only oscillations in white dwarfs and subdwarf B stars remain out of the scope of SIAMOIS, because of their too faint magnitude.

Due to the multiplex advantage of a Fourier tachometer, the radial velocity information is coded with a very limited number of pixels. This makes it possible to track on the same small CCD the signal coming from 2 fibres linked to 2 telescopes. Observing simultaneously with two dedicated small telescopes or with a 40-cm and a 2-m telescope will be possible, under the only conditions that exposure times are comparable. An automatic pipeline reduction will be developed, for real-time analysis in Europe, requiring a very limited band pass ($\simeq$ 100 kb/day for the scientific signal).

The phase A of SIAMOIS was completed in 2007. The design of the instrument takes into account the site, that is not a typical observatory offering many services for astronomy. The harsh observational conditions at Dome C impose stringent technological requirements, that translate in cost overruns: setup operation will be very limited; the instrument will operate quasi automatically. In fact, the phase A has shown that most of these requirements due to the site are similar to those of the project itself (sturdiness, high level of automation...); the cost to address the conditions in Antarctica is then limited.

A complete model of the instrument has been developed, and allows us to quantify the photon-noise limited performance. High SNR is reached for all targets presented in Fig. 2. Synthetic oscillation spectra are shown in Fig. 4 and 5. The noisy aspect of the multiplets in Fig. 4 is not due to a low SNR but to the limited lifetime of the modes in the high-frequency part of the pressure mode spectrum; fitting the Lorentzian profile of each component gives then the mode frequency, amplitude and linewidth. With a duty cycle as high as 90% during the 3-month long night, the precision on the eigenfrequency measurement will be as good as 0.2 μHz. Such simulations, compared to photometric results (Appourchaux et al 2008), shows the gain is signal-to-noise ratio when observing the Doppler signal.

Conclusion

After a score of 1-site Doppler asteroseismic observations of solar-like oscillations in solar-like stars, a limited number of two-site observations, up to date only 1 network Doppler observation, and before a possible network (Grundahl et al 2007), observing in Antarctica is required for long uninterrupted asteroseismic time series. As the observation conducted at the South Pole for helioseismology (Grec et al 1980), SIAMOIS will represent a major step in Doppler asteroseismology. With a small collector, SIAMOIS will achieve a consistent scientific program, to be conducted over more than 3 winters (ideally up to 6). According to its characteristics, SIAMOIS can be among the very first scientific projects to be conducted at Dome C.

The access to the low-frequency domain, to lower mass targets, to $\ell = 3$ modes makes Doppler observation at Dome C very competitive to space-borne observations, providing much more precise inversion results. When a photometric project such as Kepler will yield a statistical approach on a large population of faint objects, SIAMOIS will look in great detail into nearby stars. Compared to the complementary space missions, SIAMOIS can really achieve "Big Science" at Dome C, even with a small 40-cm telescope.

Acknowledgments. BM thanks warmly Annie Baglin, John Leibacher and Don Kurtz for sensible and fruitful discussions.

References

Appourchaux, T., et al. 2008, A&A, 488, 705

Arentoft, T., Kjeldsen, H., Bedding, T., et al. 2008, ApJ, 687, 1180

Baglin, A., Auvergne, M., Barge, P., et al. 2002, Stellar Structure and Habitable Planet Finding, 485, 17

Basri, G., Borucki, W.J., & Koch, D. 2005, NewAR, 49, 478

Creevey, O.L., Monteiro, M.J.P.F G., Metcalfe, T., et al. 2007, ApJ, 659, 616

Dolez, N., Vauclair, G., Kleinman, S.J., et al. (WET collaboration) 2006, A&A, 446, 237

Gabriel, A.H., et al. 1997, Proceedings of the 181st symposium of the IAU, held in Nice, France, September 30 - October 3, edited by Provost & Schmider. Dordrecht Kluwer Academic Publishers, p. 53

Grec, G., Fossat, E., & Pomerantz, M. 1980, Nature, 288, 541

Grundahl, F., Kjeldsen, H., Christensen-Dalsgaard, J., et al. 2007, CoAst, 150, 139

Bedding, T.R., & Kjeldsen, H. 2008, ASPC, 384, 21

Maillard, J.P. 1996, Applied Optics, 35, 2734

Mosser, B., Maillard, J.P., & Bouchy, F. 2003, PASP, 115, 990

Mosser, B., Bouchy, F., Catala, C., Michel, E., et al. 2005, A&A, 431, L13

Mosser, B., & Aristidi, E. 2007, PASP, 119, 127

Provost, J., Berthomieu, G., Bigot, L., & Morel, P. 2005, A&A, 432, 225

Storey, J.W.V., Ashley, M.C.B., Burton, M.G., & Lawrence, J.S. 2007, EAS Publications Series, 25, 255

Trinquet, H., Agabi, A., Vernin, J., et al. 2008, PASP, 120, 203

Vauclair, S., Laymand, M., Bouchy, F., et al. 2008, A&A, 482, L5

Comm. in Asteroseismology
Vol. 158, 2009, 38th LIAC/HELAS-ESTA/BAG, 2008
A. Noels, C. Aerts, J. Montalbán, A. Miglio and M. Briquet., eds.

Composing the verses for SONG

F. Grundahl[1], J. Christensen–Dalsgaard[1], T. Arentoft[1], S. Frandsen[1], H. Kjeldsen[1],
U.G. Jørgensen[2] and P. Kjærgaard[2]

[1] Department of Physics and Astronomy, Aarhus University,
Ny Munkegade, 8000 Aarhus C, Denmark
[2] Niels Bohr Institute, Copenhagen University, Juliane Maries Vej 30,
DK-2100 Copenhagen, Denmark

Abstract

The SONG (Stellar Observations Network Group) project aims at designing and building a global network of 1m telescopes. The primary science goals are to detect extra–solar planets (using the microlensing and radial velocity techniques) and study stellar interiors using asteroseismology. In early 2008 funding was obtained to construct a prototype network node (telescope, dome, instrumentation and software) to be ready by the end of 2011. Here we give an account of the project and the expected broad structure of the network nodes.

SONG overview

SONG is the acronym for Stellar Observations Network Group which has a primary goal of establishing a global network of 1m telescopes, dedicated to the search for extra-solar planets and the study of stars using asteroseismology.

Initial work towards reaching these goals started in 2006 and since early 2008, the project has secured funding from private and public Danish foundations which will allow the construction of a full prototype network node to be ready by the end of 2011. Currently, (autumn 2008) SONG is finalizing the optical design in preparation for ordering the major optical components.

The motivation for building a network of identical telescopes has its background in the primary science goals: doing asteroseismology and hunting for exoplanets via the microlensing method. Both of these goals require observations with a high duty cycle and over long periods of time (weeks to months) – longer than can easily be provided by current observatory organizations and single sites.

Using a network of telescopes it will be possible to obtain oscillation spectra of stars with much higher quality than currently possible using single or two-site observations (see Arentoft et al. 2008 for an example of this). Such oscillation spectra bring detailed studies of stellar interiors (convection zones, core size, ages) within reach. With the full SONG network, it will be feasible to carry out programmes which determine the ages of a large sample of stars, e.g. all planet-hosting stars brighter than $V = 6$. By September 2008, 42 such systems are known.[1] Jørgensen (2008) and Grundahl et al. (2007) have discussed the limits of planet detection for SONG in more detail.

In addition to studying stellar oscillations and exoplanets, SONG will also include a facility to carry out radial velocity observations of the Sun during daytime. This will be done by

---

[1] Radial-velocity detected planets from http://exoplanet.eu

allowing light from the blue sky to reach the spectrograph (probably through a window in the dome). Since the blue light from the sky is scattered sunlight it also carries the signal from the solar oscillations (see Kjeldsen et al 2008 for an example of such a measurement). Our purpose with these observations is twofold: it allows us to study the Sun as a star, and to monitor the instrumental behaviour by comparing results to those obtained by independent solar observations from e.g. BiSON (Chaplin et al. 1996) and the other SONG nodes.

Expected layout of a SONG node

The nodes of SONG are planned to have a high degree of robotisation such that the operational costs can be kept as low as possible. All nodes should be placed at existing observatory sites in order to avoid building major new infrastructure, and to ensure that on-site staff can react to emergencies. Our aim is to have 8 network nodes, four in each hemisphere.

Each node, will consist of an alt-az mounted telescope of 1m aperture with a Coudé feed to the instrument room which will be housed in an insulated 20 foot container next to the telescope pier. Our main motivation for choosing this configuration is to remove all instruments, except the telescope, from the dome environment. Although this causes a slight penalty in efficiency we believe that it will lead to a more stable operation of the network since the instruments will not be subjected to large variations in temperature and humidity. The telescope will be located in a standard dome, approximately 4.5m in outer diameter. In order to reduce windload on the telescope the dome shutter will have an opening which corresponds to the aperture of the telescope, and which follows the direction in which the telescope is pointing.

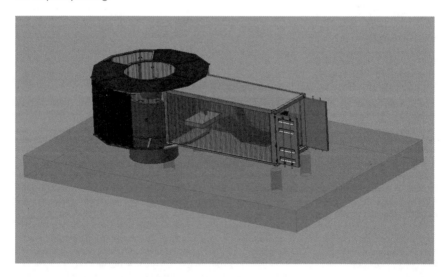

Figure 1: The footprint of a SONG node. A rough view of the location for the spectrograph and focal plane inside the container is indicated.

The instrument room will be located in the nearest half (to the pier) of the container, and will be temperature controlled to provide a stable environment for the two main instruments: a spectrograph equipped with an iodine cell to measure very precise radial velocities and imaging camera(s) for photometry in the microlensing fields. In the other half of the container, the control computers will be located. A view of the container, pier and dome support structure is shown in Fig. 1.

Requirements for the SONG instruments

There is a number of scientific requirements for a SONG node, and here we shall briefly mention the most important ones for the design.

For asteroseismology the most important requirement is to be able to reach a very high velocity precision (and accuracy), for the brightest stars in the sky, similar to the level obtained by Butler et al. (2004) on α Cen A. This implies that a velocity precision of 1m/s must be reached for stars of $V \sim 1$ in one minute of observation. The ability to measure such precise radial velocities is also an important criterion for finding exoplanets via the radial-velocity method.

For the observations and study of exoplanets the main design driver is the requirement to use the microlensing method. The microlensing fields are generally located in the direction of the Galactic Bulge region and hence the fields have a high degree of crowding – this means that high spatial resolution would be a significant advantage when aiming to do precise photometry for faint targets. At the same time only a small field need to be covered since only the microlensing target is of interest. Since many comparison stars are present in the field, a field size of $\sim$1 arcmin is sufficient for the required relative photometry. The total duration of a lensing event can be as long as a month, while a planetary anomaly during the event can last as short as $\sim$ an hour. Therefore a high-cadence as well as a long-term coverage is needed.

Focal-plane and Instruments

As part of the instrument suite, correction for atmospheric dispersion (ADC) and field rotation will be provided. For imaging, a filter wheel will be provided for the CCD camera(s). We also note that in addition to the spectrograph and imaging cameras (described below), room is also set aside which allows (later) installation of auxiliary instrumentation in the container. The focal plane assembly will also provide calibration light for the spectrograph (wavelength calibration and flat-fielding) as well as a facility to monitor the position of the star on the spectrograph slit (slit viewing) during spectroscopic observations.

Spectroscopy

The spectrograph has been designed with focus on obtaining very precise radial velocities (1m/s) for the brightest stars in the sky. To be able to do this, an iodine cell is placed in the lightpath just in front of the spectrograph slit. A working resolution of $R \sim 100.000$ is adopted (2.5 pixels per FWHM), but several (fixed) slitwidths will be available to allow for both higher and lower resolutions. The optical layout of the spectrograph is inspired by modern echelle instruments such as UVES, FEROS and HARPS.

The spectrograph is optimized for observations in the iodine wavelength range (5000Å – 6000Å), although the red limit extends to 6800Å. From 5500Å and redwards small gaps in the wavelength coverage is present; for wavelengths shorter than 6000Å the spectral coverage exceeds 86%. We are currently studying options that will allow us to cover fully the 4800–6800Å range, albeit at lower resolution.

In the design, special care has been taken to provide a PSF which varies only slowly with position across the detector; the diameter of the 80% encircled energy varies between 5 and 7 microns over the entire $2K \times 2K$ detector.

The iodine cell will be located in the beam, just before the spectrograph entrance slit for observations aiming at the highest velocity precision – it will, however, be possible to remove the cell from the light path such that "normal" spectroscopic observations can be carried out. As a further mean of improving velocity stability/precision the spectrograph will be temperature controlled. The spectrograph will only measure 50×90cm and thus fit nicely on a small optical table. We show the optical layout of the spectrograph in Fig. 2 below.

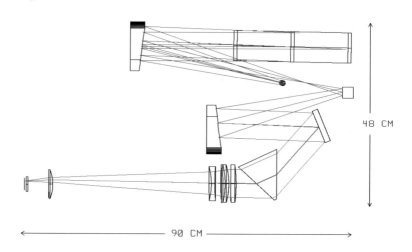

Figure 2: The layout of the spectrograph for SONG. The beam diameter is 75mm and the collimator focal ratio is 6, the echelle is an R4 echelle with 31.6 lines per mm. The optical design is by P. Spanò (INAF-Brera).

We have developed our own software package, following the strategy in Butler et al. (1996), to derive precision velocities from stellar observations obtained with an iodine cell. Currently we have reached precisions of 77cm/s per datapoint in data for αCen A obtained by Butler et al. (2004). The code and its performance is described in more detail in Grundahl et al. (2008) and in Grundahl et al. (2007) the velocity precision achievable by SONG is discussed.

Imaging

Recently a new method, lucky-imaging, to obtain high-resolution images using small and medium-sized telescopes has been introduced (Baldwin et al. 2001). Lucky-imaging is a method in which a CCD camera with fast readout (typically 10-30Hz) and low readout noise, is employed to obtain many short-exposure images. For each image the quality is evaluated and only the best (lucky) images are kept and subsequently stacked to create a final science image with higher resolution (Baldwin et al. 2001; Hormuth et al. 2008).

One of the crucial parameters governing the performance of lucky imaging is r_0, which is a measure of the typical size of the atmospheric "turbulence cells". This parameter depends on the seeing and the wavelength at which one is observing. Under typical, 1" observing conditions r_0 is $\sim$20cm in the I-band. The probability of getting a "lucky" image decreases exponentially with the ratio of the telescope diameter (D) and r_0 (Fried 1978), such that for a given value of r_0, a small telescope has a higher probability of obtaining a lucky image than a larger one.

Several groups have demonstrated that it is possible to obtain images close to the diffraction limit in I for 2m–class telescopes – this is accomplished by combining the best few % of images to a single image. For a 1m telescope the fraction of useful (high-quality) images is much higher than for a 2.5m telescope (due to the smaller value of D/r_0).

SONG will employ such a lucky-imaging camera in its focal-plane. This will have two functions: to serve as a high-quality imager for the microlensing observations and to serve as a fast guide camera which can provide correction signals to a tip-tilt mirror which feeds the spectrograph.

In the design of the focal plane unit a port for an extra lucky imager is included. With two cameras this allows dual-color simultaneous imaging over a 45×45 arcsecond field. Although this is very small for most other applications than microlensing follow-up, the design of the optical system allow fields up to 3'×3' to be observed, using either larger detectors or focal reducers.

Current work and timeline

The work for the SONG prototype is undergoing (autumn 2008) its final optical design in order to prepare for the ordering of the major optical components by late 2008 and early 2009.

During 2009 and 2010 the detailed design and production of mechanics for the spectrograph, focal plane and dome-carrying structure will take place. Our aim is for first light of the telescope in the last half of 2010. In this period a substantial effort will also be devoted to software development for robotic observations and data reduction, followed by system test and verification during 2011.

Acknowledgments. The group behind SONG gratefully acknowledges the substantial grant from the Villum Kann Rasmussen foundation which enables us to design and build the prototype node. Financial support from the Carlsberg foundation, the Danish Natural Science Research Council and Instrument centre for Danish Astronomy (IDA) is also gratefully acknowledged. We also gratefully acknowledge support from the European Helio– and Asteroseismology Network (HELAS), a major international collaboration funded by the European Commission's Sixth Framework Programme. FG wishes to thank G. Marcy for encouragement and helpful advice on the development of the iodine code.

References

Arentoft, T.A., Kjeldsen, H., Bedding, T.R., et al. 2008, ApJ, 687, 1180

Baldwin, J.E, Tubbs, R.N., Mackay, C.D., et al. 2001, A&A, 368, L1–L4

Butler, R.P., Marcy, G.W., Williams, E., et al. 1996, PASP, 108, 500

Butler, R.P., Bedding, T.R., Kjeldsen, H., et al. 2004, AJ, 600, L75

Chaplin, W. J., Elsworth, Y., Howe, R. et al. 1996, Solar Phys. 168, 1

Fried, D.L. 1978, JOSA, Vol. 68, Issue 12, 1651

Grundahl, F., Christensen–Dalsgaard, J., Arentoft, T., et al. 2008, CoAst, 157, 273

Grundahl, F., Kjeldsen, H., Christensen-Dalsgaard, J., et al. 2007, CoAst, 150, 300

Hormuth, F., Hippler, S., Brandner, W., et al. 2008, to appear in Proc. SPIE, Vol. 7014, 701448

Jørgensen, U.G. 2008, Phys. Scr., T130, 014008

Kjeldsen, H., Bedding, T.R., Arentoft, T., et al. 2008, ApJ, 682, 1370

DISCUSSION

Reese: Will SONG or SIAMOIS study line profile variations in more rapidly rotating stars?

Grundahl: SONG will be able to do it (but it doesn't seem to be the main priority).

Mosser: SIAMOIS is not designed to do this and won't be able to do it.

Comm. in Asteroseismology
Vol. 158, 2009, 38th LIAC/HELAS-ESTA/BAG, 2008
A. Noels, C. Aerts, J. Montalbán, A. Miglio and M. Briquet., eds.

BRITE-Constellation: nano-satellites for Asteroseismology

R. Kuschnig and W.W. Weiss

Institut für Astronomie, Türkenschanzstrasse 17, A-1180 Vienna, Austria

Abstract

BRITE-Constellation currently consists of two satellites, UniBRITE and BRITE-AUSTRIA (TUG-SAT1). Each will fly a CCD camera to perform high-precision two-color photometry continuously for two years or more, primarily of stars brighter than 4th magnitude (V), and with reduced accuracy also of stars down to 7th magnitude. The primary science goals are studies of massive stars in our Galactic neighbourhood, representing objects which dominate the ecology of our Universe, and also evolved stars (giants) to probe the future development of our Sun. BRITE-Constellation is made possible by innovative technology currently developed in collaboration between Canada and Austria. A launch of UniBRITE and BRITE-AUSTRIA in late 2009 is envisioned. An expansion of BRITE-Constellation may be funded by the Canadian Space Agency (CSA). The operation policy will be to observe a few fields over a long time span and possibly some short runs in between, which will assure optimum use of near polar low-earth orbits.

Satellites

Each of the 6kg BRITE satellites (Figure 1) is equipped with a small dioptric telescope. A satellite "constellation" provides improved time coverage and two-color information: two satellites have blue and two have red filters. The 20cm cube structure houses three orthogonal reaction wheels and three magnetorquer coils for three-axis attitude control and momentum dumping. Attitude determination is provided by a magnetometer, six sun sensors and a star tracker. This equipment will enable attitude determination to 10 arcseconds, attitude control accuracy to better than a degree, and attitude stability down to one arcminute rms.

Instruments

The science payload of the satellite consists of a five-lens telescope (Figure 2) with an aperture of 30mm and an interline CCD detector KAI 11002-M from Kodak with 11M pixels, along with a baffle to reduce stray light. The photometer has a resolution of 26.52 arcseconds per pixel and a field-of-view of 24 degrees. The effective wavelength range of the instrument is defined by the blue filter, which covers a wavelength range of 390-460nm, and the red filter, 550-700nm.

Science Goals

BRITE-Constellation will photometrically measure low-level oscillations and temperature variations in stars brighter than visual magnitude 4.0 (and with less accuracy also down to a visual

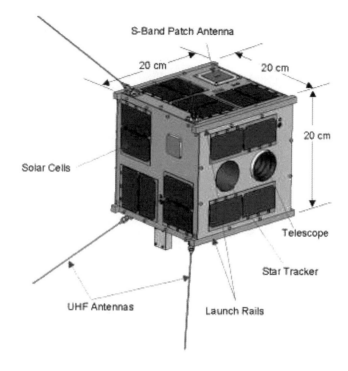

Figure 1: BRITE Satellite Model

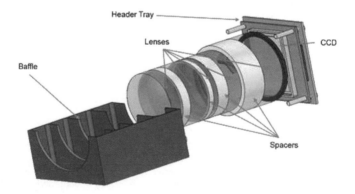

Figure 2: BRITE Telescope Model

magnitude of 7.0). There are 534 stars (Figure 3) brighter than $V = 4.0$ mag in the sky and observable at the proposed precision level with BRITE-Constellation. Considering the typical time scales for their variability ranging from an hour to several days and aiming for a frequency resolution sufficient for asteroseismology. BRITE-Constellation expects to observe on average 20 stars simultaneously.

The Figure on top shows the location of the 534 stars with V ≤ 4 mag in the Hertzsprung-Russell Diagram color-coded with the object types taken from the VISAT database. The first Announcement of Opportunity for submission of observing proposals from the community has been released (Sep. 08) at www.brite-constellation.at by the BRITE Executive Science Team. Please check for updates.

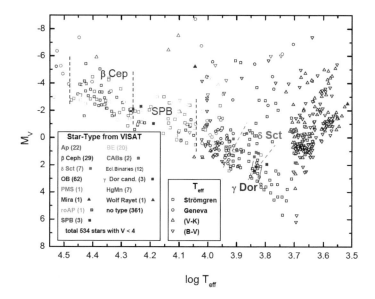

Figure 3: HRD diagram of the 534 brightest stars (V ≤ 4mag).

Acknowledgments. This project is currently funded by the University of Vienna and the Austrian Science Promotion Agency (FFG-ARL). Substantial in-kind contributions come from the Canadian Space Agency (CSA) and the Natural Sciences and Engineering Research Council (NSERC) of Canada.

Conclusions

Comm. in Asteroseismology
Vol. 158, 2009, 38th LIAC/HELAS-ESTA/BAG, 2008
A. Noels, C. Aerts, J. Montalbán, A. Miglio and M. Briquet., eds.

Concluding remarks

J. Christensen-Dalsgaard[1,2]

[1] Department of Physics and Astronomy, Aarhus University, DK-8000 Aarhus C, Denmark
[2] Danish AsteroSeismology Centre

Abstract

The 38th Liège Colloquium covered a broad range of topics within the area of massive stars on and near the main sequence. This included the physical processes in stellar interiors and modelling of stellar evolution, as well as observational aspects with emphasis on asteroseismology and the potential of recent and coming observations. This was a fitting tribute to Arlette Noels on the occasion of her retirement. Here I provide a brief summary of some of the main points of the conference.

Introduction

The Liège group has played a major role in the development of stellar astrophysics and the study of stellar pulsations, over many decades, starting with the seminal work of Paul Ledoux. In addition to his personal research, a major contribution of Ledoux was to establish a strong research group at Institut d'Astrophysique, with Arlette as an outstanding member. The work of the group has covered a large number of areas, emphasizing the physical understanding of stellar interiors and the potential for investigating stellar properties through the analysis of observations of pulsating stars. This also includes very substantial contributions to the development of helioseismology, starting from the early suggestion by Henry Hill that the Sun might be a pulsating star.

Given this background, and Arlette's important part in it, it is entirely appropriate that her retirement should be marked by one of the long series of highly successful Liège colloquia.[1] The topic of the meeting, *Evolution and Pulsation of Massive Stars on the Main Sequence and close to it*, certainly covers a substantial part of Arlette's work and includes a very broad range of stellar observations, physical processes and stellar modelling. It is evident that the present brief overview in no way can present an adequate summary of the meeting or do justice to the many excellent presentations; however, I shall try to indicate some of the main points of the meeting.

Modelling massive stars

Massive stars represent extremes of stellar internal physics, strongly affecting their evolution. Rotation is undoubtedly important, particularly in driving mixing processes in stellar interiors, and is strongly affected by angular-momentum loss from the stellar surface. An obvious first step in modelling the relevant transport processes is to determine the linear instabilities

---

[1]It was interesting to note that many of the speakers remarked that their first experience of an international conference had been at one of these colloquia; this includes the present author who took part in the 1974 colloquium as a young PhD student.

arising in the relevant situations, but it is equally obvious that, as stated, 'the World is not linear'. The mixing is likely dominated by meridional circulation, the effects of which can be modelled in the formulation going back to Zahn (1992) and assuming that turbulent diffusivity is much stronger in the horizontal than in the vertical direction. This allows a treatment of the transport of chemical elements as a diffusive process, whereas the transport of angular momentum includes an advective component. It is possible, however, that the assumption of large horizontal diffusivity becomes questionable for rapid rotation, with an angular velocity comparable to the local buoyancy frequency, where the Rossby radius becomes small (Talon 2009). Additional complications may arise from magnetic fields, either primordial or induced by a turbulent dynamo in the radiative region, which could suppress the meridional circulation.

Although the Sun is hardly a massive star, it provides a crucial test-bed for our understanding of dynamical phenomena in stellar interiors, including the evolution of rotation from a likely early state of rapid rotation to the present slowly and nearly uniformly rotating radiative interior. It is still not clear whether this has been achieved through the effect of gravity waves or through the action of a magnetic field. Recent modelling by Garaud & Garaud (2008), based on the proposal by Gough & McIntyre (1998), seems encouraging for the latter possibility.

It is evident that convection remains a major uncertainty, particularly in the mixing processes that may occur outside the normally considered unstable regions, in convective overshoot or what is usually known as 'semiconvective regions' and depending on whether the Ledoux[2] or Schwarzschild criterion is used. These processes have very substantial effects on the later evolution of stars, yet their physical nature is poorly understood.

From an observational point of view an important issue is to distinguish between mixing caused by rotation and by convective effects, given their subtly different effect on the resulting composition profile. Owing to their effect on the buoyancy frequency these differences can in principle be detected from analysis of g-mode frequencies (Miglio et al. 2009), although it is so far not clear whether the required selection of modes and frequency precision will be available in observations of any type of massive pulsating star.

In regions where there is no bulk motion the composition profile is affected by microscopic diffusion and settling. In the outer parts of massive stars with thin outer convection zones these processes are quite rapid and would lead to strong changes in composition, if not counteracted by other processes. An additional complication is the selective effect of radiation pressure, which may lead to local abundance enhancements of specific elements (Alecian 2009). Interestingly, the resulting inversion of the mean molecular weight can cause so-called 'salt-fingering instabilities' which could reduce or eliminate the effect of such radiative levitation (Vauclair 2009).

The evolution of massive stars is evidently strongly affected by mass loss which can remove a substantial fraction of the mass of a star during its evolution (Eggenberger et al. 2009). There is a close relation between rotation and mass loss: on the one hand, mass loss removes angular momentum and hence changes the rotation of the star; on the other, the near-surface structure of the star is strongly affected by rotation and hence so is the mass loss. Given that the mass loss is radiatively driven the latter effect furthermore depends on the effective temperature, and hence for main sequence stars the mass, of the star. For very hot and massive stars gravity darkening reduces the mass loss at the equator, relative to the poles; the resulting predominantly polar mass loss evidently removes relatively little angular momentum. At lower mass and effective temperature, the dominant effect is the reduction in the effective gravity at the equator; thus mass loss is mainly equatorial, leading to a very substantial loss of angular momentum. It is tempting to regard the nebulosity around η Carinae as the result of such a complex mass-loss episode.

Rotation is in most cases regarded as a perturbation on spherically symmetric evolution.

---

[2]It is interesting that Shibahashi noted that the Ledoux criterion was originally found by Karl Schwarzschild! The Schwarzschild criterion refers to his son *Martin* Schwarzschild.

However, for sufficiently rapid rotation this is clearly inadequate. Here a fully two-dimensional modelling of the hydrostatic structure is required. Such modelling is now being started (see also, for example, Roxburgh 2006; MacGregor et al. 2007). An interesting result is the appearance of a near-surface convectively unstable region near the equator (Rieutord & Espinosa Lara 2009). A further complication is obviously to combine such two-dimensional models with a treatment of the transport processes and the resulting changes in the internal angular-momentum distribution; the modelling shows a fairly complex distribution of the internal meridional circulation.

From an observational point of view, an important consequence of rapid rotation is the mixing which may dredge up products of nuclear burning to the surface. For main sequence stars the most significant signature of mixing is the increase in the surface abundance of ^{14}N, resulting from hydrogen burning through the CNO cycle (Maeder et al. 2009). With sufficiently rapid rotation mixing may become so efficient that the evolution proceeds with a nearly uniform composition. Unlike normal evolution, such homogeneous evolution does not cause the star to expand. As a result, the stars evolve on the hot side of the zero-age main sequence.

Binary stars present further challenges and opportunities (de Mink et al. 2009). The global parameters, such as mass, radius and rotation rate, are strongly constrained in well-observed eclipsing binaries; hence abundance determinations in such systems would provide excellent tests of the mixing induced by rotation. On the other hand, the evolution of the components may be affected by being in a binary system, most obviously of course in very close systems with Roch-lobe overflow of matter from one component to the other. As discussed above, the nearly homogeneous evolution in sufficiently rapidly rotating stars, as expected in very close binaries, may avoid the expansion found in normal evolution and hence the overflow.

Classical observations

The test of the theoretical models requires observations. 'Classical' observables such as effective temperature, gravity and luminosity are essential, while for massive stars also rotation rates and the strength of stellar winds are important constraints on the models. While the ideal situation is probably to match the models to the observations of an individual well-constrained star, important information can also be derived from the statistical properties of samples of stars, such as the distribution between different classes of Wolf-Rayet stars.

As already mentioned, stellar surface abundances are important diagnostics of stellar evolution and mixing processes. The abundance determination obviously requires adequate atomic data, including an understanding of the detailed microscopic processes in the stellar atmosphere; in particular, local thermodynamical equilibrium cannot be assumed (Puls 2009). Also, the basic stellar parameters must be determined. Further complications in massive stars are the likely effect of the stellar wind and the possible variations caused by pulsations or rotation, requiring an averaging over time to determine the underlying properties. Thus the determination of the composition, and other stellar parameters, depends on the underlying modelling of the stellar atmosphere. An interesting example is the possible effect of unresolved low-frequency g modes on the 'macroturbulence' affecting stellar line profiles; this may influence the determination of the rotational broadening of the lines, leading to an under-estimate of the rotation rate (Aerts et al. 2009).

An important issue is the solar composition relative to that of stars in the solar neighbourhood, particularly given the recent reduction in the inferred solar abundances of carbon, nitrogen and in particular oxygen (for a review, see Asplund 2005). Interestingly, it appears that even with this revision the Sun is somewhat overabundant in heavy elements compared with a group of nearby O and B stars (Morel 2009), although the revised solar abundances are obviously more closely in line with neighbouring objects (e.g., Turck-Chièze et al. 2004).

From stellar modelling one would expect a strong correlation between rotation rate, age and surface nitrogen abundance, resulting from dredge-up of CNO-cycle processed material. An extensive sample of observations of nitrogen abundance and projected rotational velocity obtained by Hunter et al. (2008) showed substantial departures from the expected behaviour, raising doubt about the mechanisms of rotational mixing and suggesting that other aspects, including binarity and perhaps effects of magnetic fields, must be taken into account. However, this sample contained a broad variety of stars; it appears that if more uniform sub-samples are considered the expected correlation may be recovered (Maeder et al. 2009).

Asteroseismology

Liège can properly be regarded as the birthplace of asteroseismology in a broad sense. Thus it is appropriate that asteroseismic investigations played a substantial role at the conference. An important theme was the early results from the very successful CoRoT satellite. However, there evidently also remain very significant theoretical issues in the study of stellar oscillations.

Theory

The excitation of oscillations in massive main sequence stars was for a long time a serious theoretical problem, with many imaginative suggestions for its solution. In the end, a very natural explanation was found through an increase in the opacity in the relevant temperature range, first proposed by Simon (1982) and confirmed in improved opacity calculations by Iglesias & Rogers (1991) and Seaton (1993). The revision of the opacity resulted from the full inclusion of transitions in iron-group elements which led to a very substantial opacity increase, by more than a factor of two, at temperatures around 2×10^5 K. This has successfully accounted for the observed instability regions for the slowly pulsating B stars and the β Cephei stars (see Pamyatnykh 1999 for an excellent overview). However, problems remain with the detailed interpretation of extensive observations of certain β Cephei stars where the models do not predict instability of all the modes that are observed (Dziembowski 2009). A possible solution could be the concentration of iron-group elements in the critical region through radiative levitation and settling, which seems to account for the instability of the pulsating subdwarf B stars (e.g., Fontaine et al. 2003). However, in β Cephei stars these effects would be offset by even a modest amount of meridional circulation, resulting from the typical rotation rate of the stars. Also, as mentioned above salt-finger instability could reduce the resulting concentration enhancement. Thus it cannot be excluded that further opacity revisions will be required.[3]

A potential new class of pulsating variables was identified by Lefever et al. (2007) amongst the B-type supergiants; they concluded that these stars may be pulsating as a result of opacity-driven g modes. This appeared somewhat surprising, given that nonradial modes in highly evolved stars would be expected to be heavily damped, owing to the rapidly oscillating eigenfunction in the deep interior where the mode behaves as a g mode of very high order (Dziembowski 1977). However, for suitable parameters models of these stars develop an intermediate convection zone near the hydrogen shell source which acts as a potential barrier for some g modes and effectively traps them in the outer parts of the interior where they may be excited (Dupret et al. 2009). Such models in fact predict unstable modes corresponding essentially to the observations.

While lower-mass stars, including the Sun, show intrinsically stable modes that are excited stochastically by convection, such modes have not been expected for massive stars. However, it appears that a convection zone associated with the iron opacity bump has properties which

---

[3]This may also be suggested by the difficulty in matching solar models to the helioseismic inferences, given the revised solar abundance (e.g., Montalbán et al. 2004; Bahcall et al. 2005; Christensen-Dalsgaard et al. 2009).

may lead to the stochastic excitation of higher-order acoustic modes at a level which should be detectable observationally, e.g., with the CoRoT mission (Belkacem et al. 2009). This would obviously, together with the other pulsations of the β Cephei stars, open very interesting possibilities for the asteroseismic investigation of these stars. A possibly related result is a correlation between 'microturbulence', required to model observed spectral line profiles, and the modelled properties of near-surface convection in hot stars; this suggests that the line profiles are affected by convection or perhaps by waves generated by convection (Cantiello et al. 2009). Detailed hydrodynamical modelling of these convection zones would be very interesting.

Observation of the frequency splitting induced by rotation provides a very important diagnostic of stellar internal rotation, e.g., providing evidence for radial variation in the internal angular velocity in the β Cephei star HD 129929 (Aerts et al. 2003). These effects have normally been treated through a perturbation expansion, assuming relatively slow rotation; for very slow rotation, such as in the solar case, a first-order treatment is adequate, but at more rapid rotation higher-order formulations, taking into account also near-degeneracy between the modes, may be required. However, in many of the asteroseismically relevant cases rotation is so fast that the perturbation treatment is insufficient. This is obviously closely related to the modelling of the structure of rapidly rotating and hence rotationally deformed stars, as discussed above. In this case a fully two-dimensional formulation of the pulsation equations is required (Reese et al. 2009). Extensive investigations of the resulting pulsation spectrum have been carried out by Reese et al. (2006, 2008) and Lignières & Georgeot (2008) in the case of polytropic models, demonstrating the presence of modes of oscillation which have no counterparts in the non-rotating case. It is very encouraging that these studies are now being extended to realistic stellar models. The application of the results to asteroseismic analyses of observed oscillation spectra still requires very substantial development, however.

Observations

While much attention was deservedly given to space projects, ground-based observations remain a very important part of asteroseismic investigations. These allow investigation of stars with properties making them inappropriate for the space projects, or not falling within their observing regions. Also, space observations often require supplementary ground-based studies, e.g., spectroscopic observations to secure mode identification or extend the time series. Impressive examples of the former case are the open-cluster campaigns to study β Cephei stars (Uytterhoeven 2009). Membership of clusters provides additional constraints on the stars, such as common age and initial composition, as well as independent information about cluster members from the analysis of the cluster. In particular, its age and distance can be determined from its position in colour-magnitude diagrams. Such constraints greatly improve the possibilities for asteroseismic analyses. The preliminary results, e.g., for the cluster NGC 6910, are very encouraging.

After the serendipitous use of the WIRE satellite for asteroseismology, the very successful MOST mission has been the first mission developed specifically for asteroseismic observations. Given the modest size of the mission telescope and budget, the results have been very impressive for a range of variable stars. For example, this has led to the study of the interesting class of slowly pulsating, rapidly rotating emission-line B stars (the so-called SPBe stars) which show distinct and very rich pulsation spectra, strongly influenced by the rotation of the star (Kuschnig 2009). Thus the mission has clearly demonstrated the power of space asteroseismology, without quite reaching the level required to study solar-like oscillations in main sequence stars.

This level has definitely been reached in the more ambitious CoRoT satellite, in part as a result of the very efficient baffle built in Liège which has dramatically reduced the level of scattered light. Preliminary results on solar-like oscillations have been presented by Appourchaux et al. (2008) and Michel et al. (2008). The latter paper showed that the observed amplitudes are substantially higher than for the Sun, but lower by about 50 % than the most recent theoretical predictions. The frequencies determined by Appourchaux et al. (2008), now being analyzed through comparison with stellar models, provide a very interesting opportunity to extend asteroseismic investigations to stars somewhat different from the Sun.

The asteroseismic observations from CoRoT, both from the fields dedicated to astero-seismology and from the fields optimized for the search for extra-solar planets (the so-called exofields), extend much beyond the solar-like stars. Observations of δ Scuti stars have shown a very large number of modes with low observed amplitude, consistent with theoretical ex-pectations that the number of excited modes should be far larger than the number detected in ground-based photometry. While the detection of a more complete sample of modes may in principle help the important issue of mode identification, the possible presence of modes of higher degree, detectable at the CoRoT level of precision despite the low sensitivity of integrated-disk observations, could further complicate the identification. Observations of the β Cephei star HD 180642 also show a number of additional modes which will undoubtedly provide strong asteroseismic constraints on the star (Briquet 2009). Also, the observations show tantalizing hints of what might be stochastically excited modes at rather high frequency (see above), although it may be difficult to distinguish these from the combination frequencies resulting from the larger-amplitude modes. Very detailed results have been obtained on the pulsations of emission-line B stars, likely confirming the relation between the pulsations and the ejection of the circumstellar material that causes the Be phenomena (Neiner et al. 2009). The large number of variable stars detected in the exofields allows statistical studies of their distribution in stellar parameters (Lefèvre et al. 2009).

From such an analysis in terms of surface gravity, effective temperature and characteristic frequencies two new groups of pulsating stars have been identified, one intermediate between the slowly pulsating B (SPB) stars and the δ Scuti stars, the second intermediate in frequency between the SPB and β Cephei stars (Degroote et al. 2009). This clearly shows the value of studying also stars previously thought to be nonvariable, when a new level of sensitivity is reached. Finally, a substantial number of red giants showing solar-like oscillations have been found in the exofield, spanning a range in luminosity and hence characteristic frequency. These observations will help filling the gap between the probable solar-like oscillations seen in very luminous red giants in large-scale surveys (e.g., Kiss & Bedding 2003) and stars near the main sequence. This will improve our understanding of the excitation and damping of solar-like oscillations as well as, obviously, provide important constraints on the properties of red giants.

It is clear that with these results we are just seeing the beginning of what will surely be an explosion of data on stellar variability from CoRoT which will undoubtedly revolutionize our understanding of stellar pulsations and greatly extend asteroseismic investigations.

The future

The coming years will see a continuing dramatic increase in the quantity and quality of asteroseismic data. With a launch scheduled for March 2009 the Kepler mission will yield data of a quality comparable with those of CoRoT for a large number of stars and extending over several years. This will include a very substantial number of solar-like stars observed at a 1-minute cadence, but also stars covering essentially the full range of variable stars, observed either at the rapid cadence or, for more slowly varying stars, at a 30-minute cadence. The latter category also includes very many red giants. However, since Kepler will observe a fixed field, chosen to be slightly outside the plane of the Galaxy to limit the effects of crowding, the number of massive main sequence stars that can be observed is unfortunately very limited. This limitation will to some extent be offset by the very interesting BRITE nano-satellite mission, also to be launched in 2009, which will observe a limited number of bright stars, certainly also including massive stars. Beyond Kepler the next potential major space mission with an important asteroseismic component is the PLATO mission, currently undergoing competitive assessment studies within ESA for a possible launch in 2017. While PLATO is defined predominantly as a mission to investigate extra-solar planetary systems, a crucial aspect is to make asteroseismic investigations of a very large fraction of the central stars in the systems detected.

While the photometric observations from space provide a unique opportunity to study large samples of stars they do not yield asteroseismic data of the highest possible quality. The reason is simply the intrinsic stellar 'noise': in intensity observations contributions to the observations from sources, such as granulation or activity, other than the oscillations are far higher relative to the oscillations than in velocity observations. In this sense Doppler velocity observations of carefully selected stars are an essential complement to the space observations, particularly for solar-like oscillations; the velocity observations allow detection of modes of lower frequency and hence longer lifetime and potentially higher frequency accuracy. An additional substantial advantage is the ability to detect modes of degree 3 which are barely visible in photometric observations. For other types of pulsating stars velocity observations are very important for the mode identification.

Such observations can be carried out from the ground with little degradation of the data as a result of the Earth's atmosphere. The main challenge is to secure the observing time over sufficiently long periods and ideally from several sites. In fact, observing time of up to a week has been obtained on major facilities, yielding excellent data on a few stars. One might perhaps even hope for very limited use of the coming European Extremely Large Telescope for special cases. However, it is obvious that for the extended observing periods required to obtain very high frequency precision dedicated facilities are needed.

This is the motivation for two projects to carry out Doppler-velocity observations of selected stars: the SIAMOIS project to establish a facility on Dome C in Antarctica and the SONG project to set up a network of observing stations with a suitable distribution around the World. These projects supplement each other very well: SIAMOIS has the potential for a somewhat higher duty cycle, at least for objects at high southern declination, than SONG; these objects include the important case of the α Centauri binary system. On the other hand, SIAMOIS is obviously limited to observe part of the sky, and continuous data can only be obtained during part of the year; SONG can cover the whole sky and observe throughout the year, although continuous data for any given object are typically limited to 4 – 5 months. The projects are in different stages of development but the hope is that both will be operational around 2014.

At a less ambitious, but perhaps no less important, level it should be noted that there may be a scarcity of instruments for the crucial spectroscopic follow-up observations for space asteroseismology and exoplanet investigations. This is a result of the increasing concentration on the development and use of very large telescopes, at the expense of the smaller facilities. It is important that the need for a full range of facilities, for all relevant types of observation, is kept in mind in the planning of the future of astronomical instrumentation such as is now being undertaken, e.g., in ASTRONET.

However, it is obvious that the future will see a huge increase in the data for asteroseismology, and more generally the data for studying stellar astrophysics. This represents a great challenge and opportunity also for the theoretical investigations which will need increasing sophistication in the treatment of physical effects and the modelling of stellar evolution. Much progress will be made as a result of the development of computational resources and tools for large-scale simulations which will become increasingly realistic in their representation of conditions in stellar interiors. However, the insight and understanding of the scientists involved will remain the most important aspect of these investigations. Thus I hope that Arlette, despite her no doubt busy schedule, will continue to contribute to the field for many years to come, and I wish her a very happy future.

Acknowledgments. With the concluding talk and this brief paper I wish to acknowledge the major contribution made by Arlette to the development of stellar astrophysics and, on a more personal level, to the very enjoyable and fruitful time spent by my family and me in Liège from 1978 – 1980, as well as many pleasant encounters since then. I gratefully acknowledge support from the European Helio- and Asteroseismology Network (HELAS), a major international collaboration funded by the European Commission's Sixth Framework Programme.

References

Aerts, C., Thoul, A., Daszyńska, J., et al. 2003, Science, 300, 1926

Aerts, C., Puls, J., Godart, M., & Dupret, M.-A. 2009, CoAst, 158, 66

Alecian, G. 2009, CoAst, 158, 34

Appourchaux, T., Michel, E., Auvergne, M., et al. 2008, A&A, 488, 705

Asplund, M. 2005, ARAA, 43, 481

Bahcall, J. N., Basu, S., Pinsonneault, M., & Serenelli, A. M. 2005, ApJ, 618, 1049

Belkacem, K., Goupil, M. J., Samadi, R., & Dupret, M.-A. 2009, CoAst, 158, 269

Briquet, M. 2009, CoAst, 158, 292

Cantiello, M., Langer, N., Brott, I., et al. 2009, CoAst, 158, 61

Christensen-Dalsgaard, J., Di Mauro, M. P., Houdek, G., & Pijpers, F. 2009, A&A, 494, 205

Degroote, P., Miglio, A., Debosscher, J., et al. 2009, CoAst, 158, 167

de Mink, S. E., Cantiello, M., Langer, N., & Pols, O. R. 2009, CoAst, 158, 94

Dupret, M.-A., Godart, M., Noels, A., & Lebreton, Y. 2009, CoAst, 158, 239

Dziembowski, W. 1977, Acta Astron., 27, 95

Dziembowski, W. A. 2009, CoAst, 158, 227

Eggenberger, P., Meynet, G., & Maeder, A. 2009, CoAst, 158, 87

Fontaine, G., Brassard, P., Charpinet, S., et al. 2003, ApJ, 597, 518

Garaud, P., & Garaud, J.-D. 2008, MNRAS, 391, 1239

Gough, D. O., & McIntyre, M. E. 1998, Nature, 394, 755

Hunter, I., Brott, I., Lennon, D. J., et al. 2008, ApJ, 676, L29

Iglesias, C. A., & Rogers, F. J. 1991, ApJ, 371, L73

Kiss, L. L., & Bedding, T. R. 2003, MNRAS, 343, L79

Kuschnig, R. 2009, CoAst, 158, 162

Lefever, K., Puls, J., & Aerts, C. 2007, A&A, 463, 1093

Lefèvre, L., Michel, E., Aerts, C., et al. 2009, CoAst, 158, 189

Lignières, F., & Georgeot, B. 2008, Phys. Rev. E, 78, 016215

MacGregor, K. B., Jackson, S., Skumanich, A., & Metcalfe, T. S. 2007, ApJ, 663, 560

Maeder, A., Meynet, G., S. Ekström, S., & Georgy, C. 2009, CoAst, 158, 72

Michel, E., Baglin, A., Auvergne, M., et al. 2008, Science, 322, 558

Miglio, A., Montalbán, J., Eggenberger, P., & Noels, A. 2009, CoAst, 158, 233

Montalbán, J., Miglio, A., Noels, A., et al. 2004, in Proc. SOHO 14 - GONG 2004: "Helio- and Asteroseismology: Towards a golden future", ed. Danesy, D., ESA SP-559, ESA Publication Division, p. 574

Morel, T. 2009, CoAst, 158, 122

Neiner, C., Gutiérrez-Soto, J., Floquet, M., et al. 2009, CoAst, 158, 319

Pamyatnykh, A. A. 1999, Acta Astron., 49, 119

Puls, J. 2009, CoAst, 158, 113

Reese, D., Lignières, F., & Rieutord, M. 2006, A&A, 455, 621

Reese, D., Lignières, F., & Rieutord, M. 2008, A&A, 481, 449

Reese, D., MacGregor, K. B., Jackson, S., et al. 2009, CoAst, 158, 264

Rieutord, M., & Espinosa Lara, F. 2009, CoAst, 158, 99

Roxburgh, I. W. 2006, A&A, 454, 883

Seaton, M. J. 1993, in Proc. IAU Colloq. 137: Inside the stars, eds Baglin, A. & Weiss, W. W., ASP Conf. Ser., 40, 222

Simon, N. R. 1982, ApJ, 260, L87

Talon, S. 2009, CoAst, 158, 20

Turck-Chièze, S., Couvidat, S., Piau, L., et al. 2004, Phys. Rev. Lett., 93, 211102

Uytterhoeven, K. 2009, CoAst, 158, 156

Vauclair, S. 2009, CoAst, 158, 51

Zahn, J.-P. 1992, A&A, 265, 115